Obesity and the Kidney

....
Contributions to Nephrology

Vol. 151

Series Editor

Claudio Ronco *Vicenza*

Obesity and the Kidney

Volume Editor

Gunter Wolf   Jena

36 figures, 2 in color, and 12 tables, 2006

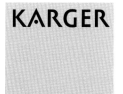

Basel · Freiburg · Paris · London · New York ·
Bangalore · Bangkok · Singapore · Tokyo · Sydney

Contributions to Nephrology

(Founded 1975 by Geoffrey M. Berlyne)

....

Gunter Wolf
Department of Internal Medicine III
University Hospital Jena
Erlanger Allee 101
DE–07740 Jena (Germany)

Library of Congress Cataloging-in-Publication Data

Obesity and the kidney / volume editor, Gunter Wolf.
   p. ; cm. – (Contributions to nephrology, ISSN 0302-5144 ; v. 151)
 Includes bibliographical references and index.
 ISBN-13: 978-3-8055-8164-6 (hard cover : alk. paper)
 ISBN-10: 3-8055-8164-5 (hard cover : alk. paper)
 1. Obesity. 2. Kidneys–Pathophysiology. I. Wolf, Gunter, 1961- II. Series.
   [DNLM: 1. Kidney Diseases–etiology. 2. Obesity–complications. W1 CO778UN v.151 2006 / WJ 300 O12 2006]
  RC628.O242 2006
  616.3′98–dc22

                                                            2006018661

Bibliographic Indices. This publication is listed in bibliographic services, including Current Contents® and Index Medicus.

Disclaimer. The statements, options and data contained in this publication are solely those of the individual authors and contributors and not of the publisher and the editor(s). The appearance of advertisements in the book is not a warranty, endorsement, or approval of the products or services advertised or of their effectiveness, quality or safety. The publisher and the editor(s) disclaim responsibility for any injury to persons or property resulting from any ideas, methods, instructions or products referred to in the content or advertisements.

Drug Dosage. The authors and the publisher have exerted every effort to ensure that drug selection and dosage set forth in this text are in accord with current recommendations and practice at the time of publication. However, in view of ongoing research, changes in government regulations, and the constant flow of information relating to drug therapy and drug reactions, the reader is urged to check the package insert for each drug for any change in indications and dosage and for added warnings and precautions. This is particularly important when the recommended agent is a new and/or infrequently employed drug.

All rights reserved. No part of this publication may be translated into other languages, reproduced or utilized in any form or by any means electronic or mechanical, including photocopying, recording, microcopying, or by any information storage and retrieval system, without permission in writing from the publisher.

© Copyright 2006 by S. Karger AG, P.O. Box, CH–4009 Basel (Switzerland)
www.karger.com
Printed in Switzerland on acid-free paper by Reinhardt Druck, Basel
ISSN 0302–5144
ISBN-10: 3–8055–8164–5
ISBN-13: 978–3–8055–8164–6

Contents

VII Foreword
Wolf, G. (Jena)

Epidemiology

1 Obesity and Chronic Kidney Disease
Kramer, H. (Maywood, Ill.)

19 Obesity and Kidney Transplantation
Srinivas, T.R.; Meier-Kriesche, H.-U. (Gainesville, Fla.)

42 Body Mass Index and the Risk of Chronic Renal Failure: The Asian Experience
Iseki, K. (Okinawa)

57 Obesity Paradox in Patients on Maintenance Dialysis
Kalantar-Zadeh, K.; Kopple, J.D. (Torrance, Calif./Los Angeles, Calif.)

Molecular Mechanisms of Obesity-Related Renal Disease

70 Adipose Tissue as an Endocrine Organ – A Nephrologists' Perspective
Chudek, J.; Adamczak, M.; Nieszporek, T.; Więcek, A. (Katowice)

91 Renal Handling of Adipokines
Kataoka, H.; Sharma, K. (Philadelphia, Pa.)

106 **Lipid Metabolism and Renal Disease**
Abrass, C.K. (Seattle, Wash.)

122 **Role of the Renin–Angiotensin–Aldosterone System in the Metabolic Syndrome**
Engeli, S. (Berlin)

135 **Functional and Structural Renal Changes in the Early Stages of Obesity**
Thakur, V.; Morse, S.; Reisin, E. (New Orleans, La.)

151 **Leptin as a Proinflammatory Cytokine**
Lord, G.M. (London)

165 **Adipose Tissue and Inflammation in Chronic Kidney Disease**
Axelsson, J.; Heimbürger, O.; Stenvinkel, P. (Stockholm)

175 **Leptin and Renal Fibrosis**
Wolf, G. (Jena); Ziyadeh, F.N. (Philadelphia, Pa.)

184 **Obesity and Renal Hemodynamics**
Bosma, R.J.; Krikken, J.A.; Homan van der Heide, J.J.; de Jong, P.E.; Navis, G.J. (Groningen)

203 **Insulin Resistance and Renal Disease**
Fliser, D.; Kielstein, J.T.; Menne, J. (Hannover)

Therapy

212 **Treatment of Obesity: A Challenging Task**
Sharma, A.M.; Iacobellis, G. (Hamilton)

221 **Weight Loss and Proteinuria**
Praga, M.; Morales, E. (Madrid)

230 **Treatment of Arterial Hypertension in Obese Patients**
Wenzel, U.O.; Krebs, C. (Hamburg)

243 **Bariatric Surgery**
Korenkov, M. (Mainz)

254 **Author Index**

255 **Subject Index**

Foreword

> *'Let me have men about me that are fat,*
> *Sleek-headed men, and such as sleep o' nights:*
> *Yond Cassius has a lean and hungry look;*
> *He thinks too much: such men are dangerous'*
>
> *Julius Caesar*, Act 1, Scene 2
> William Shakespeare (1564–1616)

Our genetic background and physiological homeostasis, orchestrated through endocrine and neuronal networks, are optimized for a world with intermittent food supply and permits us to survive periods of starvation. Although this genetic background has provided a clear survival benefit for primitive men, these systems are contraproductive in the current industrial society with fast-food restaurants, an abundance of high-energy food, and an increasingly sedentary lifestyle. The consequence is an alarming increase in obese adults, and even more disturbing of overweighed children. The ongoing epidemic of obesity is well-known in the United States and Europe, but in countries such as China, India and South America the problem has reached catastrophic dimensions. The consequences of obesity such as metabolic syndrome with the ultimate development of type 2 diabetes mellitus, cardiovascular diseases, an increased incidence of certain cancers, musculoskeletal disorder and pulmonary diseases are well-known.

Almost 30 years ago, Weisinger et al. [1] described focal-segmental glomerulosclerosis with nephrotic syndrome in four massive obese patients. Only two of them exhibited hypertension by office blood pressure measurements [1]. In the following years, several case reports describing glomerulosclerosis in very obese patients have been published, but this entity was considered as rare and rather bizarre. However, recent studies showed a dramatic increase of histological proven renal disease in obese patients in the absence of diabetes. In addition, obesity is an important risk factor for the progression of primary renal diseases and also plays a pivotal role in influencing graft function after renal transplantation.

The present volume is the first comprehensive contribution dedicated to this important topic. It brings together pathophysiological concepts describing how obesity influences renal structure and function, reviews the epidemiology of the problem and provides therapeutic suggestions. A total of 38 authors from 9 countries have contributed to this book presenting a truly international endeavor. I hope that this volume with state-of-art reviews will be a stimulating starting point for further research in this field, but may also help the clinical physician to recognize the problem and initiate appropriate therapeutic measures. Although Shakespeare was right that lean men could act dangerous, fat men certainly live dangerous, partially because of kidney disease.

Professor Dr. med. Gunter Wolf

Reference

1. Weisinger JR, Kempson RL, Eldrige FL, Swenson RS: The nephrotic syndrome: a complication of massive obesity. Ann Intern Med 1974;81:440–447.

Obesity and Chronic Kidney Disease

Holly Kramer

Loyola University Medical Center, Maywood, Ill., USA

Abstract

Background/Aims: The prevalence of obesity among U.S. adults has doubled within the past two decades, and if trends continue, over one-third of U.S. adults may be obese by the year 2008. Concurrent with the rising prevalence of obesity is an epidemic of chronic kidney disease (CKD) with an estimated 18 million U.S. adults currently affected. This review discusses the strong and consistent association between CKD risk and increasing body mass index noted in several observational studies. Potential mechanisms for obesity's role in the development and progression of CKD and secondary focal segmental glomerulosclerosis are also discussed. **Methods:** Literature review. **Results:** Although obesity is an important risk factor for diabetes and hypertension, the two most common etiologies of kidney failure, obesity itself may increase CKD risk by increasing the metabolic demands on the kidney, which leads to higher glomerular capillary pressures and glomerular hypertrophy. The hyperinsulinemia frequently linked with obesity may also accelerate structural damage by interacting with angiotensin II and increasing collagen production and deposition. The histologic changes in the kidney noted in some obese, especially morbidly obese, adults frequently mimic those changes associated with secondary focal segmental glomerulosclerosis, which may occur in disease states such as severely reduced nephron mass and hemodynamic stress. Given the presence of genetic susceptibility and/or reduced nephron mass, obesity may potentiate the development and progression of secondary focal segmental glomerulosclerosis. **Conclusions:** Obesity is an important risk factor for CKD. Treatment plans for obese adults with CKD should include weight loss and exercise because these interventions may simultaneously reduce the metabolic demands on the kidney, lower systemic and glomerular pressures, and improve insulin sensitivity. However, more studies are needed to further optimize the treatment and prevention of CKD associated with obesity.

Copyright © 2006 S. Karger AG, Basel

Epidemiology of Obesity

The U.S. and most other developed countries are facing a growing epidemic of obesity. Within the past 20 years, the prevalence of obesity among U.S. adults

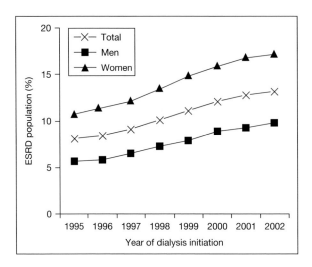

Fig. 1. Prevalence of obesity among incident dialysis patients by year of dialysis initiation. Adapted from reference [6].

has doubled from 15 to 30.5% [1]. If trends continue as projected, 39% of U.S. adults may be obese by the year 2008 [2]. Data from the 1999 to 2000 National Health and Nutrition Examination Survey show that approximately two-thirds of U.S. adults are now overweight or obese [1], a substantial increase from the 56% prevalence of overweight and obesity noted in the survey conducted during 1988–1994 period. The number of morbidly obese U.S. adults, defined as a body mass index (BMI) $\geq 40 \, kg/m^2$ has quadrupled from 1 in 200 adults to 1 in 50 [3]. The prevalence of obesity worldwide has increased sharply from approximately 200 million adults in 1995 to over 300 million adults in 2000 [4]. Currently, the number of overweight or obese adults worldwide exceeds one billion [4]. Among developing nations, overnutrition now frequently coexists with undernutrition, especially in urban areas [5], and over 115 million people in developing countries now face obesity-related health problems [4].

Epidemiology of CKD in Overweight and Obese Adults

Evidence of the obesity epidemic may be witnessed in the changing demographics of the U.S. incident dialysis population. From 1995 to 2002, the rate of increase in mean BMI among the incident end-stage renal disease (ESRD) population was approximately 2-fold higher compared to the U.S. population and these findings were consistent across all age groups [6]. The increasing prevalence of obesity among incident U.S. dialysis patients is shown in figure 1.

In 2002, 13% of all patients initiating dialysis had a BMI $\geq 35\,kg/m^2$, an obesity level which may preclude transplantation. By 2007, this may increase to 18% [6]. Currently, 60% of all ESRD patients who receive a kidney transplant are either overweight or obese at the time of transplantation [7]. Not only is the average body habitus of the ESRD population expanding, but the total number of prevalent ESRD patients in the U.S. is expected to increase by 70% over the next decade exceeding 700,000 patients by the year 2015 [8].

In order to address the expanding number of patients with kidney disease, the National Kidney Foundation created guidelines for the detection and assessment of chronic kidney disease (CKD) [9]. These guidelines define CKD as the presence of kidney damage and/or a glomerular filtration rate (GFR) $< 60\,ml/min/1.73\,m^2$ body surface area for ≥ 3 months. Kidney damage may be indicated by increased urine albumin excretion, histological changes or abnormalities in the urine sediment and/or imaging tests. CKD has been divided into 5 stages depending on the estimated GFR. Stages 1 and 2 are defined as normal ($\geq 90\,ml/min/1.73\,m^2$) or mildly decreased ($60-89\,ml/min/1.73\,m^2$) GFR, respectively, given the presence of other markers of kidney damage such as increased urine albumin excretion. Stages 3–5 are classified as a GFR $< 60\,ml/min/1.73\,m^2$ regardless of other markers of kidney damage. Stage 5 includes patients with kidney failure or GFR < 15 [9].

The majority of information on CKD prevalence in the U.S. is based on data from national multistage complex probability health surveys called the National Health and Nutrition Examination Surveys (NHANES). These surveys provide estimates of health indicators and disease prevalence as if the entire non-institutionalized population had been surveyed. NHANES participants provide blood and spot urine samples which are analyzed for serum creatinine and albumin/creatinine ratios, respectively. GFR is then estimated using the modified Modification of Diet in Renal Disease (MDRD) GFR prediction formula [10] after calibrating serum creatinine values with the MDRD laboratory. Prevalence estimates of stages 1 and 2 CKD are based on persistent albuminuria (spot urine albumin/creatinine ratio $\geq 30\,mg/g$) in the setting of a GFR $\geq 60\,ml/min/1.73\,m^2$. Due to sampling estimates, the number of adults with stage 5 CKD is extremely small and unreliable; thus, survey estimates focus on stages 1–4 CKD.

According to the NHANES completed during 1999–2000, stage 3–4 CKD (estimated GFR $15-59\,ml/min/1.73\,m^2$) is present among 3.8% of U.S. adults with a population estimate of seven million [11]. An additional 11 million U.S. adults have stages 1–2 CKD [11]. CKD prevalence is substantially higher among adults with hypertension or diabetes compared to healthy adults. In fact, only 1.4% of adults without hypertension or diabetes have stages 3–4 CKD. Prevalence of CKD also varies by BMI categories. For example, prevalence of GFR $< 60\,ml/min/1.73\,m^2$ increases from 2.9% among adults with an ideal BMI

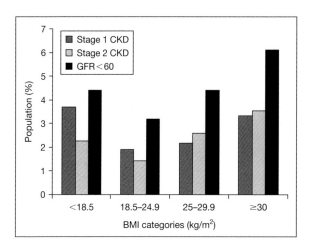

Fig. 2. Prevalence of CKD by BMI categories. National Health and Nutrition Examination Survey 1999–2000. Data are age adjusted for U.S. 2000 census. Unpublished data.

(18.5–24.9 kg/m^2) to 4.5% among obese adults (BMI ≥ 30 kg/m^2) [12]. Even after age adjustment, the prevalence of GFR < 60 ml/min/1.73 m^2 remains higher among obese adults compared to adults with an ideal BMI (fig. 2).

Using data from the Hypertension Detection and Follow-Up Program (HDFP), we tested the hypothesis that overweight and obesity are associated with incident CKD in hypertensive adults [13]. The HDFP is a multi-center, 5 year randomized trial comparing stepped care (SC) vs. referred care (RC) for the treatment of hypertension in approximately 10,000 White and African-American adults. Serum and spot urine samples were collected at baseline and at year five. CKD was defined as the presence of ≥1+ proteinuria on routine urinalysis and/or an estimated GFR < 60 ml/min/1.73 m^2 using the MDRD GFR prediction equation. After excluding participants with CKD at baseline, overweight (OR: 1.21; 95% CI: 1.05, 1.41) and obesity (OR: 1.40; 95% CI: 1.20, 1.63) were both associated with increased odds of incident CKD at year five after adjustment for covariates including diabetes mellitus (DM), blood pressure and hypertension treatment. Results did not change substantially after exclusion of participants with diabetes [13].

Similar results were noted in a cohort of approximately 11,000 healthy male physicians followed for 14 years in the Physicians Health Study, a randomized trial of aspirin and beta-carotene for primary prevention of cardiovascular disease and cancer completed in the 1980s [14]. CKD in this study was defined as an estimated GFR < 60 ml/min/1.73 m^2 using the MDRD GFR prediction equation. Participants in the highest BMI quintile (>26.6 kg/m^2) at baseline had a 45% increased odds (95% CI: 1.19, 1.76) of CKD at study end compared to participants in the lowest quintile (BMI < 22.7 kg/m^2) after adjustment for age, smoking, exercise, alcohol consumption, and family history of premature cardiovascular disease [14]. The association between BMI and risk of CKD was not modified by physical activity, but further adjustment for baseline hypertension and diabetes decreased the odds of CKD to 1.26 (95% CI: 1.03, 1.54) [14]. A population based prospective study of Japanese adults without baseline proteinuria or renal insufficiency, defined as a serum creatinine >1.2 and >1.0 mg/dl in men and women, respectively, reported a 45% increased risk (95% CI: 1.13–1.86) of developing proteinuria over a 2-year period in obese adults compared to lean adults after adjustment for hypertension and DM [15].

Few studies have examined whether obesity increases risk of ESRD due to the large sample size needed to study this uncommon outcome. However, a study by Hsu et al. showed a very strong biologic gradient whereby increasing levels of BMI were associated with increasing risk of ESRD. The study cohort consisted of 320,252 adults who volunteered for a screening health checkup between 1964 and 1985 and were followed until death, ESRD or December 31, 2000. Cases of ESRD including transplantation, hemodialysis and peritoneal dialysis were ascertained by linking the Kaiser Permanente data with the U.S. Renal Data System registry. Compared to patients with an ideal BMI of (18.5–24.9 kg/m^2), relative risk of ESRD was 3.57 (95% CI: 3.05, 4.18) for those with BMI of 30–34.9 kg/m^2, 6.12 (95% CI: 4.97, 7.54) for those with BMI 35–39.9 kg/m^2 and 7.07 (95% CI: 5.37, 9.31) for those with BMI $\geq$ 40 kg/m^2. Controlling for baseline blood pressure and presence of diabetes did attenuate the associations but the strong gradient between increasing body size and ESRD risk remained. Among 100,753 Japanese adults followed for 17 years, the cumulative incidence of ESRD was over 2-fold higher in adults with baseline BMI $\geq$ 25.5 kg/m^2 compared to adults with baseline BMI < 21.0 kg/m^2 [16]. However, after stratification by sex, the association between BMI and risk of ESRD was only noted in the men.

Obesity Is a Risk Factor for CKD Risk Factors

Obesity is a well-recognized risk factor for diabetes [17–19] and hypertension [20–22], thus, the global obesity epidemic translates into substantially heightened CKD risk factors worldwide. In fact, the number of adults with diabetes worldwide is expected to exceed 300 million by the year 2030 [4]. According to the World Health Organization, over half of the 177 million cases of DM worldwide in 2000 may be attributed to overweight and obesity [4].

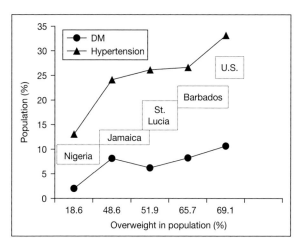

Fig. 3. Prevalence of diabetes and hypertension across the African diaspora by prevalence of overweight in the respective countries. Adapted from references [25, 26].

Among U.S. women, the prevalence of type 2 diabetes ranges from 2.4% with an ideal BMI to 7.1% in the overweight to almost 20% in the morbidly obese [23]. Prevalence rates of type 2 diabetes among men also increase steeply across BMI categories as well, but remain below levels noted among women [23]. Diabetes and hypertension remain the two most common etiologies of ESRD accounting for approximately 45 and 25% of all prevalent ESRD cases [24]. Incidence of ESRD secondary to diabetes has grown substantially over the past two decades with rates increasing from 12.5 per million population in 1980 to 146.8 per million population in 2000 [24]. Although incidence rates appear to have plateaued over the past several years [24], it is doubtful we will see rates of diabetic kidney failure decline substantially anytime within the next decade.

In the U.S., prevalence of hypertension, diabetes and kidney disease are substantially higher among African-Americans and Native Americans compared to whites, suggesting the role of genetic factors. However, international studies of populations of African origin have shown a strong gradient of hypertension and diabetes prevalence across the African Diaspora. This spectrum of risk is positively correlated with BMI and varies consistently with environmental factors, especially obesity and intake of sodium and potassium [25, 26] (fig. 3). Similar findings have been found among Pima Indians with diabetes prevalence 6-fold higher among those living in the U.S. compared to those living in Mexico [27]. Higher blood pressures are also noted among U.S. Pima Indians compared to Mexican Pima Indians [27]. Most of these differences can be explained by diet, physical activity and obesity, which is 4-fold more prevalent among Pima Indians living in the U.S. compared to those living in Mexico [27]. These findings do not rule out genetic susceptibility among these populations, but they do demonstrate the importance of environmental factors, especially obesity, in the development of CKD risk.

Obesity, especially abdominal obesity, not only increases the risk of hypertension, but also makes hypertension more resistant to treatment [28]. The higher blood pressures associated with overweight and obesity are probably due to multiple factors and include activation of the sympathetic nervous and renin–angiotensin systems [19, 29], increased serum leptin levels [30–33], volume expansion [19, 29], and sleep apnea [34, 35]. Uncontrolled hypertension in obese adults may certainly accelerate loss of kidney function over time, especially when compounded by the additional CKD risks, which accompany obesity as detailed below.

Obesity as an Independent Risk Factor for CKD

Obesity may be the number one preventable risk factor for CKD due to its strong link with diabetes and hypertension, the two primary causes of CKD and kidney failure. Aside from its link with traditional CKD risk factors, obesity itself may increase susceptibility to CKD via several potential mechanisms. The fact that central adiposity modifies the association between overweight and obesity and measures of CKD point to the potential role of the metabolic syndrome [36–39]. The Prevention of Renal and Vascular End-Stage Disease (PREVEND) study collected two 24-hour urine samples and one serum sample from 8,592 adults between the ages of 28 and 75 living in Groningen, The Netherlands who attended an outpatient screening program. The prevalence of microalbuminuria, defined as an average urine albumin excretion $\geq$30 mg/24 h, not only increased with increasing BMI, but was also higher in adults with central adiposity (waist to hip ratio $\geq$0.9 and $\geq$0.8 in men and women, respectively) in all BMI categories [40]. After adjusting for confounders such as age, sex, blood pressure and medication use, central adiposity among obese adults was associated with a 70% increased odds of microalbuminuria compared to lean participants. Odds of reduced creatinine clearance increased by 2.6-fold among obese adults with central adiposity compared to lean adults after adjustment for confounders [40].

Central adiposity is frequently accompanied not only by hypertension but also hypertriglyceridemia, low HDL cholesterol, inflammation, and a prothrombotic state. These metabolic changes reflect a state of insulin resistance and its interaction with obesity [41]. The clustering of these traits defines the metabolic syndrome which is delineated by the presence of three of the

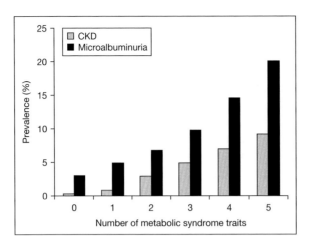

Fig. 4. Prevalence of CKD (estimated GFR < 60 ml/min/1.73) and microalbuminuria (spot urine albumin/creatinine ratio ≥ 30 mg/g) by number of metabolic syndrome traits in the non-diabetic U.S. population. Adapted from reference [43].

following five traits: abdominal obesity, impaired fasting glucose, hypertension, hypertriglyceridemia, and low HDL cholesterol [42]. Cross-sectional studies have documented a strong and consistent association between the metabolic syndrome and measures of CKD [43–45]. For example, in a population survey of non-diabetic Native Americans participating in the Inter-Tribal Heart Project, the odds of microalbuminuria increased by 80% in the presence of one metabolic syndrome trait to 230% in the presence of 3 or more traits compared to no traits [44]. Similar results were reported in a study using data from a nationally representative sample of U.S. adults without DM [43]. Compared to presence of zero–one metabolic syndrome trait, presence of two metabolic syndrome traits was associated with a 2-fold increased odds of CKD, defined as an estimated GFR < 60 ml/min/1.73 m^2. Presence of five metabolic syndrome traits was associated with a 5-fold increased odds of CKD compared to presence of zero–one trait [43]. Prevalence of microalbuminuria was also positively correlated with number of metabolic syndrome traits (fig. 4) [43].

Metabolic syndrome represents hyperinsulinemia which may lead to structural changes via mechanisms not yet established but several theories have been postulated [46]. Renal mesangial cells cultured in the presence of insulin demonstrate increased synthesis of type 1 and III collagen [47, 48]. In dogs fed a high fat diet for 7 weeks with 64% increases in body weight, plasma renin activity and insulin concentrations increased by over 2-fold. Histologic kidney specimens showed mesangial expansion in obese but not lean dogs. No difference was noted in glomerulosclerosis score between obese and lean dogs, but specimens from obese dogs exhibited 3-fold higher TGF-β1 staining compared to lean dogs [49]. The heightened activity of TGF-β1 may be due to the additive effects of insulin and angiontensin II [50], which could potentially promote production of extracellular matrix and accelerate glomerular damage [51–55]. Leptin levels, which are strongly and positively correlated with degree of adiposity, may also interact with TGF-β1 and amplify its effects leading to increased collagen production [33].

Increased metabolic demands on the kidney may also mediate increased CKD risk in overweight and obesity. Given that nephron number is fixed at birth, weight gain increases the work of each individual nephron. Body size is positively correlated with glomerular size [56], and studies have demonstrated higher GFR and effective renal plasma flow (ERPF) in overweight and obese adults compared to lean adults [57–60]. These differences in GFR and ERPF are eliminated after correction for body size [57, 59, 61, 62]. Ribstein et al. [58] estimated GFR and ERPF by measuring clearances of technetium-labeled diethylene triaminopentaacetic acid and ^{131}iodine-orthoiodohippurate, respectively, in 40 normotensive adults and 80 never-treated hypertensives matched for age and sex. After stratifying normotensive and hypertensive study participants by presence of overweight, defined in this study as a BMI > 27 kg/m^2, GFR and ERPF were higher among overweight participants compared to lean participants but not after normalization for BSA. Higher filtration fraction (GFR/ERPF) was noted among hypertensive overweight participants compared to hypertensive lean participants, while no difference in filtration fraction was noted between overweight and lean participants who were normotensive [58].

Increased filtration fraction in overweight and obese individuals in the presence of increased blood flow suggests the potential existence of increased glomerular capillary pressure. As an individual gains weight, single nephron GFR must increase and this occurs at the expense of increased capillary pressures. Significant increases in GFR and renal plasma flow along with higher kidney weight and expansion of Bowman's capsule have been documented in dogs fed a high fat diet and gain weight compared to lean dogs [49]. Because increases in arterial blood pressure accompany the increases in renal plasma flow with weight gain, these hemodynamic changes result in increased glomerular capillary pressures. Experimental animal models have demonstrated that both glomerular hypertrophy and increased glomerular capillary pressures are risk factors for glomerulosclerosis [63–67]. In fact, acute increases in glomerular volume and capillary pressures are associated with increased TGF-β1 expression [68, 69]. As discussed earlier, increased TGF-β1 expression increases matrix production and deposition. The glomerular hypertrophy itself may also predispose the glomerular capillary wall to hemodynamic injury because capillary wall tension is a direct function of its diameter, which could potentially increase with glomerular hypertrophy [70]. Because podocytes do not undergo adaptive replication in the setting of glomerular hypertrophy, the density of podocytes for a given glomerular surface area decreases. These podocytes then enlarge to compensate for the reduced density potentially leading to structural changes and detachment of the foot processes from the basement membrane [71]. Thus, in the setting of obesity, there are multiple stimulants of TGF-β1, including increased insulin and angiotensin II levels and heightened glomerular volume and capillary pressures. These factors all interact and promote structural changes and glomerular damage.

Individuals with reduced nephron mass possess a high risk for CKD in the setting of overweight and obesity. The compensatory glomerular hypertrophy among individuals with reduced nephron mass is compounded by the increased metabolic load imposed on the kidney by overweight and obesity. These individuals are also at high risk for subsequent development of other CKD risk factors including hypertension [72, 73] and diabetes [74]. However, this cascade of CKD risk factors can be avoided if these individuals with reduced nephron mass maintain an ideal body weight.

Obesity and Focal Segmental Glomerulosclerosis

Obesity, especially morbid obesity, has been linked with focal segmental glomerulosclerosis (FSGS) since the 1970s when case reports of morbidly obese adults with proteinuria were first published [75, 76]. Several subsequent reports documented FSGS in patients with morbid obesity and sleep apnea [77, 78]. Investigators have also noted differences in the clinical and histologic features between patients with idiopathic FSGS and obese (particularly morbidly obese) patients with FSGS [70, 79]. Obese patients with FSGS frequently have higher serum albumin levels and less proteinuria and edema compared to non-obese patients with FSGS [78, 79]. Histologic differences most notably include larger glomerular size among the obese patients [79]. Heavy proteinuria has also been documented in morbidly obese adults in the absence of other glomerular abnormalities aside from glomerulomegaly [80, 81], although sampling error must be considered.

Obesity alone does not appear to be the sole mediator of the development of glomerulosclerosis as body size does not appear to predict percent sclerosed glomeruli [56]. However, body size is positively and directly correlated with glomerular size [56], and some investigators view glomerulomegaly accompanied by obesity and proteinuria as a specific disease entity termed obesity-related glomerulopathy [81]. In a single center study of 6,818 kidney biopsies collected over a 14 year period, obesity-related glomerulopathy was defined as presence of glomerulomegaly and/or FSGS in a patient with a BMI $\geq 30\,\text{kg/m}^2$

in the absence of other primary and secondary causes of FSGS such as HIV, heroin abuse, and reduced renal mass. All patients who fit these criteria were referred for kidney biopsy due to proteinuria [81]. Glomerulomegaly was noted in 10% of idiopathic FSGS cases vs. 100% of obesity associated FSGS cases. Foot process effacement was also significantly lower among patients with obesity associated FSGS vs. idiopathic FSGS [81]. The histologic differences may not necessarily indicate a new disease but rather evidence of secondary FSGS in these adults with glomerular hypertrophy, obesity and proteinuria.

Glomerular hypertrophy frequently antedates development of several experimental models of glomerular capillary hypertension such as the 5/6 nephrectomy animal model [65] and diabetic glomerulosclerosis [70, 71, 82]. In these models, structural alterations include increase in glomerular size, mesangial matrix expansion, and increased basement membrane thickness [71]. These same structural changes were noted in dogs fed a high fat diet for 7–9 weeks [49]. Histologic changes in some obese adults with FSGS may actually reflect structural damage due to increased glomerular capillary pressures and other factors which culminate in glomerulosclerosis and secondary FSGS. Secondary FSGS is the end result of adaptations to reduced functioning nephron population (e.g. primary kidney disease, reduced kidney mass from congenital or surgical means) [83–85] or hemodynamic stresses such as cyanotic congenital heart disease [86], pulmonary hypertension [87] or morbid obesity [56, 76, 78] accompanied by sleep apnea [77]. Clinically, secondary FSGS is frequently marked by absence of hypoalbuminemia, edema and high lipid levels and may be differentiated histologically from primary FSGS by increased diameter of non-sclerosed glomeruli, mild segmental foot process fusion over non-sclerosed segments and less visceral cell hypertrophy [70, 71].

Obesity itself is unlikely to be the sole mediator of secondary FSGS, given the rarity of this disease in the general population contrasted with the high prevalence of obesity. However, within a background of genetic susceptibility and perhaps other clinical risk factors including sleep apnea and reduced nephron number, obesity could potentiate development of secondary FSGS. These changes may begin with increased GFR and renal plasma flow in order to meet the increased metabolic demands associated with obesity. Glomerular capillary pressures and glomerular volume increase leading to a cascade of growth factors including TGF-β1 and angiotensin II and subsequent increased matrix production and deposition. The hemodynamic changes associated with obesity damage visceral glomerular epithelial cells, which do not undergo cell division in this setting [71]. The same number of podocytes must now cover a larger surface area due to glomerular hypertrophy. Foot processes eventually become distorted and detached leading to increased hydraulic conductivity and proteinuria [71, 88]. These structural abnormalities may be compounded by insulin resistance and hypertension [19]. Thus, obesity should be considered a 'high-risk state' for development of CKD and secondary FSGS as it represents the interaction of hemodynamic and metabolic abnormalities, which may lead to structural damage, matrix deposition and eventual glomerulosclerosis.

Interventions

Hypertension treatment with a blood pressure goal <130/80 mm Hg remains one of the most important interventions for any patient with CKD. Lifestyle modifications including maintaining a normal body weight, engaging in regular physical exercise and following a diet low in fat and sodium and high in potassium [89], should be included in all treatment plans. Among obese patients, weight loss should be viewed as an indispensable treatment for the prevention and treatment of CKD. Serial direct GFR measurements in normotensive morbidly obese adults who underwent gastric bypass surgery show that even moderate weight loss decreases GFR and renal plasma flow [61]. Among eight morbidly obese adults without hypertension (mean BMI $48 \pm 2.4\,kg/m^2$), directly measured GFR decreased from 145 ± 14 to $101 \pm 4\,ml/min$ ($p < 0.01$), while renal plasma flow decreased from 803 ± 39 to $698 \pm 42\,ml/min$ ($p < 0.02$). Urine albumin excretion decreased from 16 to 5 µg/min [61]. These changes were noted despite the fact mean BMI remained at obesity levels ($>30\,kg/m^2$) after weight loss. Thus, an obese individual who loses even a moderate amount of weight will dramatically reduce the work of the kidney, and perhaps decrease his/her risk of CKD. Weight loss may also be beneficial in obese patients with established CKD. In a small study of 30 patients with a BMI $> 27\,kg/m^2$ with established CKD and proteinuria due to either diabetic or non-diabetic causes, a 4.1% weight loss in the absence of oral protein restriction was associated with significant decreases in proteinuria from 2.8 to 1.9 g/24 h ($p < 0.005$) after 5 months. Patients assigned to the control group actually gained weight and showed increases in proteinuria during the 5-month period [90].

The renal benefits of weight loss in obese patients may be mediated not only by reduced systemic and glomerular [71] pressures but also by ameliorating insulin resistance. In the Diabetes Prevention Program Randomized Trial, lifestyle intervention, which aimed for a 7% weight loss and 150 min of exercise per week, decreased the 3-year cumulative incidence of metabolic syndrome by 25 and 40% compared to the metformin and placebo arms, respectively [91]. Compared to the baseline prevalence of metabolic syndrome, at study end the overall prevalence of metabolic syndrome actually decreased among participants in the lifestyle intervention group from 51 to 43% but increased in both the placebo and metformin arms [91]. This trial did not collect information on kidney disease measures, but it seems intuitive that prevention and regression of metabolic syndrome with weight loss and exercise would decrease risk and progression of CKD.

Aside from weight loss, medications which block the renin–angiotensin system may provide additional benefits aside from lowering blood pressure. By binding to the angiotensin I receptor, angiotensin II inhibits the recruitment and differentiation of adipocytes [92]. Lack of adipocyte differentiation promotes storage of excess calories as fat in many tissues such as the liver, muscle and pancreas [93]. Drugs which inhibit the renin–angiotensin system will improve insulin resistance by promoting the recruitment and differentiation of adipocytes [93]. Several large randomized trials have shown significantly lower incidence rates of new-onset type 2 diabetes among participants assigned to medications which block the renin–angiotensin system compared to calcium channel blockers [94, 95], diuretics [94], β-blockers [96] or placebo [97]. Although angiotensin converting enzyme inhibitors and angiotensin II receptor blockers may provide protection from development of type 2 diabetes, the most common cause of ESRD, physicians should not overlook the benefits of weight loss and exercise, which may provide even greater payback than drugs and at lower cost.

Conclusions

Obesity is the number one risk factor for CKD risk factors, especially type 2 diabetes and hypertension, and may be the number one preventable risk factor for CKD. The development of CKD is rarely a 'one-hit' phenomena and is usually the culminating result of the interaction of multiple risk factors. Obesity represents one example of a 'multi-hit' state and given the background of genetic susceptibility and/or reduced nephron number, obesity may initiate and/or accelerate kidney damage. Because weight loss has been shown to improve glomerular hemodynamics and reduce urine albumin excretion [61], obese patients with CKD should be counseled on the benefits of weight loss. We are just now stepping on the iceberg of obesity and its link with CKD. Clearly, more studies are needed on the potential renal benefits of lifestyle interventions and medications for obese patients with all forms of kidney disease.

References

1 Flegal KM, Carroll MD, Ogden CL, Johnson CL: Prevalence and trends in obesity among US adults, 1999–2000. JAMA 2002;288:1723–1727.
2 Hill JO, Wyatt HR, Reed GW, Peters JC: Obesity and the environment: where do we go from here? Science 2003;299:853–855.

3   Sturm R: Increases in clinically severe obesity in the United States, 1986–2000. Arch Intern Med 2003;163:2146–2148.
4   World Health Organization: The World Health Report 2002: Reducing Risk, Promoting Healthy Life. 2002.
5   Luke A, Cooper RS, Prewitt TE, Adeyemo AA, Forrester TE: Nutritional consequences of the African diaspora. Annu Rev Nutr 2001;21:47–71.
6   Kramer H, Saranathan A, Luke A, Durazo-Arvizu R, Hou S, Cooper R: Increasing BMI and obesity in the incident end-stage renal disease population. J Am Soc Nephrol 2006;17:1453–1459.
7   Friedman AN, Miskulin DC, Rosenberg IH, Levey AS: Demographics and trends in overweight and obesity in patients at time of kidney transplantation. Am J Kidney Dis 2003;41:480–487.
8   Gilbertson DT, Liu J, Xue JL, Louis TA, Solid CA, Ebben JP, Collins AJ: Projecting the number of patients with end-stage renal disease in the United States to the year 2015. J Am Soc Nephrol 2005;16:3736–3741.
9   National Kidney Foundation: K/DOQI clinical practice guidelines for chronic kidney disease: evaluation, classification and stratification. Part 4. Definition and classification of stages of chronic kidney disease. Am J Kidney Dis 2002;39:S72–S75.
10 Levey A, Bosch J, Breyer Lewis J, Greene T, Rogers N, Roth D: A more accurate method to estimate glomerular filtration rate from serum creatinine. Ann Intern Med 1999;139:461–470.
11 Coresh J, Byrd-Holt D, Astor BC, Briggs JP, Eggers PW, Lacher DA, Hostetter TH: Chronic kidney disease awareness, prevalence, and trends among U.S. adults, 1999 to 2000. J Am Soc Nephrol 2005;16:180–188.
12 National Health and Nutrition Examination Survey 1999–2000. U.S. Department of Health and Human Services. Centers for Disease Control and Prevention. National Center for Health Statistics, Hyattsville, MD, 1999–2000.
13 Kramer H, Luke A, Bidani A, Cao G, Cooper R, McGee D: Obesity and prevalent and incident chronic kidney disease: The Hypertension Detection and Follow-up Program. Am J Kidney Dis 2005;46:587–594.
14 Gelber RP, Kurth T, Kausz AT, Manson JE, Buring JE, Levey AS, Gaziano JM: Association between body mass index and CKD in apparently healthy men. Am J Kidney Dis 2005;46:871–880.
15 Tozawa M, Iseki K, Iseki C, Oshiro S, Ikemiya Y, Takishita S: Influence of smoking and obesity on the development of proteinuria. Kidney Int 2002;62:956–962.
16 Iseki K, Ikemiya Y, Kinjo K, Inoue T, Iseki C, Takishita S: Body mass index and the risk of development of end-stage renal disease in a screened cohort. Kidney Int 2004;65:1870–1876.
17 Chan JM, Rimm EB, Colditz GA, Stampfer MJ, Willett WC: Obesity, fat distribution, and weight gain as risk factors for clinical diabetes in men. Diabetes Care 1994;17:961–969.
18 He J, Whelton PK, Appel LJ, Charleston J, Klag MJ: Long-term effects of weight loss and dietary sodium reduction on incidence of hypertension. Hypertension 2000;35:544–549.
19 Hall JE, Kuo JJ, da Silva AA, de Paula RB, Liu J, Tallam L: Obesity-associated hypertension and kidney disease. Curr Opin Nephrol Hypertens 2003;12:195–200.
20 Liu K, Ruth KJ, Flack JM, Jones-Webb R, Burke G, Savage PJ, Hulley SB: Blood pressure in young blacks and whites: relevance of obesity and lifestyle factors in determining differences. The CARDIA Study. Coronary Artery Risk Development in Young Adults. Circulation 1996;93:60–66.
21 Wilson PW, D'Agostino RB, Sullivan L, Parise H, Kannel WB: Overweight and obesity as determinants of cardiovascular risk: the Framingham experience. Arch Intern Med 2002;162:1867–1872.
22 Kaufman JS, Durazo-Arvizu RA, Rotimi CN, McGee DL, Cooper RS: Obesity and hypertension prevalence in populations of African origin. The Investigators of the International Collaborative Study on Hypertension in Blacks. Epidemiology 1996;7:398–405.
23 Must A, Spadano J, Coakley EH, Field AE, Colditz G, Dietz WH: The disease burden associated with overweight and obesity. JAMA 1999;282:1523–1529.
24 U.S. Renal Data System, USRDS 2005 Annual Data Report: Atlas of End-Stage Renal Disease in the United States, National Institutes of Health, National Institute of Diabetes and Digestive and Kidney Diseases, Bethesda, MD., 2005.
25 Cooper R, Rotimi C, Ataman S, McGee D, Osotimehin B, Kadiri S, Muna W, Kingue S, Fraser H, Forrester T, et al: The prevalence of hypertension in seven populations of west African origin. Am J Public Health 1997;87:160–168.

26 Cooper RS, Rotimi CN, Kaufman JS, Owoaje EE, Fraser H, Forrester T, Wilks R, Riste LK, Cruickshank JK: Prevalence of NIDDM among populations of the African diaspora. Diabetes Care 1997;20:343–348.
27 Valencia ME, Weil EJ, Nelson RG, Esparza J, Schulz LO, Ravussin E, Bennett PH: Impact of lifestyle on prevalence of kidney disease in Pima Indians in Mexico and the United States. Kidney Int Suppl 2005:S141–S144.
28 Sharma AM, Engeli S: Managing big issues on lean evidence: treating obesity hypertension. Nephrol Dial Transplant 2002;17:353–355.
29 Hall JE, Brands MW, Dixon WN, Smith MJ Jr: Obesity-induced hypertension. Renal function and systemic hemodynamics. Hypertension 1993;22:292–299.
30 Haynes WG: Role of leptin in obesity-related hypertension. Exp Physiol 2005;90:683–688.
31 Hall JE, Hildebrandt DA, Kuo J: Obesity hypertension: role of leptin and sympathetic nervous system. Am J Hypertens 2001;14:103S–115S.
32 Aizawa-Abe M, Ogawa Y, Masuzaki H, Ebihara K, Satoh N, Iwai H, Matsuoka N, Hayashi T, Hosoda K, Inoue G, et al: Pathophysiological role of leptin in obesity-related hypertension. J Clin Invest 2000;105:1243–1252.
33 Wolf G, Chen S, Han DC, Ziyadeh FN: Leptin and renal disease. Am J Kidney Dis 2002;39:1–11.
34 Peppard PE, Young T, Palta M, Skatrud J: Prospective study of the association between sleep-disordered breathing and hypertension. N Engl J Med 2000;342:1378–1384.
35 Nieto FJ, Young TB, Lind BK, Shahar E, Samet JM, Redline S, D'Agostino RB, Newman AB, Lebowitz MD, Pickering TG: Association of sleep-disordered breathing, sleep apnea, and hypertension in a large community-based study. Sleep Heart Health Study. JAMA 2000;283:1829–1836.
36 Cirillo M, Senigalliesi L, Laurenzi M, Alfieri R, Stamler J, Stamler R, Panarelli W, De Santo NG: Microalbuminuria in nondiabetic adults: relation of blood pressure, body mass index, plasma cholesterol levels, and smoking: The Gubbio Population Study. Arch Intern Med 1998;158: 1933–1939.
37 Liese AD, Hense HW, Doring A, Stieber J, Keil U: Microalbuminuria, central adiposity and hypertension in the non-diabetic urban population of the MONICA Augsburg survey 1994/95. J Hum Hypertens 2001;15:799–804.
38 Kim YI, Kim CH, Choi CS, Chung YE, Lee MS, Lee SI, Park JY, Hong SK, Lee KU: Microalbuminuria is associated with the insulin resistance syndrome independent of hypertension and type 2 diabetes in the Korean population. Diabetes Res Clin Pract 2001;52:145–152.
39 Metcalf P, Baker J, Scott A, Wild C, Scragg R, Dryson E: Albuminuria in people at least 40 years old: effect of obesity, hypertension, and hyperlipidemia. Clin Chem 1992;38:1802–1808.
40 Pinto-Sietsma SJ, Navis G, Janssen WM, de Zeeuw D, Gans RO, de Jong PE: A central body fat distribution is related to renal function impairment, even in lean subjects. Am J Kidney Dis 2003;41:733–741.
41 Grundy SM: Obesity, metabolic syndrome, and cardiovascular disease. J Clin Endocrinol Metab 2004;89:2595–2600.
42 Third Report of the National Cholesterol Education Program (NCEP): Expert Panel on Detection, Evaluation, and Treatment of High Blood Cholesterol in Adults (Adult Treatment Panel III) final report. Circulation 2002;106:3143–3421.
43 Chen J, Muntner P, Hamm LL, Jones DW, Batuman V, Fonseca V, Whelton PK, He J: The metabolic syndrome and chronic kidney disease in U.S. adults. Ann Intern Med 2004;140:167–174.
44 Hoehner CM, Greenlund KJ, Rith-Najarian S, Casper ML, McClellan WM: Association of the insulin resistance syndrome and microalbuminuria among nondiabetic native Americans. The Inter-Tribal Heart Project. J Am Soc Nephrol 2002;13:1626–1634.
45 Chen J, Muntner P, Hamm LL, Fonseca V, Batuman V, Whelton PK, He J: Insulin resistance and risk of chronic kidney disease in nondiabetic US adults. J Am Soc Nephrol 2003;14:469–477.
46 Bagby SP: Obesity-initiated metabolic syndrome and the kidney: a recipe for chronic kidney disease? J Am Soc Nephrol 2004;15:2775–2791.
47 Abrass CK, Spicer D, Raugi GJ: Induction of nodular sclerosis by insulin in rat mesangial cells in vitro: studies of collagen. Kidney Int 1995;47:25–37.
48 Berfield AK, Raugi GJ, Abrass CK: Insulin induces rapid and specific rearrangement of the cytoskeleton of rat mesangial cells in vitro. J Histochem Cytochem 1996;44:91–101.

49 Henegar JR, Bigler SA, Henegar LK, Tyagi SC, Hall JE: Functional and structural changes in the kidney in the early stages of obesity. J Am Soc Nephrol 2001;12:1211–1217.
50 Anderson PW, Zhang XY, Tian J, Correale JD, Xi XP, Yang D, Graf K, Law RE, Hsueh WA: Insulin and angiotensin II are additive in stimulating TGF-beta 1 and matrix mRNAs in mesangial cells. Kidney Int 1996;50:745–753.
51 Liboi E, Di Francesco P, Gallinari P, Testa U, Rossi GB, Peschle C: TGF beta induces a sustained c-fos expression associated with stimulation or inhibition of cell growth in EL2 or NIH 3T3 fibroblasts. Biochem Biophys Res Commun 1988;151:298–305.
52 Yamamoto T, Noble NA, Miller DE, Border WA: Sustained expression of TGF-beta 1 underlies development of progressive kidney fibrosis. Kidney Int 1994;45:916–927.
53 Park IS, Kiyomoto H, Abboud SL, Abboud HE: Expression of transforming growth factor-beta and type IV collagen in early streptozotocin-induced diabetes. Diabetes 1997;46:473–480.
54 Sharma K, Ziyadeh FN: The emerging role of transforming growth factor-beta in kidney diseases. Am J Physiol 1994;266:F829–F842.
55 Kagami S, Border WA, Miller DE, Noble NA: Angiotensin II stimulates extracellular matrix protein synthesis through induction of transforming growth factor-beta expression in rat glomerular mesangial cells. J Clin Invest 1994;93:2431–2437.
56 Kasiske B, Napier J: Glomerular sclerosis in patients with massive obesity. Am J Nephrology 1985;5:45–50.
57 Chagnac A, Weinstein T, Korzets A, Ramadan E, Hirsch J, Gafter U: Glomerular hemodynamics in severe obesity. Am J Physiol Renal Physiol 2000;278:F817–F822.
58 Ribstein J, du Cailar G, Mimran A: Combined renal effects of overweight and hypertension. Hypertension 1995;26:610–615.
59 Porter LE, Hollenberg NK: Obesity, salt intake, and renal perfusion in healthy humans. Hypertension 1998;32:144–148.
60 Reisin E, Messerli FG, Ventura HO, Frohlich ED: Renal haemodynamic studies in obesity hypertension. J Hypertens 1987;5:397–400.
61 Chagnac A, Weinstein T, Herman M, Hirsh J, Gafter U, Ori Y: The effects of weight loss on renal function in patients with severe obesity. J Am Soc Nephrol 2003;14:1480–1486.
62 Praga M, Hernandez E, Herrero JC, Morales E, Revilla Y, Diaz-Gonzalez R, Rodicio JL: Influence of obesity on the appearance of proteinuria and renal insufficiency after unilateral nephrectomy. Kidney Int 2000;58:2111–2118.
63 Brenner BM: Hemodynamically mediated glomerular injury and the progressive nature of kidney disease. Kidney Int 1983;23:647–655.
64 Hostetter TH, Olson JL, Rennke HG, Venkatachalam MA, Brenner BM: Hyperfiltration in remnant nephrons: a potentially adverse response to renal ablation. Am J Physiol 1981;241:F85–F93.
65 Hostetter TH, Rennke HG, Brenner BM: Compensatory renal hemodynamic injury: a final common pathway of residual nephron destruction. Am J Kidney Dis 1982;1:310–314.
66 Brenner BM, Meyer TW, Hostetter TH: Dietary protein intake and the progressive nature of kidney disease: the role of hemodynamically mediated glomerular injury in the pathogenesis of progressive glomerular sclerosis in aging, renal ablation, and intrinsic renal disease. N Engl J Med 1982;307:652–659.
67 Purkerson ML, Hoffsten PE, Klahr S: Pathogenesis of the glomerulopathy associated with renal infarction in rats. Kidney Int 1976;9:407–417.
68 Cortes P, Zhao X, Riser BL, Narins RG: Regulation of glomerular volume in normal and partially nephrectomized rats. Am J Physiol 1996;270:F356–F370.
69 Cortes P, Riser B, Narins RG: Glomerular hypertension and progressive renal disease: the interplay of mesangial cell stretch, cytokine formation and extracellular matrix synthesis. Contrib Nephrol 1996;118:229–233.
70 D'Agati V: The many masks of focal segmental glomerulosclerosis. Kidney Int 1994;46:1223–1241.
71 Rennke HG, Klein PS: Pathogenesis and significance of nonprimary focal and segmental glomerulosclerosis. Am J Kidney Dis 1989;13:443–456.
72 Woods LL, Ingelfinger JR, Nyengaard JR, Rasch R: Maternal protein restriction suppresses the newborn renin-angiotensin system and programs adult hypertension in rats. Pediatr Res 2001;49: 460–467.

73 Vehaskari VM, Aviles DH, Manning J: Prenatal programming of adult hypertension in the rat. Kidney Int 2001;59:238–245.
74 Holemans K, Aerts L, Van Assche FA: Lifetime consequences of abnormal fetal pancreatic development. J Physiol 2003;547:11–20.
75 Weisinger JR, Kempson RL, Eldridge FL, Swenson RS: The nephrotic syndrome: a complication of massive obesity. Ann Intern Med 1974;81:440–447.
76 Warnke RA, Kempson RL: The nephrotic syndrome in massive obesity: a study by light, immunofluorescence, and electron microscopy. Arch Pathol Lab Med 1978;102:431–438.
77 Jennette JC, Charles L, Grubb W: Glomerulomegaly and focal segmental glomerulosclerosis associated with obesity and sleep-apnea syndrome. Am J Kidney Dis 1987;10:470–472.
78 Kasiske BL, Crosson JT: Renal disease in patients with massive obesity. Arch Intern Med 1986;146:1105–1109.
79 Praga M, Hernandez E, Morales E, Campos AP, Valero MA, Martinez MA, Leon M: Clinical features and long-term outcome of obesity-associated focal segmental glomerulosclerosis. Nephrol Dial Transplant 2001;16:1790–1798.
80 Wesson DE, Kurtzman NA, Frommer JP: Massive obesity and nephrotic proteinuria with a normal renal biopsy. Nephron 1985;40:235–237.
81 Kambham N, Markowitz GS, Valeri AM, Lin J, D'Agati VD: Obesity-related glomerulopathy: an emerging epidemic. Kidney Int 2001;59:1498–1509.
82 Seyer-Hansen K, Hansen J, Gundersen HJ: Renal hypertrophy in experimental diabetes. A morphometric study. Diabetologia 1980;18:501–505.
83 Bohle A, Biwer E, Christensen JA: Hyperperfusion injury of the human kidney in different glomerular diseases. Am J Nephrol 1988;8:179–186.
84 Zucchelli P, Cagnoli L, Casanova S, Donini U, Pasquali S: Focal glomerulosclerosis in patients with unilateral nephrectomy. Kidney Int 1983;24:649–655.
85 McGraw M, Poucell S, Sweet J, Baumal R: The significance of focal segmental glomerulosclerosis in oligomeganephronia. Int J Pediatr Nephrol 1984;5:67–72.
86 Hida K, Wada J, Yamasaki H, Nagake Y, Zhang H, Sugiyama H, Shikata K, Makino H: Cyanotic congenital heart disease associated with glomerulomegaly and focal segmental glomerulosclerosis: remission of nephrotic syndrome with angiotensin converting enzyme inhibitor. Nephrol Dial Transplant 2002;17:144–147.
87 Faustinella F, Uzoh C, Sheikh-Hamad D, Truong LD, Olivero JJ: Glomerulomegaly and proteinuria in a patient with idiopathic pulmonary hypertension. J Am Soc Nephrol 1997;8:1966–1970.
88 Kriz W, Elger M, Nagata M, Kretzler M, Uiker S, Koeppen-Hageman I, Tenschert S, Lemley KV: The role of podocytes in the development of glomerular sclerosis. Kidney Int Suppl 1994;45: S64–S72.
89 Chobanian AV, Bakris GL, Black HR, Cushman WC, Green LA, Izzo JL Jr, Jones DW, Materson BJ, Oparil S, Wright JT Jr, et al: The Seventh Report of the Joint National Committee on Prevention, Detection, Evaluation, and Treatment of High Blood Pressure: the JNC 7 report. JAMA 2003;289:2560–2572.
90 Morales E, Valero MA, Leon M, Hernandez E, Praga M: Beneficial effects of weight loss in overweight patients with chronic proteinuric nephropathies. Am J Kidney Dis 2003;41:319–327.
91 Orchard TJ, Temprosa M, Goldberg R, Haffner S, Ratner R, Marcovina S, Fowler S: The effect of metformin and intensive lifestyle intervention on the metabolic syndrome: the Diabetes Prevention Program randomized trial. Ann Intern Med 2005;142:611–619.
92 Janke J, Engeli S, Gorzelniak K, Luft FC, Sharma AM: Mature adipocytes inhibit in vitro differentiation of human preadipocytes via angiotensin type 1 receptors. Diabetes 2002;51:1699–1707.
93 Sharma AM, Janke J, Gorzelniak K, Engeli S, Luft FC: Angiotensin blockade prevents type 2 diabetes by formation of fat cells. Hypertension 2002;40:609–611.
94 Sierra C, Ruilope LM: New-onset diabetes and antihypertensive therapy: comments on ALLHAT trial. J Renin Angiotensin Aldosterone Syst 2003;4:169–170.
95 Julius S, Kjeldsen SE, Weber M, Brunner HR, Ekman S, Hansson L, Hua T, Laragh J, McInnes GT, Mitchell L, et al: Outcomes in hypertensive patients at high cardiovascular risk treated with regimens based on valsartan or amlodipine: the VALUE randomised trial. Lancet 2004;363: 2022–2031.

96 Dahlof B, Devereux RB, Kjeldsen SE, Julius S, Beevers G, de Faire U, Fyhrquist F, Ibsen H, Kristiansson K, Lederballe-Pedersen O, et al: Cardiovascular morbidity and mortality in the Losartan Intervention For Endpoint Reduction in Hypertension Study (LIFE): a randomised trial against atenolol. Lancet 2002;359:995–1003.

97 Jacobs DR Jr, Murtaugh MA, Steffes M, Yu X, Roseman J, Goetz FC: Gender- and race-specific determination of albumin excretion rate using albumin-to-creatinine ratio in single, untimed urine specimens: the Coronary Artery Risk Development in Young Adults Study. Am J Epidemiol 2002;155:1114–1119.

Holly Kramer, MD, MPH
Loyola University Medical Center
Departments of Preventive Medicine and Medicine
Division of Nephrology and Hypertension
2160 First Avenue
Maywood, IL 60153 (USA)
Tel. +1 708 327 9039, Fax +1 708 327 9009, E-Mail HKramer@lumc.edu

Obesity and Kidney Transplantation

Titte R. Srinivas, Herwig-Ulf Meier-Kriesche

University of Florida, Gainesville, Fla., USA

Abstract

Obesity is an increasingly common comorbidity at the time of kidney transplantation. Obesity is also a frequent complication of kidney transplantation. The obese transplant recipient suffers adverse metabolic and cardiovascular consequences of obesity. These metabolic and cardiovascular sequelae of obesity are major contributors to morbidity and mortality after kidney transplantation. We discuss the epidemiology and pathophysiology of obesity in the kidney transplant recipient and provide an overview of clinical studies on obesity in kidney transplantation. An approach to obesity in renal transplantation, based on the evidence to date, is presented.

Copyright © 2006 S. Karger AG, Basel

Introduction

Obesity is an increasingly common co-morbidity in patients with chronic renal diseases with up to 60% of patients being obese or overweight at the time of transplant [1].

Obesity is also extremely common post-transplantation, the average weight gain after successful renal transplantation is about 10% in the first year [2]. The impact of obesity in the recipient on graft and patient survival after transplantation is not straightforward. Some studies show that pre-transplant obesity negatively impacts graft and patient survival, whereas others show no impact of obesity on outcomes after transplantation [3]. While certain generalizations are simple, such as pre-transplant obesity being usually followed by its persistence post-transplantation, whether pre-transplant obesity that persists after transplantation carries the same import as that developing de novo in the post-transplant period is not clear.

In this article, we discuss the impact of both pre-transplant obesity and post-transplant obesity on outcomes following renal transplantation. We shall review key short-term and long-term studies on obesity in kidney transplantation.

We offer general guidelines that may help the treating physician manage the obesity in the context of optimizing graft and patient survival.

Definitions

Obesity is a pathophysiologic state characterized by excessive adipose tissue mass. In the medical context, obesity may be defined as a state of increased body weight, more specifically adipose tissue, of sufficient magnitude to impact adversely on health [4]. The most widely accepted classification of obesity is based on the body mass index (BMI), defined as body weight in kilograms (kg) divided by height in meters squared (m^2). BMI's between 18.5 and 24.9 kg/m^2 represent normal weight, BMI's between 25 and 29.9, overweight and BMI's > 30 are classified as obesity. Large scale epidemiologic studies suggest that cause specific mortality such as cardiovascular death start rising with BMIs > 25 [5]. Much of the literature examining the relationship between BMI and mortality supports the hypothesis of a curvilinear relation where risk is highest among the most heavy and the leanest [5]. The term *overweight* is used by some to describe individuals with BMIs between 25 and 30. In general, a BMI falling between 25 and 30 must be considered medically significant and targeted for treatment especially if accompanied by significant comorbidities such as hypertension (HTN) and insulin resistance/glucose intolerance.

Abdominal obesity, defined by a waist–hip ratio >0.9 in women and >1.0 in men connotes the preponderance of intra-abdominal fat in the adiposity. This phenotype is associated with a greater prevalence of the most important complications of obesity, namely HTN, insulin resistance, diabetes, hyperlipidemia in men and women and in addition hyperandrogenism in women. Use of the waist–hip ratio to classify the state of adiposity has not been studied systematically in large numbers of transplant patients though.

Epidemiology

The prevalence of obesity continues to increase globally and in developed countries it can be as high as 20% [6]. Sedentary lifestyles, cultural and behavioral factors and dietary indiscretion are the two most commonly utilized explanations for the burgeoning epidemic of obesity [6]. In the U.S., data from the National Health and Nutrition Examination Surveys (NHANES) that the proportion of American adults with obesity (BMI > 30) has increased from 14.5% (1976–1980) to 32.2% (2003–2004). About two-thirds of U.S. adults were overweight (BMI: 25–30) between the years 2003 and 2004 [7]. Extreme obesity (BMI > 40) is an increasingly common problem and affects 2.8% of U.S. men and 6.9% of U.S. women [8, 9]. Obesity is more common among women and the poor and is increasing at an alarming rate in children [9]. The prevalence of obesity in the kidney transplant population has increased; the proportion of renal transplant recipients classified as obese increased by 116% between 1987 and 2001 [1]. The increasing prevalence of obesity in the renal transplant population has been attributed to many factors including: (1) increased transplantation among older type 2 diabetics, most of whom are obese; (2) improved management of ESRD with better preservation of nutrition and hence weight gain during dialysis; (3) improved sense of well being after transplantation with an increase in appetite and hence food intake and weight and (4) declining levels of physical activity as a secular trend [3]. In the chronic renal failure and dialysis populations from whom transplant candidates are selected obese patients in contrast to their leaner counterparts exhibit fewer manifestations of protein energy malnutrition when on dialysis [10, 11]. Indeed, Kalantar-Zadeh et al. [12] have shown that hemodialysis patients with decreased baseline fat and decreasing body fat over time have increased mortality. This translates into better survival in this population where malnutrition and its inevitable accompaniment, infection is one of the major contributors to death. This stands in direct opposition to the situation in the general population, where obesity is clearly a risk factor for excess cardiovascular mortality. This phenomenon has been termed the risk factor paradox or reverse epidemiology in uremia [10]. In contrast to the impact of BMI in the ESRD population and similar to the general population, higher BMIs have been shown to be associated with increased patient mortality in transplantation.

Pathophysiologic Considerations

In the general population, obesity portends poor health and longevity as it is both an independent risk factor for cardiovascular disease and also clusters almost inextricably with other risk factors for cardiovascular disease and progressive renal insufficiency such as HTN, hyperlipidemia and insulin resistance [5]. Obesity is also independently associated with abnormalities of coronary circulation in the population at large [13]. Obesity has also been linked to the development of focal and segmental glomerulosclerosis and obesity related glomerulopathy characterized by glomerulomegaly with proteinuria and progressive loss of renal function. In addition obesity has been implicated in the progression of primary renal disease in the general population [14, 15]. Recently, proteinuria and progressive renal insufficiency have been reported in obese patients undergoing nephrectomy for reasons other than kidney donation [16]. Given that in the transplant patient, hyperfiltration is a necessary and given accompaniment of functional adaptation, obesity could potentially be implicated in post-transplant renal dysfunction. A number of factors can mediate decreased long-term renal allograft survival [17]. Such variables as have been studied in detail include: living vs. cadaveric donor status, delayed graft function (DGF), rejection episodes (both early and late), HLA matching, donor-recipient size mismatches, and specific metabolic factors [18–22]. Increased BMI has been linked to decreased short-term graft survival and an increase in recipient mortality. Increased BMI at transplantation could potentially decrease long-term allograft survival by adverse effects on patient survival or by independent effects on the allograft itself. Independent effects of obesity on the allograft could in turn be due to either increased metabolic demand (donor/recipient size disparity) [23, 24] or potentially due to effects on immunosuppressant drug concentrations [25]. In the succeeding sections, we examine the relationship between obesity in the renal transplant recipient and its effects on short- and long-term outcomes.

Obesity and Its Impact on Short-Term Sequelae of Renal Transplantation

Perioperative Complications
In the general population, obesity does increase the risk of perioperative complications such as deep vein thrombosis, delayed wound healing, wound infections and hernias [26]. These risks can be logically expect to transfer to the CRF patients with the reasonable expectation that malnutrition, immunosuppression and co-existing cardiovascular morbidity could compound the overall risk.

Humar et al. [27] studied 2,013 renal transplant recipients at the University of Minnesota and found a 4.4-fold risk for wound infections in recipients with BMIs in excess of 30 ($p < 0.001$). In another study of 493 recipients transplanted between 1994 and 2000, Johnson et al. noted that 14% of obese subjects suffered superficial wound breakdown vs. 4% in non-obese ($p < 0.01$) and complete wound dehiscence (3 vs. 0%). Halme et al. [28] have shown a significant graft loss risk in obese subjects primarily due to technical vascular complications; an increased incidence of re-intubation, intensive care unit admission and increased length of hospital stay in obese subjects have been reported. In another study, only female obese patients had an increased length of stay in the hospital. In general, most studies show that obese recipients incur more postoperative wound complications and that these are usually of minor significance [29, 30].

Early Complications

DGF and acute rejection are two significant complications in the early post-transplant period.

The role of obesity in the development of DGF (usually defined as requirement for dialysis in the first week after renal transplantation) is uncertain. A number of single-center studies of cadaveric renal transplantation showed an increased DGF risk with BMIs exceeding 36 [31–35]. Explanations for the association between recipient obesity and DGF include prolonged total operating time and prolonged revascularization due to anatomic constraints [36, 37]. More recent studies do not report an association between obesity and DGF [29, 38]. Whether this reflects only improvements in the technical aspects of revascularization or overall improvements such as the use of polyclonal antibody induction for transplantation is not entirely clear. In an analysis of 51,927 primary adult renal transplants reported the U.S. Renal Data System (USRDS), a BMIs in excess of 36 conferred and increased risk for DGF (relative risk (RR) = 1.51; 95% confidence interval (CI) = 1.27–1.85). No significant associations between BMI and acute rejection episodes were noted in this study, however [39]. In summary, recipient obesity may pre-dispose to DGF especially at the upper extreme of obesity.

Most studies, however, do not demonstrate associations between obesity and acute rejection episodes both early and late [28, 29, 40, 41]. It appears, therefore that any impact of recipient obesity on transplant outcomes is likely independent of acute rejection.

Post-transplant HTN complicates the majority of transplants (up to 90%) in the calcineurin inhibitor era [42, 43]. In the general population, HTN is twice as common in those who are obese than in the non-obese [44]; mortality risk increases with increasing elevations of blood pressure even in patients within the normal range weight range [45]. Pre-transplant obesity is a major risk factor for the development and persistence of HTN post-transplantation [42, 43]. Kasiske et al. studied the correlates of post-transplant HTN in 1,666 consecutive recipients of renal transplants who were transplanted in two different time periods: 1976–1992 and 1993–2002. In this study, the BMI of the transplanted subject increased significantly over the two eras ($23.5 \pm 4.5\,kg/m^2$ (1976–1992) vs. $26.6 \pm 5.8\,kg/m^2$ (1993–2002); $p < 0.001$). An increasing proportion of obese patients was transplanted in the latter era (8.5% (1976–1992) vs. 20.8% (1993–2002). Pre-transplant BMI remained a strong correlate of post-transplant HTN even after adjustment for other covariates such as calcineurin inhibitor use, pre-transplant HTN and CAN [42]. Given that HTN is closely tied to graft survival its occurrence with greater frequency in the obese recipient deserves close attention.

In a more recent study, Kasiske et al. studied more than 53,000 Medicare beneficiaries who were placed on the deceased donor waiting list and evaluated the effects of transplantation on the risk of development of an acute myocardial infarction. Overall, transplantation reduced the risk for development of an acute MI. However, these risk reductions conferred by transplantation were lesser for those within the first 3 months post-transplantation, those older than 65 years, the diabetic and pre-transplantation time in excess of 1 year. However, the risk for AMI after listing was no different for individuals who were either underweight (BMI = 18.5 kg/m^2) or overweight (BMI > 30 kg/m^2), compared with individuals with normal BMI. The risk for AMI for BMI < 30 kg/m^2 (vs. BMI > 30 kg/m^2 was similar on the waiting list) and early or late (0.92; 95% CI = 71.2–1.18) after transplantation. Thus regardless of recipient BMI, older individuals, diabetics, recipients with dialysis exceeding a year in duration and all subjects within the first 3 months post-transplantation deserve aggressive management of cardiovascular risk factors [46].

Post-Transplant Hyperlipidemia

In the general population, hyperlipidemia is strongly associated with obesity with hyperlipidemia being one of the defining lesions of the metabolic syndrome [44]. Post-transplant hyperlipidemia is extremely common and is reported in the majority of patients in the calcineurin-inhibitor era. Its etiology is multifactorial and includes calcineurin inhibitors, corticosteroid use, genetic pre-disposition, uncontrolled diabetes mellitus and other immunosuppressants such as sirolimus [47, 48]. One of the early descriptions of hyperlipidemia in renal transplant patients and those on dialysis and uremic subjects noted the contribution of obesity to hyperlipidemia [49]. Carvalho and Soares studied the relationships between BMI and lipid levels in 72 renal transplant recipients. Pre-transplant BMIs were greater in the hypercholesterolemic recipients (25.6 ± 4.1 vs. 23.8 ± 3.2 kg/m^2; p = 0.04). No such differences in BMI were demonstrable over different triglyceride levels. Furthermore, even in this small study, the authors were able to show that the risk of doubling of serum creatinine was significantly higher in the high cholesterol group; no differences in graft loss were noted in the reported follow-up period across lipid levels and BMI strata [50, 51]. It has also been shown that the use of lipid lowering therapy can lower the incidence of cardiac events in the transplant recipient [52]. Based on the foregoing, it is probably reasonable to surmise that one must at the very least expect hyperlipidemia in the obese renal transplant recipient and treat it with appropriate measures.

Insulin Resistance and Post-Transplant Diabetes Mellitus

Obesity is a recognized contributor to the development of insulin resistance and is strongly associated with type II diabetes mellitus in the general population [44].

The links between obesity and the metabolic syndrome remain demonstrable across varying levels of creatinine clearances as demonstrated in an analysis of NHANES 3 data [53]. The association between insulin resistance, obesity and diabetes mellitus persist after successful transplantation with the added effects of glucocorticoids and calcineurin inhibitors. The incidence of post-transplant diabetes mellitus (PTDM) may be higher in patients treated with tacrolimus than with cyclosporine. Both impaired insulin secretion and decreased sensitivity to insulin are the central underpinnings of post-transplantation diabetes mellitus [54]. Numerous studies have demonstrated that obesity is an independent risk factor in the development of PTDM and that this risk increases with time in the obese transplant recipient. In a recent analysis of USRDS data obesity was an independent risk factor for the development of PTDM (RR = 1.73; (p < 0.0011) [55]. Cosio et al. [56] reported that 20% of 1,811 renal transplant recipients followed 8 years on average developed PTDM. Those developing PTDM had significantly higher BMIs at time of transplantation (PTDM: 29 ± 7; No-PTDM: 25.9 ± 6.5; p < 0.001). More recently, Cosio et al. [57] studied 490 kidney transplant recipients who were not diagnosed with diabetes pre-transplant transplanted between 1998 and 2003 who were followed for up to 4 years. Within a week post-transplant, 45% of recipients had abnormal fasting blood sugars between 100 and 125 mg/dl), and 21% had frank PTDM (fasting blood glucose ≥126). At 1-year post-transplant, 33% of patients had impaired fasting glycemia, and 13% had PTDM. Significant risk factors for hyperglycemia at 1 year included along with higher BMI – older recipient, male gender, higher pre-transplant blood glucose, and higher blood glucose 1 week post-transplant (p < 0.002 for each of the risk factors by multivariate analyses). The occurrence of post-transplant hyperglycemia in this study was not an innocuous accompaniment. During a follow-up period of up to 4 years, 12% of recipients suffered cardiovascular events. Increasing fasting glucose levels (>100 mg/dl) at 1, 4, and/or 12 months post-transplant were significantly related to CV events (post-transplant cardiac [p = 0.001] and peripheral vascular disease events [p = 0.003]). Furthermore, this relationship between fasting blood sugar and CV events was independent of all other CV risk factors, such as older age, pre-transplant CV disease or event, male gender, dyslipidemia, and transplant vintage. Studies such as this further go on to show the complex interplay between CV risk factors in the obese transplant recipient.

Impact of Obesity on Graft and Patient Survival after Kidney Transplantation

As discussed previously, the effects of mild obesity and overweight have been less well studied. Drafts et al. [33] reported that mild obesity did not correlate

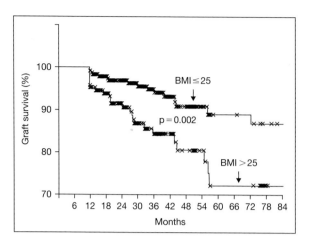

Fig. 1. Kaplan-Meier analysis of graft survival. Individuals with BMI ≤ 25 had a superior 7-year graft survival compared to those with BMIs < 25 (88 vs. 72%, respectively) [reproduced with permission from 39].

with long-term graft outcome. Bumgardner et al. [58] showed that mild obesity (BMI > 27) adversely affected long-term graft survival of both pancreas and renal allografts in patients receiving simultaneous transplants of both organs.

Does mild obesity affect long-term outcome of solitary renal transplants?

This question was also addressed in a single-center study of renal transplant recipients performed in the U.S. [39]. Four hundred and five consecutive recipients of renal allografts transplanted between 1990 and 1997 were included. Immunosuppression consisted of cyclosporine (Sandimmune) or Cyclosporine microemulsion (Neoral) with azathioprine or mycophenolate mofetil and corticosteroids. Graft survival and patient survival were analyzed in patients with BMIs below 25 and equal to or above 25 by the Kaplan-Meier method. A significantly greater 7-year graft survival was evident in the individuals with BMI ≤ 25 (88%) as opposed to those with BMI > 25 (72%) ($p = 0.002$) (fig. 1). Seven-year patient survival was inferior in the group with BMIs > 25 (BMI ≤ 25: 92%; BMI > 25: 81%; $p = 0.01$). Cox regression was then used to analyze the effects of various covariates on graft and patient survival. Significant risk factors associated with graft loss were BMI > 25, cadaveric donor status, acute rejection episode and use of azathioprine as opposed to mycophenolate mofetil. These data are shown in table 1. Reasons for graft loss in the two groups were then analyzed. The major cause of death in both BMI groups was infection followed by cardiovascular death with a trend towards more infectious death in the BMI > 25 group. When censored for death, the major cause of graft loss was chronic rejection. The proportion of

Table 1. Cox proportional hazard model (after backward elimination) demonstrating the effect of clinically relevant covariates on graft and patient survival [reproduced with permission from 39].

Variable	Graft survival			Patient survival		
	Risk ratio	95% CI	p value	risk ratio	95% CI	p value
BMI > 25	2.0	1.1–3.6	<0.05	2.0	1.1–4.6	<0.05
Cadaveric donor	13.3	1.8–15.4	<0.01	2.6	1.1–5.2	<0.05
Rejection	2.5	1.3–4.6	<0.01	2.8	1.3–6.3	<0.05
Azathioprine vs. mycophenolate mofetil	2.5	1.2–8.5	<0.05	NS	–	–

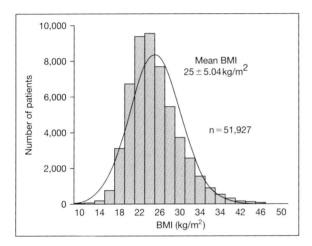

Fig. 2. Distribution of BMI data.

patients with chronic rejection trended higher in the BMI > 25 group and was just barely short of statistical significance (p = 0.06). In this study, BMI > 25 conferred a 2-fold RR of long-term graft loss. Furthermore, BMI > 25 remained a significant and independent risk factor for graft loss in the multivariate analyses. As noted, death censored graft loss trended higher in the BMI > 25 group and was independent of race, acute rejection, serum cholesterol, CsA levels or other covariates. Mycophenolate mofetil appeared to confer a relative benefit on graft survival and was only later on shown to have independent effects on patient survival and death censored graft survival (fig. 2).

This study raised the question as to whether obesity exerts effects on renal allograft survival independent of its effects on patient death. Single-center centers often lack the power to demonstrate effects of risk factors on death-censored

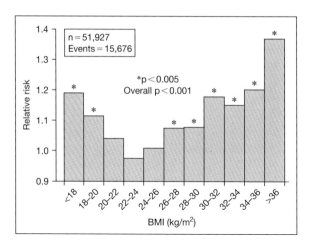

Fig. 3. Relative risk for graft loss by BMI [reproduced with permission from 40].

graft survival as noted above. Similarly, single-center studies typically do not afford the sample sizes needed to investigate the effects of specific risk factors such as BMI on cause specific mortality. In the U.S., the United Network for Organ sharing (UNOS) maintains a large database which is distributed, enriched with social security and Center for Medicare and Medicaid Services (CMS) data by the United States Renal Data System (USRDS). The UNOS data is also distributed to the Scientific Registry of Renal Transplant Recipients (SRTR) a unique database that contains all Organ Procurement and Transplant Network (OPTN) data since 1987 [59, 60]. Endpoints like graft loss and patient death are reported with surprising accuracy in these databases, when both CMS and social security data is integrated [59]. Importantly, it should be noted that retrospective analyses provide evidence of associations between risk factors (e.g. treatment regimen or race) and outcomes (e.g. time to graft loss or death) and do yield valuable measures of the strength and significance of such associations [61]. However, proof of causality relationships between a risk factor and outcome should necessarily include among other lines of evidence, prospective experimental evidence [62].

In order to test the hypothesis that obesity affected long-term graft survival independent of patient death, and to investigate the effects of obesity on cause-specific mortality in renal transplantation, Meier-Kriesche et al. [40] performed an analysis of USRDS data.

Data from 51,927 first solitary adult renal transplant recipients registered in the USRDS database between 1988 and 1997 were used in this analysis. Follow-up data starting from date of transplant to graft loss, death or study end date in mid 1998 were utilized. Primary endpoints in this study were patient

Table 2. Cox proportional hazard model for graft loss by categorized BMI

BMI (kg/m^2)	RR	95% CI	p Value
<18	1.213	1.110–1.326	<0.001
18–20	1.114	1.044–1.189	0.001
20–22	1.034	0.976–1.094	0.252
22–24	0.963	0.912–1.017	0.117
24–26	1.000	Reference	0.117
26–28	1.071	1.008–1.136	0.026
28–30	1.073	1.004–1.140	0.047
30–32	1.181	1.098–1.271	<0.001
32–34	1.151	1.055–1.257	0.002
34–36	1.205	1.084–1.339	0.001
>36	1.385	1.300–1.551	<0.001

Model corrected for donor and recipient age, race, gender, and CMV IgG antibody status; primary cause of ESRD; donor source; cold ischemia times; HLA mismatch; presensitization (PRA); cyclosporine vs. tacrolimus vs. neoral treatment; induction vs. no induction treatment; mycophenolate mofetil vs. azathioprine, and delayed graft function [adapted from reference 40].

death, death censored graft loss, chronic allograft failure (CAF) and overall graft loss (composite of patient death and graft loss). CAF was defined as graft loss due to chronic allograft nephropathy or graft loss after 6 months post-transplant censored for patient death, graft loss secondary to acute rejection, graft thrombosis, infection, surgical complications or recurrent disease. Other endpoints were acute rejection in the first 6 months post-transplant and DGF. The relationships between BMI and study endpoints were analyzed using Cox regression after correcting for confounding variables. Logistic regression models were also used to investigate relationships between pre-transplant BMI and DGF and acute rejection. For the purposes of analysis, BMI values were stratified into 11 categories: <18, from 18 to 36 in 2 unit increments and >36 kg/m^2.

In this study, BMIs were distributed normally as depicted in figure 3. For purposes of analysis, the BMI = 24–26 kg/m^2 category was used as the reference group. When compared to this category, BMIs of 20–22 and 24–26 kg/m^2 were not associated with significantly different relative risks for graft loss (table 2). However, the next lower BMI category, 18–20 kg/m^2 was associated with a 11% relative increase in risk for graft loss (RR = 1.114; 95% CI = 1.044–1.1189). Importantly, the lowest BMI category, <18 kg/m^2 exhibited the highest elative risk for graft loss (RR = 1.213; 95% CI = 1.110–1.326) among those patients with BMIs falling below the reference range.

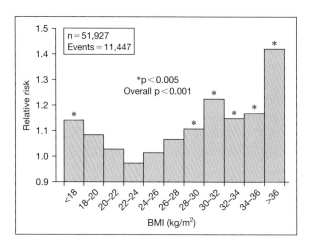

Fig. 4. Relative risk for death censored graft loss by BMI [reproduced with permission from 40].

All BMI categories above the reference range exhibited significant and progressively higher RR for graft loss. For instance, in the lower risk category, BMI = 26–28 kg/m², the increase in RR was mild albeit significant (RR = 1.071: 95% CI = 1.008–1.136). However as one proceeded towards the higher BMI categories, RR for graft loss increased exponentially, the highest risk being observed in the most obese subjects (BMI > 36 kg/m², RR = 1.385; 95% CI = 1.300–1.551). These results are depicted in graphic fashion in figure 3. The relationship between BMI and graft loss was U-shaped with the lowest risk being incurred by subjects with BMI = 22–26 kg/m² and significant and progressive increases to either side of that reference range.

When death censored graft loss was analyzed, a similar picture emerged (fig. 4). Subjects with BMI < 18 kg/m² exhibited significantly higher risk. On the other end of the BMI spectrum, BMIs in excess of 28 kg/m² were associated with increasingly higher risk for death censored graft loss.

Next, the impact of BMI on CAF was addressed. This relationship again exhibited a U-shape (fig. 5) with the exception being that increased RR for CAF in the lower BMI categories failed to attain statistical significance. In contrast, BMIs in excess of 30 kg/m² were associated with significant and progressively higher risk for CAF. When death with a functioning graft was addressed, the U-shaped relationship again emerged (fig. 6). However, the range of BMIs with relative freedom from increased risk was wider, i.e. 22–32 kg/m². The shape of the curve was very similar when the model was applied only to patients who had been on dialysis for at least 2 years. With regard to BMI's impact on cardiovascular mortality, the curve differed from that for overall mortality in that the interval of optimal risk close to the

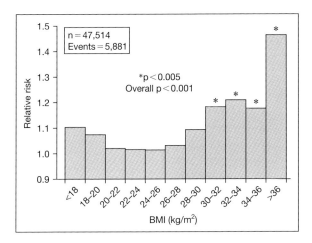

Fig. 5. Relative risk for CAF by BMI [reproduced with permission from 40].

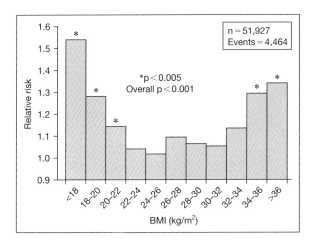

Fig. 6. Relative risk for death with functioning graft by BMI [reproduced with permission from 40].

reference range was very narrow (fig. 7). While not reaching statistical significance, tendencies to increasing cardiovascular risk with BMI $< 22 \text{ kg/m}^2$ and above 26 kg/m^2 were noted. As one proceeded from the lowest RR category to the highest risk group of BMI $> 36 \text{ kg/m}^2$, RR for cardiovascular death increased exponentially. In contrast, Howard et al. reporting on a single-center study did not show any effects of recipient obesity at the time of transplantation on graft or patient survival; only a higher incidence of post-transplant diabetes was noted in recipients with

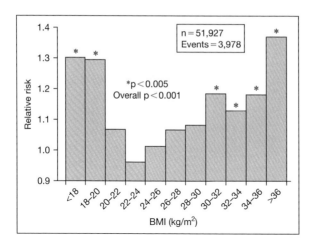

Fig. 7. Relative risk for cardiovascular death by BMI [reproduced with permission from 40].

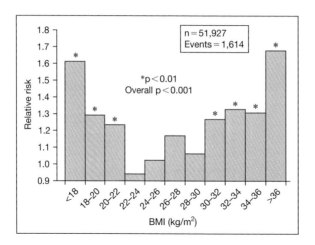

Fig. 8. Relative risk for infectious death by BMI [reproduced with permission from 40].

BMIs exceeding 35. In this study, it should be pointed out that only 98 patients had BMIs in excess of 30 as opposed to 457 with BMIs < 25 and 278 with BMIs between 25 and 30 [38]. These findings could again represent lack of power.

With regard to infectious death, there was a very strong association of the RR for this endpoint both with very low and elevated BMIs. Again, the relationship was a curvilinear U-shaped one with the interval free of significant risk between BMIs of 22 and 30 kg/m² (fig. 8).

Interestingly, BMI was significantly associated with the RR for DGF. The RR for DGF increased gradually with increasing BMI and was greatest for BMIs exceeding $36\,kg/m^2$ (RR = 1.51; 95% CI = 1.27–1.85). For DGF, the BMI < $18\,kg/m^2$ group exhibited a 22% lower RR when compared to the reference group (RR = 0.78; 95% CI = 0.60–0.98). In contrast, no significant association between BMI and the odds ratio for acute rejection episodes within the first 6 months post-transplantation were demonstrable.

This study was thus able to show that recipient BMI is a strong independent risk factor both for patient mortality and graft failure independent of patient death after renal transplantation. As described in the preceding section, the relationship between BMI and mortality after renal transplantation exhibits a U-shaped pattern that is similar to that observed in the general population. Recently, Gore et al. [63] using the UNOS database have also shown that obesity in the renal transplant recipient is associated with decreased allograft survival.

These relationships stand in contrast to the finding in dialysis patients that higher BMIs are associated with lower mortality risk. The most plausible explanation that is put forth to explain this apparent paradox is that subjects on dialysis with higher BMIs are those very patients with better nutrition. This could explain why those patients at the lowest BMI strata experience a higher risk for infectious death after renal transplantation; malnutrition being a risk factor for infection in most populations. In addition, subjects with lower BMIs could potentially receive inappropriately high levels of immunosuppression especially when one uses agents such as mycophenolate mofetil and corticosteroids which are not monitored therapeutically and are not dosed by body weight. The other possibility is that pharmacologic immunosuppression is additive to the anergic states that usually accompany malnutrition. It should be noted that none of these considerations take into account the tremendous changes in body composition that necessarily occur when a patient transits the hemodialysis milieu for the transplanted state. Certainly, one must keep in mind the selection bias that is necessarily inherent to kidney transplantation as transplant is offered and then performed only in the fittest of carefully screened dialysis or CRF patients. To the extent that the unfavorable metabolic milieu of uremia is largely removed by successful transplantation, the transplant patient seems to behave much more like a member of the general population when obesity is evaluated as a risk factor for mortality. In that respect, the same risk factors that operate to mediate cardiovascular risk in the general population: hyperlipidemia, diabetes mellitus and insulin resistance and HTN are both strongly associated with obesity and compound the mortality risk of obesity in the transplant recipient.

Death censored graft survival is a parameter that measures graft survival independent of patient death. Thus, relationships of BMI with this parameter do not reflect the overall mortality risk conferred by a mere increase in mortality risk. Given the intrinsic proclivity of the obese to proteinuria, glomerulosclerosis and progressive loss of renal function, it is quite possible that obesity could be a significant mediator of CAF. However, given the close clustering of obesity with comorbidities that can themselves affect the allograft such as diabetes mellitus or HTN, 'purer' effects of obesity that mediate graft loss such as hyperfiltration and a consequent glomerulopathy need to be explored in greater depth.

Recently, Armstrong et al. [64] have shown in a prospective study of 90 renal transplant recipients followed to 2 years that the prevalence of central obesity, impaired glucose tolerance, hyperlipidemia and HTN at 2 years increased with increasing BMI. Importantly, progression of proteinuria over time was strongly associated with increased BMI. Interestingly, in their study progressive decline in eGFR (MDRD) was not associated with BMI in the 2-year follow-up. These findings are of considerable interest in that one would expect progression of proteinuria to precede declining GFR if indeed hyperfiltration is a significant mediator of obesity induced attrition of renal function.

Weight Gain after Transplantation

Thus, far we have discussed the effects of BMI at transplant on outcomes after transplantation. However, weight gain following successful kidney transplantation is frequent and occurs in up to 50% of recipients. In the emaciated recipient it may represent the restoration of appetite and nutrition; in others the effects of steroid induced obesity. Steroids have been blamed for most obesity that follows transplantation. However, pre-transplant obesity is the major precedent to post-transplant obesity. Other risk factors include younger age, African-American race and female gender [2]. Weight gain post-transplant is in the first year independent of cumulative steroid dose, donor source, rejection episodes or post-transplant renal function; after the first year cumulative steroid dose is the primary determinant of weight gain [65, 66].

While obesity and overweight at the time of transplantation affect outcome, less is known about outcomes when weight gain follows transplantation. Feldman et al. [67], in a cohort study conducted between 1985 and 1990 with up to 5 years of follow-up reported that the adjusted RR of allograft failure for a 15-kg increase in recipient body weight was 1.47, $p < 0.0001$ (95% CI = 1.21–1.78) and the adjusted RR for a 10 unit increase in recipient BMI was 2.34, $p < 0.0001$ (95% CI = 1.53–3.58) [67].

El-Agroudy et al. [68] studied 650 non-diabetic live donor kidney recipients, all with BMI at transplant less than 25 kg/m² and followed them for up to

10 years. They then looked at the impact of obesity at 6 months post-transplantation on several outcomes. Post-transplantation obesity was associated with higher incidence of each of chronic allograft nephropathy, post-transplant HTN, post-transplant hyperlipidemia, PTDM and ischemic heart disease. Furthermore, with increasing post-transplant obesity patient death from cardiovascular disease increased and graft function was decreased (increased creatinine) in the patients with the highest BMI.

In another study, Ducloux et al. [69] studied 292 renal transplant recipients and obtained anthropometric measurements and biochemical data such as fasting glucose, lipids and CRP as a measure of systemic inflammation at baseline and at 1 year post-transplantation. One year graft loss was 10.3%. In multivariate analyses, BMI increase greater than 5% at 1 year was associated with almost trebling of the RR for death censored graft loss (HR = 2.82; 95% CI = 1.11–7.44; p = 0.015) [69]. Low creatinine clearance, increasing proteinuria and the occurrence of DGF were other independent predictors of graft loss in this study [69].

We have evaluated patients' BMI levels at ESRD onset, time of wait listing and transplant from the national USRDS database for patients wait listed after 1995 (unpublished observations). We then calculated changes in BMI over these time periods and generated multivariate survival models to examine the association of changes in BMI on outcomes of death on the waiting list and patient and graft survival after transplantation. Of all wait listed patients on dialysis, patients with low BMI (<20) and 'normal' BMI (20–25) had the highest risk for death prior to transplant (AHR = 1.45, 95% CI = 1.34–1.56) and (AHR = 1.17, 95% CI = 1.12, 1.23) relative to patients with BMI levels 25–30. The change in BMI from wait listing to transplant had no association with outcomes following transplant, which was also demonstrated when stratifying the analysis by obese/non-obese patients. Thus, even though BMI is a strong risk factor for graft loss, there is no evidence to suggest that weight loss prior to renal transplantation has a beneficial impact on graft survival. It does remain possible that complication rates can be positively affected by pre-transplant weight loss. However, given the negative impact of extended waiting time on dialysis, and the potential nutritional hazard of significant weight loss on dialysis, aggressive weight loss programs may not be justified prior to renal transplantation.

An Approach to Obesity in the Transplant Recipient

Given the foregoing discussion, the obese ESRD patient presenting for transplantation presents a profile that is prone to cardiovascular risk if not actually already burdened with cardiovascular disease. With this in mind, the pre-transplant evaluation should include a diligent search for and appropriate management of coronary artery disease. Most transplant institutions screen for coronary artery disease with non-invasive tests such as radionuclide stress tests and may proceed directly to coronary angiography in high risk subjects such as diabetics and those with previous history of coronary events. Optimal management of hyperlipidemia, HTN and diabetes should be instituted and maintained through the time that the patient waits for the transplant. The surgical evaluation is focused on aspects that can impinge on the occurrence of post-operative complications such as hernias and wound dehiscence. This may play a role in regimen selection since certain drugs such as sirolimus may be associated with a higher rate of poor wound healing. Pre-anesthesia screening should focus on risks for venous thromboses and ventilatory failure such as pre-existing sleep apnea.

Whether patients should be made to lose weight before transplantation is certainly not clear. Despite the unfavorable effects of obesity on transplant outcome, a high BMI in the dialysis patient is not associated with a universally poor outcome. Importantly, this 'risk factor paradox' needs to be kept in mind when one takes into account that most transplant centers either arbitrarily exclude recipients with BMIs in excess of 30 or certainly delay a transplant until a recipient can reach some pre-defined (oftentimes arbitrary) weight target despite overwhelming and incontrovertible evidence demonstrating the survival advantage of transplantation over dialysis even in the obese. That renal transplantation continues to confer a survival advantage even in the obese patient was reported by Glanton et al. using USRDS data pertaining to 7,521 patients who presented with ESRD between 1995 and 1999 and later enrolled on the renal transplantation waiting list with BMI $\geq 30 \text{ kg/m}^2$ at the time of presentation to ESRD. Follow-up data was available for up to 5 years. They excluded recipients of preemptive renal transplantation or organs other than kidneys and calculated adjusted, time-dependent hazard ratios (HR) for time to death in a given patient during the study period, controlling for renal transplantation, demographics and comorbidities. The incidence of mortality in this study was was 3.3 episodes per 100 patient-years in cadaveric renal transplantation and 1.9/100 patient-years in living donor renal transplantation compared with 6.6 episodes/100 patient-years in all patients on the transplant waiting list. In comparison to maintenance dialysis, both recipients of solitary cadaveric kidneys (HR = 0.39, 95% CI = 0.33–0.47), and recipients of living donor kidneys (HR = 0.23, 95% CI = 0.16–0.34) had statistically significant improved survival. However, these benefits did not apply to those patients with BMI $\geq 41 \text{ kg/m}^2$ (HR = 0.47, 95% CI = 0.17–1.25, p = 0.13).

Furthermore, drastic weight loss goals may be unattainable given the limited exercise capacity of the dialysis patient. All of this causes considerable frustration to the recipients, transplant centers and the treating physicians [38]. Importantly, the negative impact of forced starvation on a nutritionally stressed ESRD patient could well tilt the delicate nutritional balance to a less advantaged infection prone mortality pre-disposed phenotype. Indeed the Canadian Society of Transplantation in its guidelines states: '(1) few data exist to suggest which, if any, obese (BMI $\geq 30\,\text{kg/m}^2$) patients should be denied transplantation based on obesity; (2) supervised weight-loss therapy is recommended for obese candidates, with target BMI $< 30\,\text{kg/m}^{2}$'.

We submit that every transplant center has had experience with obese candidates and that certainly successful transplantation has been reported in morbidly obese subjects.

If dietary interventions for weight loss are to be instituted, these should be done in a supervised setting with monitoring of nutritional status using markers such as pre-albumin under the guidance of a physician and the active participation of an experienced dietician. Case reports document successful transplantation is subjects who have had bariatric surgery and this should be considered in the appropriate candidate who has been adequately worked for absence of cardiovascular disease [70, 71].

In the immediate post-operative period, close attention needs to be paid to prevention of wound infections and deep vein thromboses. Early ambulation with active institution of physiotherapy at the bedside should be utilized.

The successfully transplanted patient is justifiably elated with the lifting of the dietary sanctions engendered by uremia. In addition, high-dose steroids induce a hyperphagic state and promote the development of obesity. Most patients can be counseled about these changes pre-transplant to prevent inordinate gains of weight.

Management of obesity after transplantation is best carried out using a multidisciplinary approach that includes behavior modification, exercise counseling and extensive and continuing nutritional counseling with appropriate use of group support techniques.

Drugs used to study obesity in the general population have not been well studied in the transplant setting. Sibutramine is of concern given its propensity to raise blood pressure. Orlistat can interfere with intestinal absorption of cyclosporine [72].

Non-insulin dependent diabetes associated with obesity can be managed by using sulfonylureas and thiazolidinediones. The propensity for fluid retention with the latter must be kept in mind when appropriate insulin should be used. These interventions should best be managed in close consultation with or by an endocrinologist experienced in the care of the transplant recipient.

Hyperlipidemia is managed in the usual fashion with statins or fibric acid derivatives and dietary intervention. Given the overall cardiovascular risk in these patients, non-selective beta-blockers are used to treat HTN. ACE inhibitors are introduced as early as feasible in these subjects and may well forestall progression of proteinuria [73].

Management of immunosuppression should pay diligent attention to adherence to tapering of corticosteroids as per protocol. If subjects have had bariatric surgery, absorption of immunosuppressants may be erratic and may require parenteral administration guided by levels [70].

In the longstanding obese patient with a transplant, the same multidisciplinary approach should be instituted. Patients desiring to come off steroids should be counseled on the lack of evidence supporting considerable weight loss with this as the sole approach and the definite evidence of increased rejection risk with steroid withdrawal.

Conclusion

In conclusion, the obese renal transplant patient presents a considerable clinical challenge. Obesity compounds perioperative risk and is associated with or may lead to many medical complications of transplantation such as obesity, diabetes and HTN. Obesity compounds the cardiovascular risk accrued by the renal transplant recipient. Obesity is also associated with chronic graft loss, the mechanisms underlying which are not entirely clear. Whether obese patients on the transplant waiting list should lose weight prior to being transplanted or listed for transplantation is a subject of controversy. The optimal management of obesity in renal transplantation and the definition of optimal pathways to minimize risk in this population are works in progress.

References

1 Friedman AN, Miskulin DC, Rosenberg IH, Levey AS: Demographics and trends in overweight and obesity in patients at time of kidney transplantation. Am J Kidney Dis 2003;41:480–487.
2 Johnson CP, Gallagher-Lepak S, Zhu YR, Porth C, Kelber S, Roza AM, et al: Factors influencing weight gain after renal transplantation. Transplantation 1993;56:822–827.
3 Armstrong KA, Campbell SB, Hawley CM, Johnson DW, Isbel NM: Impact of obesity on renal transplant outcomes. Nephrology (Carlton) 2005;10:405–413.
4 Flier JS: Obesity wars: molecular progress confronts an expanding epidemic. Cell 2004;116: 337–350.
5 Calle EE, Thun MJ, Petrelli JM, Rodriguez C, Heath CW Jr: Body-mass index and mortality in a prospective cohort of U.S. adults. N Engl J Med 1999;341:1097–1105.
6 Stamler J: Epidemic obesity in the United States. Arch Intern Med 1993;153:1040–1044.
7 Ogden CL, Carroll MD, Curtin LR, McDowell MA, Tabak CJ, Flegal KM: Prevalence of overweight and obesity in the United States, 1999–2004. JAMA 2006;295:1549–1555.
8 Ogden CL, Carroll MD, Curtin LR, McDowell MA, Tabak CJ, Flegal KM: Prevalence of overweight and obesity in the United States, 1999–2004. JAMA 2006;295:1549–1555.

9   Ogden CL, Carroll MD, Curtin LR, McDowell MA, Tabak CJ, Flegal KM: Prevalence of overweight and obesity in the United States, 1999–2004. JAMA 2006;295:1549–1555.
10 Fleischmann EH, Bower JD, Salahudeen AK: Risk factor paradox in hemodialysis: better nutrition as a partial explanation. ASAIO J 2001;47:74–81.
11 Hakim RM, Lowrie E: Obesity and mortality in ESRD: is it good to be fat? Kidney Int 1999;55:1580–1581.
12 Kalantar-Zadeh K, Kuwae N, Wu DY, Shantouf RS, Fouque D, Anker SD, et al: Associations of body fat and its changes over time with quality of life and prospective mortality in hemodialysis patients. Am J Clin Nutr 2006;83:202–210.
13 Schindler TH, Cardenas J, Prior JO, Facta AD, Kreissl MC, Zhang XL, et al: Relationship between increasing body weight, insulin resistance, inflammation, adipocytokine leptin, and coronary circulatory function. J Am Coll Cardiol 2006;47:1188–1195.
14 Kambham N, Markowitz GS, Valeri AM, Lin J, D'Agati VD: Obesity-related glomerulopathy: an emerging epidemic. Kidney Int 2001;59:1498–1509.
15 Bonnet F, Deprele C, Sassolas A, Moulin P, Alamartine E, Berthezene F, et al: Excessive body weight as a new independent risk factor for clinical and pathological progression in primary IgA nephritis. Am J Kidney Dis 2001;37:720–727.
16 Praga M, Hernandez E, Herrero JC, Morales E, Revilla Y, Diaz-Gonzalez R, et al: Influence of obesity on the appearance of proteinuria and renal insufficiency after unilateral nephrectomy. Kidney Int 2000;58:2111–2118.
17 Schweitzer EJ, Matas AJ, Gillingham KJ, Payne WD, Gores PF, Dunn DL, et al: Causes of renal allograft loss. Progress in the 1980s, challenges for the 1990s. Ann Surg 1991;214:679–688.
18 Hamar P, Muller V, Kohnle M, Witzke O, Albrecht KH, Philipp T, et al: Metabolic factors have a major impact on kidney allograft survival. Transplantation 1997;64:1135–1139.
19 Moreso F, Seron D, Anunciada AI, Hueso M, Ramon JM, Fulladosa X, et al: Recipient body surface area as a predictor of posttransplant renal allograft evolution. Transplantation 1998;65:671–676.
20 Matas AJ, Gillingham KJ, Payne WD, Najarian JS: The impact of an acute rejection episode on long-term renal allograft survival (t1/2). Transplantation 1994;57:857–859.
21 Cecka JM: Outcome statistics of renal transplants with an emphasis on long-term survival. Clin Transplant 1994;8(pt 2):324–327.
22 Kerman RH, Van Buren CT, Lewis RM, DeVera V, Baghdahsarian V, Gerolami K, et al: Improved graft survival for flow cytometry and antihuman globulin crossmatch-negative retransplant recipients. Transplantation 1990;49:52–56.
23 Tullius SG, Tilney NL: Both alloantigen-dependent and -independent factors influence chronic allograft rejection. Transplantation 1995;59:313–318.
24 Brenner BM, Cohen RA, Milford EL: In renal transplantation, one size may not fit all. J Am Soc Nephrol 1992;3:162–169.
25 Yee G, Kennedy MS: Cyclsporine; in Evans W, Shentag J, Jusko WS (eds): Applied Pharmacokinetics. Spokane, Applied Therapeutics, 1986.
26 Pasulka PS, Bistrian BR, Benotti PN, Blackburn GL: The risks of surgery in obese patients. Ann Intern Med 1986;104:540–546.
27 Humar A, Ramcharan T, Denny R, Gillingham KJ, Payne WD, Matas AJ: Are wound complications after a kidney transplant more common with modern immunosuppression? Transplantation 2001;72:1920–1923.
28 Halme L, Eklund B, Kyllonen L, Salmela K: Is obesity still a risk factor in renal transplantation? Transpl Int 1997;10:284–288.
29 Johnson DW, Isbel NM, Brown AM, Kay TD, Franzen K, Hawley CM, et al: The effect of obesity on renal transplant outcomes. Transplantation 2002;74:675–681.
30 Bennett WM, McEvoy KM, Henell KR, Valente JF, Douzdjian V: Morbid obesity does not preclude successful renal transplantation. Clin Transplant 2004;18:89–93.
31 Holley JL, Shapiro R, Lopatin WB, Tzakis AG, Hakala TR, Starzl TE: Obesity as a risk factor following cadaveric renal transplantation. Transplantation 1990;49:387–389.
32 Blumke M, Keller E, Eble F, Nausner M, Kirste G: Obesity in kidney transplant patients as a risk factor. Transplant Proc 1993;25:2618.

33 Drafts HH, Anjum MR, Wynn JJ, Mulloy LL, Bowley JN, Humphries AL: The impact of pre-transplant obesity on renal transplant outcomes. Clin Transplant 1997;11(pt 2):493–496.
34 Orofino L, Pascual J, Quereda C, Burgos J, Marcen R, Ortuno J: Influence of overweight on survival of kidney transplant. Nephrol Dial Transplant 1997;12:855.
35 Pirsch JD, Armbrust MJ, Knechtle SJ, D'Alessandro AM, Sollinger HW, Heisey DM, et al: Obesity as a risk factor following renal transplantation. Transplantation 1995;59:631–633.
36 Gill IS, Hodge EE, Novick AC, Steinmuller DR, Garred D: Impact of obesity on renal transplantation. Transplant Proc 1993;25(pt 2):1047–1048.
37 Modlin CS, Flechner SM, Goormastic M, Goldfarb DA, Papajcik D, Mastroianni B, et al: Should obese patients lose weight before receiving a kidney transplant? Transplantation 1997;64:599–604.
38 Howard RJ, Thai VB, Patton PR, Hemming AW, Reed AI, Van der Werf WJ, et al: Obesity does not portend a bad outcome for kidney transplant recipients. Transplantation 2002;73:53–55.
39 Meier-Kriesche HU, Vaghela M, Thambuganipalle R, Friedman G, Jacobs M, Kaplan B: The effect of body mass index on long-term renal allograft survival. Transplantation 1999;68:1294–1297.
40 Meier-Kriesche HU, Arndorfer JA, Kaplan B: The impact of body mass index on renal transplant outcomes: a significant independent risk factor for graft failure and patient death. Transplantation 2002;73:70–74.
41 Merion RM, Twork AM, Rosenberg L, Ham JM, Burtch GD, Turcotte JG, et al: Obesity and renal transplantation. Surg Gynecol Obstet 1991;172:367–376.
42 Kasiske BL, Anjum S, Shah R, Skogen J, Kandaswamy C, Danielson B, et al: Hypertension after kidney transplantation. Am J Kidney Dis 2004;43:1071–1081.
43 First MR, Neylan JF, Rocher LL, Tejani A: Hypertension after renal transplantation. J Am Soc Nephrol 1994;4(suppl):S30–S36.
44 Gregg EW, Cheng YJ, Cadwell BL, Imperatore G, Williams DE, Flegal KM, et al: Secular trends in cardiovascular disease risk factors according to body mass index in US adults. JAMA 2005;293:1868–1874.
45 Vasan RS, Larson MG, Leip EP, Evans JC, O'Donnell CJ, Kannel WB, et al: Impact of high-normal blood pressure on the risk of cardiovascular disease. N Engl J Med 2001;345:1291–1297.
46 Kasiske BL, Maclean JR, Snyder JJ: Acute myocardial infarction and kidney transplantation. J Am Soc Nephrol 2006;17:900–907.
47 Rao KV, Andersen RC: Long-term results and complications in renal transplant recipients. Observations in the second decade. Transplantation 1988;45:45–52.
48 Kasiske BL: Clinical practice guidelines for managing dyslipidemias in kidney transplant patients. Am J Transplant 2005;5:1576.
49 Ibels LS, Alfrey AC, Huffer WE, Craswell PW, Anderson JT, Weil R III: Arterial calcification and pathology in uremic patients undergoing dialysis. Am J Med 1979;66:790–796.
50 Ibels LS, Simons LA, King JO, Williams PF, Neale FC, Stewart JH: Studies on the nature and causes of hyperlipidaemia in uraemia, maintenance dialysis and renal transplantation. Q J Med 1975;44:601–614.
51 Carvalho MF, Soares V: Hyperlipidemia as a risk factor of renal allograft function impairment. Clin Transplant 2001;15:48–52.
52 Holdaas H, Fellstrom B, Cole E, Nyberg G, Olsson AG, Pedersen TR, et al: Long-term cardiac outcomes in renal transplant recipients receiving fluvastatin: the ALERT extension study. Am J Transplant 2005;5:2929–2936.
53 Beddhu S, Kimmel PL, Ramkumar N, Cheung AK: Associations of metabolic syndrome with inflammation in CKD: results from the Third National Health and Nutrition Examination Survey (NHANES III). Am J Kidney Dis 2005;46:577–586.
54 Hjelmesaeth J, Jenssen T, Hagen M, Egeland T, Hartmann A: Determinants of insulin secretion after renal transplantation. Metabolism 2003;52:573–578.
55 Kasiske BL, Snyder JJ, Gilbertson D, Matas AJ: Diabetes mellitus after kidney transplantation in the United States. Am J Transplant 2003;3:178–185.
56 Cosio FG, Pesavento TE, Kim S, Osei K, Henry M, Ferguson RM: Patient survival after renal transplantation: IV. Impact of post-transplant diabetes. Kidney Int 2002;62:1440–1446.

57 Cosio FG, Kudva Y, van der Velde M, Larson TS, Textor SC, Griffin MD, et al: New onset hyperglycemia and diabetes are associated with increased cardiovascular risk after kidney transplantation. Kidney Int 2005;67:2415–2421.
58 Bumgardner GL, Henry ML, Elkhammas E, Wilson GA, Tso P, Davies E, et al: Obesity as a risk factor after combined pancreas/kidney transplantation. Transplantation 1995;60:1426–1430.
59 Dickinson DM, Bryant PC, Williams MC, Levine GN, Li S, Welch JC, et al: Transplant data: sources, collection, and caveats. Am J Transplant 2004;4(suppl 9):13–26.
60 Merion RM: 2004 SRTR Report on the State of Transplantation. Am J Transplant 2005;5 (pt 2):841–842.
61 Kaplan B, Schold J, Meier-Kriesche HU: Overview of large database analysis in renal transplantation. Am J Transplant 2003;3:1052–1056.
62 Hill AB: Statistical evidence and inference; in Hill AB (ed): Principles of Medical Statistics. London, The Lancet Limited, 1971, pp 309–323.
63 Gore JL, Pham PT, Danovitch GM, Wilkinson AH, Rosenthal JT, Lipshutz GS, et al: Obesity and outcome following renal transplantation. Am J Transplant 2006;6:357–363.
64 Armstrong KA, Campbell SB, Hawley CM, Nicol DL, Johnson DW, Isbel NM: Obesity is associated with worsening cardiovascular risk factor profiles and proteinuria progression in renal transplant recipients. Am J Transplant 2005;5:2710–2718.
65 van den Ham EC, Kooman JP, Christiaans MH, Nieman FH, van Hooff JP: Weight changes after renal transplantation: a comparison between patients on 5-mg maintenance steroid therapy and those on steroid-free immunosuppressive therapy. Transpl Int 2003;16:300–306.
66 Baum CL, Thielke K, Westin E, Kogan E, Cicalese L, Benedetti E: Predictors of weight gain and cardiovascular risk in a cohort of racially diverse kidney transplant recipients. Nutrition 2002;18:139–146.
67 Feldman HI, Fazio I, Roth D, Berlin JA, Brayman K, Burns JE, et al: Recipient body size and cadaveric renal allograft survival. J Am Soc Nephrol 1996;7:151–157.
68 El Agroudy AE, Wafa EW, Gheith OE, Shehab el-Dein AB, Ghoneim MA: Weight gain after renal transplantation is a risk factor for patient and graft outcome. Transplantation 2004;77:1381–1385.
69 Ducloux D, Kazory A, Simula-Faivre D, Chalopin JM: One-year post-transplant weight gain is a risk factor for graft loss. Am J Transplant 2005;5:2922–2928.
70 Marterre WF, Hariharan S, First MR, Alexander JW: Gastric bypass in morbidly obese kidney transplant recipients. Clin Transplant 1996;10:414–419.
71 Bennett WM, McEvoy KM, Henell KR, Valente JF, Douzdjian V: Morbid obesity does not preclude successful renal transplantation. Clin Transplant 2004;18:89–93.
72 Evans S, Michael R, Wells H, Maclean D, Gordon I, Taylor J, et al: Drug interaction in a renal transplant patient: cyclosporin-Neoral and orlistat. Am J Kidney Dis 2003;41:493–496.
73 Praga M, Hernandez E, Andres A, Leon M, Ruilope LM, Rodicio JL: Effects of body-weight loss and captopril treatment on proteinuria associated with obesity. Nephron 1995;70:35–41.

Herwig-Ulf Meier-Kriesche, MD
Transplant Nephrologist, University of Florida
1600 SW Archer Road, Box 100224
Gainesville, FL 32610–0224 (USA)
Tel. +1 352 846 2692, Fax +1 352 392 3581, E-Mail meierhu@medicine.ufl.edu

Body Mass Index and the Risk of Chronic Renal Failure: The Asian Experience

Kunitoshi Iseki

Dialysis Unit, University Hospital of The Ryukyus, Okinawa, Japan

Abstract

Background: There is a worldwide obesity epidemic, and the number of patients requiring dialysis because of obesity-related renal disease such as diabetes mellitus and hypertension is increasing. Obesity increases the risk of cardiovascular disease and premature death due to chronic kidney disease (CKD) and end-stage renal disease (ESRD). Although the effect of obesity might differ among races, obesity has a significant impact on CKD and ESRD. **Methods:** We examined the relationship between obesity (i.e., body mass index [BMI]) and CKD or ESRD using a community based screening registry in Okinawa, Japan. For this purpose, we used a general screening registry (1983, 1993, and 2003), the ESRD patient registry (1971–2000), a hospital-based cardiovascular disease registry (1988–1990), and a hospital-based screening registry (2003). **Results:** The prevalence of obesity, based on a BMI $\geq 30\,\text{kg/m}^2$, was 3.5% (1983), 4.7% (1993), and 6.2% (2003) in the general adult population. The incidence of ESRD increased when BMI increased, particularly in men. In the hospital-based screening study, the number of components of metabolic syndrome was significantly related with the prevalence of CKD. The relationship was linear when the modified National Cholesterol Education Program criteria were used to define abdominal obesity as a waist circumference of 85 cm or more in men and 90 cm or more in women. **Conclusion:** Although prospective studies are needed, our findings indicate that obesity, including metabolic syndrome, is a potential treatable cause of CKD and ESRD. Rigorous efforts should be made to optimize weight to reduce the risk of CKD and ESRD by a judicious combination of diet, exercise, and psychologic therapies.

Copyright © 2006 S. Karger AG, Basel

Introduction

The number of patients with end-stage renal disease (ESRD) is increasing worldwide [1–3]. Therefore, early detection and treatment are important for people who are at risk [4, 5]. Chronic kidney disease (CKD) is a cause of ESRD, but is also a recently recognized predictor of cardiovascular disease (CVD) [6–8]. Established CVD risk factors are associated with the development of new-onset CKD [9]. The recent increase in the number of ESRD patients with multiple comorbid conditions might reflect a mutual association between CVD and CKD [6, 7]. Metabolic syndrome, which is mostly associated with obesity, is a common phenotype of patients at increased risk for CKD among adults in the US [10].

The effect of obesity on CKD might differ among races. Asians and Japanese are more vulnerable than whites and blacks [11]. The prevalence of obesity-related glomerulopathy is increasing in renal biopsy patients [12]. For the Japanese, there are only a few epidemiologic studies on this important issue [13]. Japan is a rapidly aging society and the relationship between CVD and CKD has a large impact not only for nephrologists, but also other practicing clinicians. Using the community-based screening registries of both the general population and patients with ESRD, we have been studying the significance of obesity on renal diseases. In contrast to CKD patients, obesity is not common in ESRD patients in Japan [1]. More studies of the role of obesity in CKD and ESRD are needed.

Methods

Okinawa comprises subtropical islands that are separated from mainland Japan, so there is relatively little migration of patients, particularly those with ESRD and CVD. The current population is approximately 1.34 million. We accessed data from three independent registries covering the entire area of Okinawa: the registry of the screening program of the Okinawa General Health Maintenance Association (OGHMA) [14–16], the Okinawa Dialysis Study (OKIDS) registry [17, 18], and the hospital-based registry of the Cooperative Study Group of Morbidity and Mortality of CVD in Okinawa (COSMO) [19]. Subjects who had been among the screening participants and those individuals who had had either a stroke or acute myocardial infarction (AMI) or were entering a dialysis program were identified using the screening registry. The identity of these patients was verified using multiple sources of information that included name, sex, birth date, and medical chart review. Only standard analysis files of the OGHMA, OKIDS, and COSMO were used. Approval from the ethics committee or equivalent permission was obtained from each organization. Furthermore, we accessed a hospital-based screening registry to examine the relationship between metabolic syndrome and CKD.

Community-Based Screening Program in Okinawa, Japan

The OGHMA, a non-profit organization founded in 1972, conducts an annual large community-based health examination. Once each year, the staff, doctors, and nurses visit residences and work places throughout the prefecture to perform health examinations. The OGHMA personnel provide mass screening, inform the participants of the results, and, when necessary, recommend further evaluation or treatment. This process includes an interview concerning health status, a physical examination, and urine and blood tests. A nurse or doctor measures blood pressure using a standard mercury sphygmomanometer with the subject in the sitting position. Dipstick testing using an Ames dipstick (Tokyo, Japan) is performed in spontaneously voided fresh urine. Computer-based registry data for standard analysis are available for the 1983 (n = 106,171), 1993 (n = 143,948), and 2003 (n = 154,019) screenings. Approximately 14% of the total adult population participates in each screening registry. OGHMA is the largest provider of health screening in Okinawa. There are other organizations, both profit and non-profit, that also provides screening programs. Subjects already on chronic dialysis were excluded from the data. All subjects voluntarily participated in the screening. Waist circumference is not measured in the OGHMA screening. Therefore, obesity was defined as body mass index (BMI) $\geq 30.0 \text{ kg/m}^2$.

A subgroup of the screening participants who visited the central OGHMA clinic was examined further. These subjects answered questions about various lifestyle habits, including smoking, alcohol consumption, and exercise, as well as their medical history, current medications, and whether or not they had been diagnosed with diabetes mellitus (DM) [20].

ESRD Patient Registry in Okinawa

The details of every ESRD patient treated in Okinawa since 1971 are maintained in an independent community-based registry, the OKIDS studies. All chronic dialysis patients residing in the prefecture who survived for at least 1 month on scheduled dialysis are included in the registry. By the end of 2000, there were 46 dialysis units in Okinawa: 9 in the public sector, 17 in private hospitals, and 20 in clinics. All patients (n = 5,246) were followed-up until the occurrence of a major medical event or until January 2001, whichever occurred first, and all outcomes were verified.

Hospital-Based CVD Registry

All patients who suffered stroke and AMI in Okinawa were registered in the COSMO registry, which filed all hospital cases of stroke and AMI that occurred in Okinawa during the 3-year period from April 1, 1988, to March 31, 1991. The study physicians regularly visited these medical facilities where the patients were supposed to be admitted, and investigated the medical documents. Stroke cases were identified according to the criteria of the Ministry of Health and Welfare, Japan. Criteria for identification of AMI were similar to those used in the MONICA project, and both definite and possible AMI were included. A total of 747 patients with AMI and 3,809 patients with stroke were eligible for the study.

Hospital-Based Screening Registry

To examine the relationship between the number of components of metabolic syndrome and the prevalence of CKD, we analyzed a hospital-based screening registry [21]. Metabolic syndrome was defined using the criteria recommended in the National Cholesterol Education Program (NCEP) Adult Treatment Panel III (ATP III) guidelines [22]. Specifically, elevated blood pressure was defined as systolic or diastolic blood pressure of 130/85 mm Hg or higher; low HDL cholesterol level was defined as less than 1.036 mmol/l (40 mg/dl) in men or less than 1.295 mmol/l (50 mg/dl) in women; high serum triglyceride levels were defined as 1.695 mmol/l (150 mg/dl) or more; and elevated fasting plasma glucose level was defined as 6.10 mmol/l (110 mg/dl) or more. Finally, abdominal obesity was defined as a waist circumference of 102 cm or greater in men or 88 cm or greater in women. Metabolic syndrome was defined as the presence of three or more of these components. Data

Table 1. Demographic changes in the screenees in Okinawa, Japan

Year of screening	1983	1993	2003
Number of screenees	106,171	143,948	154,019
Men, %	47.6	47.6	48.0
Mean age, years	49.1	49.5	49.8
Mean BMI, kg/m²	23.4	23.9	24.1
BMI ≥ 30 kg/m²			
Total, %	3.5	4.7	6.2
Men, %	2.7	4.7	6.9
Women, %	4.3	4.8	5.7
Age, years			
20, %	2.6	3.8	6.0
40, %	4.7	5.9	6.5
60, %	2.9	4.1	6.1

were also analyzed according to the modified NCEP criteria defining abdominal obesity as a waist circumference of 85 cm or greater in men and 90 cm or greater in women. Diabetes was defined as a self-reported history of a previous diagnosis of diabetes or a fasting plasma glucose level of 7.0 mmol/l (126 mg/dl) or higher. Hypertension was defined as systolic blood pressure ≥ 140 mm Hg or diastolic blood pressure ≥ 90 mm Hg. Estimated glomerular filtration rate (GFR) was calculated using the abbreviated Modification of Diet in Renal Disease formula [23]. CKD was defined as a GFR < 60 ml/min/1.73 m² or dipstick proteinuria (≥1+) [24]. Proteinuria was defined by a dipstick urinalysis score of 1+ or more.

Results

Increasing Prevalence of Obesity

The mean BMI of the general population in the US is 26.3 kg/m² [25]; the corresponding figure in Okinawa was 23.4 kg/m². The prevalence of obesity in Okinawa was 3.5% in 1983, 4.7% in 1993, and 6.2% in 2003 (table 1); this is similar to the prevalence of 5.7% in Singapore [26]. Okinawa was under US control from the end of the World War II until 1972. Since their return to Japanese control, the lifestyle on the islands has changed, as Okinawa has become more a part of the industrialized world. Individuals now do less walking and are more likely to become overweight with a BMI of at least 25.0 kg/m². The mean BMI was 24.1 kg/m² in the 2003 screening.

The prevalence of metabolic syndrome in Okinawa (12.4%) [21] is approximately half that reported in the US (24.7%) [10]. When we used the modified

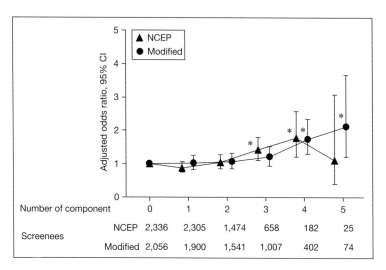

Fig. 1. Relationship between the number of metabolic syndrome components and the prevalence of CKD [21]. *p < 0.01.

NCEP criteria, however, the prevalence of metabolic syndrome increased to 21.2%. Metabolic syndrome was more prevalent in men and elderly people. In younger women (<50 years), the prevalence of metabolic syndrome was lower than that in men; whether this is related to differences in the hormonal environment is not known.

Obesity as a Risk Factor of CKD in the General Population

Obesity has a negative impact on renal disease, and is closely associated with hypertension, hyperlipidemia, and microalbuminuria [27, 28]. In a large cross-sectional study of middle-aged men and women, there was a significant correlation between obesity and urinary albumin excretion [29]. We recently reported that obesity is a significant predictor of developing proteinuria [20] (fig. 1). Furthermore, we demonstrated a strong, positive relationship between the number of metabolic syndrome components and the prevalence of CKD (fig. 2) [21]. The duration of metabolic syndrome might be longer in older participants (≥60 years) than in younger participants (<60 years), however, and this might partly explain the higher prevalence of CKD in the older participants.

To our knowledge, this is the first study of the association between CKD and metabolic syndrome in the Japanese. The cross-sectional study design makes it difficult, however, to infer causality between metabolic syndrome and risk for CKD or proteinuria. A recent follow-up study from Thailand demonstrated that systolic hypertension, hyperuricemia, and elevated BMI ($>24.0 \, kg/m^2$) were

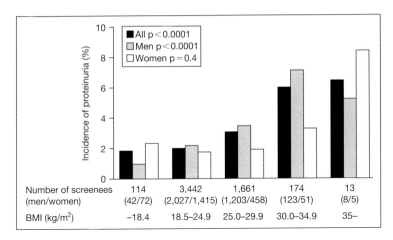

Fig. 2. Relationship between baseline BMI and the development of proteinuria [20].

significant predictors of decreased renal function [30]. We used serum creatinine and calculated GFR to define CKD [24]. Although inulin or iothalamate clearance techniques might provide a more sensitive estimate of renal function, serum creatinine is widely used in large epidemiologic studies and in clinical practice for estimating renal function. Our findings are applicable to clinical and public health practice settings. Several studies demonstrate that the Japanese have a lower GFR than whites or blacks, which is not necessarily due to renal disease [31–33]. Therefore, our results might have underestimated the prevalence of CKD. In contrast to our cohort, CKD was uncommon in those with no metabolic syndrome components in the National Health and Nutrition Examination Survey III [34]. Other parameters associated with obesity and glucose metabolism, such as physical activity [35] and alcohol consumption [36], were not examined in the present study. There was a graded linear relationship between fasting plasma glucose and the prevalence of CKD [37]. High fasting plasma glucose (>126 mg/dl) is a predictor of ESRD [38].

The significance of insulin resistance and metabolic syndrome was recently raised as a factor in renal dysfunction [10, 39]. Kubo et al. [39] demonstrated that hyperinsulinemia is a significant risk factor for renal dysfunction. The results support the notion that hyperinsulinemia has a significant role in renal dysfunction, and therefore it is important to recognize the risk of end organ complications [40]. In our cohort, the initial event in the development of metabolic syndrome seemed to be related to high blood pressure and abdominal obesity [21]. The impact of high triglyceride levels or abdominal obesity seemed to be larger in younger participants. Triglyceride-rich apo B-containing lipoproteins clearly promote the progression of human renal insufficiency [41].

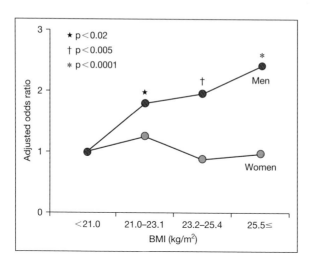

Fig. 3. Baseline BMI and the risk of ESRD in the screenees in Okinawa, Japan [15].

High triglyceride levels are a risk factor for developing proteinuria [42]. As the GFR falls, protein and energy intake might decline in moderate and advanced CKD [43]. The relationship between obesity and metabolic impairment might differ according to ethnicity [44, 45]. Ramirez et al. [26] reported a J-shaped function between proteinuria and BMI; patients with a very low BMI had an increased risk of proteinuria. They also demonstrated racial differences in the relationship between BMI and proteinuria. The factors related to these racial differences, however, need to be determined.

Another potential component of metabolic syndrome is hyperuricemia. Hyperuricemia is often associated with metabolic syndrome and other conventional risk factors of CVD [46–48], and is a risk factor of renal dysfunction [49]. Uric acid might be a cause of metabolic syndrome, possibly due to its ability to inhibit endothelial function. The ingestion of fructose might have a major role in the epidemic of metabolic syndrome and obesity due to its ability to increase uric acid [50].

Evidence for the Role of Obesity in Increasing ESRD

There are two papers in which the contribution of obesity to the risk of ESRD was examined using a large population [15, 51]. The results indicated that an association between an increased BMI and an increased risk for developing ESRD in men (fig. 3) [15]. These data suggest that the maintenance of an optimal body weight might reduce the risk of ESRD, independent of the effect of blood pressure and proteinuria. Recently, Hsu et al. [51] extended our observation. Given the increased morbidity and mortality of obese patients, our data might underestimate the true importance of BMI to the risk of ESRD. Both studies measured BMI on only one occasion; the effects of weight loss remain to be shown in a prospective study.

Obese patients are at higher risk for glomerulomegaly and focal segmental glomerulosclerosis [12]. Among patients who underwent uninephrectomy or have IgA nephropathy, obesity is associated with more rapid loss of renal function [52]. There are several plausible mechanisms for the association between BMI and ESRD. High BMI is a risk factor for hypertension, DM, and hyperlipidemia, all of which increase the risk for ESRD. Whether the effect of increased BMI is dependent on its identification as a risk factor for DM or hypertension was not fully excluded in our study. The number of glomeruli is dependent on birth weight, and does not increase after birth. Patients with hypertension [53] or low birth weight have a low number of nephrons. The incidence of low birth weight (defined as weight $< 2,500\,g$) in Okinawa is 7.4–9.3%, and is higher than the US national average of 5.7% [24]. Low birth weight is linked to the subsequent development of hypertension and renal failure [54]. This is more evident in Okinawa, where the prevalence of obesity has recently increased. Interestingly, although the number of glomeruli is different for each individual, the average number of glomeruli per person does not differ by sex [55].

There are several ethnic groups with a low nephron number [56]. Being overweight leads to renal hyperperfusion and glomerular hyperfiltration, which in turn causes proteinuria and focal segmental glomerulosclerosis [57]. Leptin, which is produced from adipose tissue, might directly lead to renal fibrosis. It is also possible that the risk conferred by being overweight or obese might be different in those with normal BMI. Body weight is related to current health status or health practices. For example, smoking or chronic disease might lead to weight loss.

Increasing Incidence of ESRD (Data from the Japanese Society for Dialysis Therapy)

The number of patients with ESRD requiring chronic dialysis therapy is increasing worldwide, and therefore the medical cost is escalating. The prevalence of ESRD is close to 2,000 per million population in Japan, and DM nephropathy as a cause of ESRD was more than 40% of the incident dialysis patients in 2004. The number is still increasing linearly, because the acceptance policy for ESRD is quite open in Japan. The sex difference in ESRD incidence is getting larger (fig. 4) [58, 59]. Sex differences in the progression of various renal diseases are reported. Men are at higher risk of developing ESRD, and tend to develop ESRD earlier in life than women [60–62]. This might be partly due to the influence of testosterone and other sex hormones on the risk of proteinuria and glomerular sclerosis [63]. In addition, differences in the risk

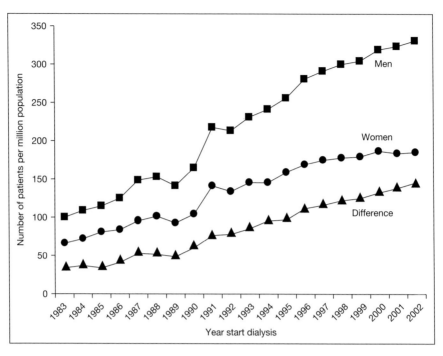

Fig. 4. Incidence of ESRD in Japan [59].

conferred by being overweight or obese contribute to the sex differences in the risk for ESRD [15].

In Japan, Okinawa has the highest incidence of ESRD [1, 64]. Factors explaining such phenomena are likely targets for intervention to reduce ESRD [65].

CKD as a Risk Factor of CVD and ESRD

CKD is a new risk factor of CVD. Few epidemiologic studies, however, have examined the relationship between CKD and the risk of developing CVD in a general population. We evaluated the significance of CKD on the incidence of CVD using large community-based mass screening registries in Okinawa, Japan [66]. Screenees of the 1983 OGHMA were investigated to determine whether they also registered in the COSMO for CVD and OKIDS for ESRD. All relative risks of CVD and dialysis were adjusted for sex, blood pressure, and the presence of CKD at the time of screening. Among the 13,983 screenees, there were 7,477 subjects with CKD. Of these subjects, there were 121 CVD patients and 116 ESRD patients. The presence of CKD is a significant predictor of CVD, with an adjusted hazard ratio of 2.650 and a 95% confidence interval of 1.693–4.148, $p < 0.0001$. CKD was defined as the presence of proteinuria or an estimated creatinine clearance of <60 ml/min. The results support the notion that the presence of CKD is an important predictor of CVD. Recently, another study from Japan demonstrated that CKD is a significant risk factor for CVD [67]. On the other hand, patients with CVD have a higher risk for developing ESRD, particularly in patients under 60 years of age [68].

There are several biases inherent in the screened population [66]. Subjects who chose to participate in the screening were generally healthy individuals who were interested in their health. Individuals who were already diagnosed with renal failure or cancer might have been less likely to participate in the screening. The incidence of primary renal diseases in the screened ESRD patients was slightly different than in the entire ESRD Okinawa population during the study period. ESRD was caused by DM in 24.3% of screened ESRD patients, but in 33.7% of ESRD patients in the general Okinawa population during this study period. Low income and education, potential risk factors for the development of CKD [24], were also not measured in our study. There are ethnic differences in DM-induced organ damage and in the number of affected nephrons. Asians tend to have more renal damage than whites [69].

Prevention of CKD and ESRD: Early Detection and Treatment
Obesity is a major health issue, including CKD and ESRD, in the developed countries and now also in developing countries. In Asia, there are many countries in which renal replacement therapy is not readily available. Therefore, when ESRD develops, the prognosis is dismal. The prevention and treatment of obesity, at least in men, is an important strategy for reducing the burden of ESRD. Early detection of metabolic syndrome or high triglyceride levels might be beneficial if accompanied by early intervention, such as statins to lower triglyceride levels and suppress the pathways for renal injury. Weight reduction is associated with a decrease in proteinuria [29, 70]. While screening for early identification of metabolic syndrome might improve health and prevent the progression of renal disease; it could also lead to unnecessary harm and excessive cost. Whether the measurement of abdominal circumference in addition to body weight and height is superior to current screening practices in Japan [5, 21] remains to be examined.

Obesity is a profoundly socio-economic issue, and medical approaches might not be effective [71]. Political authorities should recognize the immense values of screening programs [5].

Reverse Epidemiology: Obesity and Survival in
Chronic Hemodialysis Patients
Obesity has a significant impact on survival in both dialysis and renal transplant patients [72, 73]. The prognosis is somewhat better in those with higher BMI than those with normal to low BMI. Several epidemiologic studies

demonstrate a close relationship between these variables (blood pressure and BMI) and survival, but it is quite different from that of the general population (reverse epidemiology) [74–77]. The untoward effect of obesity, which is a conventional risk factor for CVD in the general population, might take a long time, years and decades, to manifest; therefore, it might not significantly impact survival in chronic dialysis patients. The effect of obesity on survival might differ among races [78]. Asians and Japanese are more vulnerable than whites and blacks. In contrast to HD patients, obesity is common in CKD patients [79, 80].

The mechanisms of this phenomenon are not known, however, the mortality rate is high even in those with high BMI when compared to the normal population. In the general population, the mortality reaches its highest rate among people in the top BMI categories [81]. The reverse epidemiology is also seen in several conditions such as rheumatoid arthritis, congestive heart failure, recovery from heart surgery, and also among the aged [82]. Patients with malnutrition–inflammation complex (cachexia) syndrome or malnutrition–inflammation atherosclerosis syndrome are at high risk of death, therefore should be under vigorous scrutiny. These results suggest that we need to explore specific risk factors or treatment modalities in such HD patients.

Growing confusion has developed among physicians, some of whom are no longer confident about whether to treat obesity in ESRD patients. At least, it would be advisable to avoid malnutrition or too much calorie restriction for patients at the pre-ESRD and ESRD stages. Protein restriction might be effective to retard the progression of CKD with a combination of renin–angiotensin suppression [83].

Conclusion

Analysis of the community-based screening program demonstrated that obesity has a significant role in the development and survival in CKD and ESRD, particularly in men. Differences in lifestyle between men and women might underlie the sex difference. Asians are more vulnerable than whites and others. Therefore, more research is needed on this important public problem. In developing countries, people might not have proper access to healthier foods and information [84, 85]. People in Okinawa, which has changed from a developing area to a developed area in just a half century, have experienced rapid changes in lifestyle and economics. The Okinawa experience might help us to understand the effect of such social changes in renal disease [5]. From a public health perspective, encouraging healthy weight and physical activity behaviors to control the obesity epidemic will probably be the greatest benefit of the current focus on metabolic syndrome.

Finally, the author would like to express sincere gratitude to the colleagues in Okinawa who supported the epidemiologic studies, in particular the staff of the OGHMA and dialysis units.

References

1. Nakai S, Shinzato T, Sanaka T, et al: The current state of chronic dialysis treatment in Japan. J Jpn Soc Dial Ther 2002;35:1155–1184.
2. US Renal Data System: Excerpts from the USRDS 2001 Annual Data Report: Atlas of End-Stage Renal Disease in the United States. Am J Kidney Dis 2001;38:(suppl 3):S1–S248.
3. Schena FP: Epidemiology of end-stage renal disease: international comparisons of renal replacement therapy. Kidney Int 2000;57:(suppl 74):S39–S45.
4. El Nahas M: The global challenge of chronic kidney disease. Kidney Int 2005;68:2918–2929.
5. Iseki K: Screening for renal disease – what can be learned from Okinawa experience. Nephrol Dial Transplant 2006;21:839–843.
6. Sarnak MJ, Levey AS, Schoolwerth AC, Coresh J, Culleton, Hamm LL, et al: Kidney disease as a risk factor for development of cardiovascular disease: a statement from the American Heart Association Councils on Kidney in Cardiovascular Disease, High Blood Pressure research, Clinical Cardiology, and Epidemiology and Prevention. Circulation 2003;108:2154–2169.
7. Go AS, Chertow GM, Fan D, et al: Chronic kidney disease and the risks of death, cardiovascular events, and hospitalization. N Engl J Med 2004;351:1296–1305.
8. Anavekar NS, McMurray JJV, Velazquez EJ, et al: Relation between renal dysfunction and cardiovascular outcomes after myocardial infarction. N Engl J Med 2004;351:1285–1295.
9. Fox CS, Larson MG, Leip EP, Culleton B, Wilson PWF, Levy D: Predictors of new-onset kidney disease in a community-based population. JAMA 2004;291:844–850.
10. Chen J, Muntner P, Hamm LL, Jones DW, Batuman V, Fonseca V, Whelton PK, He J: The metabolic syndrome and chronic kidney disease in U.S. adults. Ann Intern Med 2004;140:167–174.
11. Johansen K, Young B, Kaysen GA, Chertow GM: Association of body size with outcomes among patients beginning dialysis. Am J Clin Nutr 2004;80:324–332.
12. Kambham N, Markowitz GS, Valeri AM, et al: Obesity-related glomerulopathy: an emerging epidemic. Kidney Int 2001;59:1498–1509.
13. Iseki K, Miyasato F, Uehara H, Tokuyama K, Toma S, Nishime K, Yoshi S, Shiohira Y, Oura T, Tozawa M, Fukiyama K: Outcome study of renal biopsy patients in Okinawa, Japan. Kidney Int 2004;66:914–919.
14. Iseki K, Iseki C, Ikemiya Y, Fukiyama K: Risk of developing end-stage renal disease in a cohort of mass screening. Kidney Int 1996;49:800–805.
15. Iseki K, Ikemiya Y, Kinjo K, Inoue T, Iseki C, Takishita S: Body mass index and the risk of development of end-stage renal disease in a screened cohort. Kidney Int 2004;65:1870–1876.
16. Iseki K, Ikemiya Y, Fukiyama K: Risk factors of end-stage renal disease and serum creatinine in a community-based mass screening. Kidney Int 1997;51:850–854.
17. Iseki K, Kawazoe N, Osawa A, Fukiyama K: Survival analysis of dialysis patients in Okinawa, Japan (1971–1990). Kidney Int 1993;43:404–409.
18. Iseki K, Tozawa M, Iseki C, Takishita S, Ogawa Y: Demographic trends in the Okinawa Dialysis Study (OKIDS) registry (1971–2000). Kidney Int 2002;61:668–675.
19. Kinjo K, Kimura Y, Shinzato Y, Tomori M, Komine Y, Kawazoe N, Takishita S, Fukiyama K, COSMO Group: An epidemiological analysis of cardiovascular diseases in Okinawa, Japan. Hypertens Res 1992;15:111–119.
20. Tozawa M, Iseki K, Iseki C, Oshiro S, Ikemiya Y, Takishita S: Influence of smoking and obesity on the development of proteinuria. Kidney Int 2002;62:956–962.
21. Tanaka H, Shiohira Y, Uezu Y, Higa A, Iseki K: Metabolic syndrome and chronic kidney disease in Okinawa, Japan. Kidney Int 2006;69:369–374.

22 National Cholesterol Education Program: Executive Summary of the Third Report of National Cholesterol Education Program (NCEP): expert panel on detection, evaluation, and treatment of high blood cholesterol in adults (Adults Treatment Panel III). JAMA 2001;285:2486–2497.

23 Levey AS, Bosch JP, Lewis JB, et al: A more accurate method to estimate glomerular filtration rate from serum creatinine: a new prediction equation. Ann Intern Med 1999;130:461–470.

24 National Kidney Foundation: K/DOQI Clinical Practice Guidelines for Chronic Kidney Disease: Evaluation, Classification and Stratification. Am J Kidney Dis 2002;39(suppl 1):S1–S266.

25 Kuczmarski RJ, Flegal KM, Campbell SM, Johnson CL: Increasing prevalence of overweight among US adults. JAMA 1994;272:205–211.

26 Ramirez SP, McClellan W, Port FK, Hsu SIH: Risk factors for proteinuria in a large, multiracial, southeast Asian population. J Am Soc Nephrol 2002;13:1907–1917.

27 Spangler JG, Konen JC: Hypertension, hyperlipidemia, and abdominal obesity and the development of microalbuminuria in patients with non-insulin-dependent diabetes mellitus. J Am Board Fam Pract 1996;9:1–6.

28 Reid M, Bennett F, Wilks R, Forrester T: Microalbuminuria, renal function and waist: hip ratio in black hypertensive Jamaicans. J Hum Hypertens 1998;12:221–227.

29 Metcalf P, Baker J, Scott A, et al: Albuminuria in people at least 40 years old: effect of obesity, hypertension, and hyperlipidemia. Clin Chem 1992;38:1802–1808.

30 Domrongkitchaiporn S, Sritara P, Kitiyakara C, Stitchantrakul W, Krittaphol V, Lolekha P, Cheepudomwit S, Yipintsoi T: Risk factors for development of decreased kidney function in a Southeast Asian population: a 12-year cohort study. J Am Soc Nephrol 2005;16:791–799.

31 Orita Y, Okada M, Harada S, Horio M: Skim soy protein enhances GFR as much as beefsteak protein in healthy human subjects. Clin Exp Nephrol 2004;8:103–108.

32 Hosoya T, Toshima R, Icida K, et al: Changes in renal function with aging among Japanese. Intern Med 1995;34:520–527.

33 Coresh J, Astor BC, Greene T, et al: Prevalence of chronic kidney disease and decreased kidney function in the adult US population: Third National Health and Nutrition Examination Survey. Am J Kidney Dis 2006;41:1–12.

34 Must A, Spadano J, Coakley EH, Field AE: The disease burden associated with overweight and obesity. JAMA 1999;282:1523–1529.

35 Hu G, Lindstrom J, Valle TT, et al: Physical activity, body mass index, and risk of type 2 diabetes in patients with normal or impaired glucose regulation. Arch Intern Med 2004;164:892–896.

36 Howard AA, Amsten JH, Gourevitch MN: Effect of alcohol consumption on diabetes mellitus. Ann Intern Med 2004;140:211–219.

37 Iseki K, Oshiro S, Tozawa M, Ikemiya Y, Fukiyama K, Takishita S: Prevalence and correlates of diabetes mellitus in a screened cohort in Okinawa, Japan. Hypertens Res 2002;25:185–190.

38 Iseki K, Ikemiya Y, Kinjo K, Iseki C, Takishita S: Prevalence of high fasting plasma glucose and risk of developing end-stage renal disease in a screened cohort. Clin Exp Nephrol 2004;8:250–256.

39 Kubo M, Kiyohara Y, Kato I, et al: Effect of hyperinsulinemia on renal function in a general Japanese population: the Hisayama study. Kidney Int 1999;55:2450–2456.

40 Diabetes Prevention Program Research Group: Reduction in the incidence of type 2 Diabetes with lifestyle intervention or metformin. N Engl J Med 2002;346:393–403.

41 Samuelsson O, Attman PO, Knight-Gibson C, et al: Complex apolipoprotein B-containing lipoprotein particles are associated with a higher rate of progression of human chronic renal insufficiency. J Am Soc Nephrol 1998;9:1482–1488.

42 Tozawa M, Iseki K, Iseki C, et al: Triglyceride, but not total cholesterol or low-density lipoprotein levels, predict development of proteinuria. Kidney Int 2002;62:1743–1749.

43 Kopple JD, Greene T, Chumlea WC, et al: Relationship between nutritional status and the glomerular filtration rate: Results from the MDRD study. Kidney Int 2000;57:1688–1703.

44 Simmons D, Williams DRR, Powell MJ: Prevalence of diabetes in a predominantly Asian community: preliminary findings of the Coventry diabetes study. Br Med J 1989;298:18–21.

45 Samanta A, Burden AC, Jagger C: A comparison of the clinical features and vascular complications of diabetes between migrant Asians and Caucasians in Leicester, U.K. Diabetes Res Clin Pract 1991;14:205–214.

46   Iseki K, Oshiro S, Tozawa M, Iseki C, Ikemiya Y, Takishita S: Significance of hyperuricemia on the early detection of renal failure in a cohort of screened subjects. Hypertens Res 2001;24: 691–697.
47   Nagahama K, Iseki K, Inoue T, Touma T, Ikemiya Y, Takishita S: Hyperuricemia and cardiovascular risk factor clustering in a screened cohort in Okinawa, Japan. Hypertens Res 2004;27:227–234.
48   Nagahama K, Inoue T, Iseki K, Touma T, Kinjo K, Ohya Y, Takishita S: Hyperuricemia as a predictor of hypertension in a screened cohort in Okinawa, Japan. Hypertens Res 2005;27:835–841.
49   Johnson RJ, Kivlighn SD, Kim YG, Suga S, Fogo AB: Reappraisal of the pathogenesis and consequences of hyperuricemia in hypertension, cardiovascular disease, and renal disease. Am J Kidney Dis 1999;33:225–234.
50   Nakagawa T, Hu H, Zharikov S, et al: A causal role for uric acid in fructose-induced metabolic syndrome. Am J Physiol Renal Physiol 2006;290:F625–F631.
51   Hsu CY, McCulloch CE, Iribarren C, Darbinian J, Go AS: Body mass index and risk for end-stage renal disease. Ann Intern Med 2006;44:21–28.
52   Brenner BM, Cherton GM: Congenital oligonephropathy and the etiology of adult hypertension and progressive renal injury. Am J Kidney Dis 1994;23:171–175.
53   Keller G, Zimner G, Mall G, et al: Nephron number in patients with primary hypertension. N Engl J Med 2003;348:101–108.
54   Luft FC: Food intake and the kidney: the right amounts at the right times. Am J Kidney Dis 2000;37:629–631.
55   Hopper J Jr, Trew PA, Biava CG: Membranous nephropathy: its relative benignity in women. Nephron 1981;29:18–24.
56   Hughson M, Farris AB 3rd, Douglas-Denton R, Hoy WE, Bertram JF: Glomerular number and size in autopsy kidneys: the relationship to birth weight. Kidney Int 2003;63:2113–2122.
57   Brenner BM, Chertow GM: Congenital oligonephropathy and the etiology of adult hypertension and progressive renal injury. Am J Kidney Dis 1994;23:171–175.
58   Wakai K, Nakai S, Kikuchi K, Iseki K, Miwa N, Masakane I, Wada A, Shinzato T, Nagura Y, Akiba T: Trends in incidence of end-stage renal disease in Japan, 1983–2000: age-adjusted and age-specific rates by gender and cause. Nephrol Dial Transplant 2004;16:2044–2052.
59   Iseki K, Nakai S, Shinzato T, Nagura Y, Akiba T: Increasing difference between sexes in the incidence of chronic dialysis patients in Japan. Ther Aphe Dial 2005;9:407–411.
60   Donadio JV Jr, Torres VE, Velosa JA, et al: Idiopathic membranous nephropathy: the natural history of untreated patients. Kidney Int 1988;33:708–715.
61   Gretz N, Zeier M, Gerberth S, Strauch M, Ritz E: Is gender a determinant for evolution of renal disease? A study in autosomal dominant polycystic kidney disease. Am J Kidney Dis 1989;14: 178–183.
62   Sakemi T, Toyoshima H, Morito F: Testosterone eliminates the attenuating effect of castration on the progressive glomerular injury in hypercholesterolemic male Imai rats. Nephron 1994;67: 469–476.
63   Horner D, Fliser D, Klimm HP, Ritz E: Albuminuria in normotensive and hypertensive individuals attending offices of general practitioners. J Hypertens 1996;14:655–660.
64   Usami T, Koyama K, Takeuchi O, et al: Regional variation in the incidence of end-stage renal failure in Japan. JAMA 2000;284:2622–2624.
65   Iseki K: Factors influencing development of end-stage renal disease. Clin Exp Nephrol 2005;9:5–14.
66   Iseki K, Iseki C, Okumura K, Kinjo K, Takishita S: Chronic kidney disease and the risk of developing cardiovascular disease events: a community-based approach. Vasc Dis Prev 2006, in press.
67   Ninomiya T, Kiyohara Y, Kubo M, et al: Chronic kidney disease and cardiovascular disease in a general Japanese population: The Hisayama Study. Kidney Int 2005;68:228–236.
68   Iseki K, Wakugami K, Maehara A, Tozawa M, Muratani H, Fukiyama K: Evidence for high incidence of end-stage renal disease in patients after stroke and acute myocardial infarction at age 60 or younger. Am J Kidney Dis 2001;38:1235–1239.
69   Mandavilli A, Cyranoski D: Asia's big problem. Nat Med 2004;10:325–327.
70   Praga M, Hernandez E, Andres A, et al: Effects of body-weight loss and captopril treatment on proteinuria associated with obesity. Nephron 1995;70:35–41.

71 McCarthy M: The economics of obesity. Lancet 2004;364:2169–2170.
72 Kenchaiah S, Evans JC, Levy D, et al: Obesity and the risk of heart failure. N Engl J Med 2002;347:305–313.
73 Jindal RM, Zawada ET Jr: Review. Obesity and kidney transplantation. Am J Kidney Dis 2004;43:943–952.
74 Iseki K, Miyasato F, Tokuyama K, et al: Low diastolic blood pressure, hypoalbuminemia, and risk of death in a cohort of chronic hemodialysis patients. Kidney Int 1997;51:1212–1217.
75 Kalantar-Zadeh K, Block G, Humphreys MH, Kopple JD: Reverse epidemiology of cardiovascular risk factors in maintenance dialysis patients. Kidney Int 2003;63:793–808.
76 Kalantar-Zadeh K, Kilpatrick RD, McAllister CJ, et al: Reverse epidemioogy and cardiovascular death in the hemodialysis population. Hypertension 2005;45:811–817.
77 Kalantar-Zadeh K, Abbott KC, Salahudeen AK, et al: Survival advantages of obesity in dialysis patients. Am J Clin Nutr 2005;81;543–554.
78 Wong JS, Port FK, Hulbert-Shearon TE, et al: Survival advantage in Asian American end-stage renal disease patients. Kidney Int 1999;55:2515–2523.
79 Hayashi T, Boyko EJ, Leonetti DL, et al: Visceral adiposity is an independent predictor of incident hypertension in Japanese Americans. Ann Intern Med 2004;140:992–1000.
80 Meigs JB, Wilson PWF, Nathan DM, et al: Prevalence and characteristics of the metabolic syndrome in the San Antonio Heart and Framingham Offspring Studies. Diabetes 2003;52:2160–2167.
81 Fontaine KR, Redden DT, Wang C, Westfall AD, Allison DB: Years of life lost due to obesity. JAMA 2003;289:187–193.
82 Escalante A, Haas RW, del Rincon I: Paradoxical effect of body mass index on survival in rheumatoid arthritis. Arch Intern Med 2005;165:1624–1629.
83 Nakao N, Yoshimura A, Morita H, Takada M, Kayano T, Ideura T: Combination treatment of angiotensin-II receptor blocker and angiotensin-converting-enzyme inhibitor in non-diabetic renal disease (COOPERATE): a randomised controlled trial. Lancet 2003;361:117–124.
84 Jafar TH: The growing burden of chronic kidney disease in Pakistan. N Engl J Med 2006;354: 995–997.
85 Barsoum RS: Chronic kidney disease in the developing world. N Engl J Med 2006;354:997–999.

Kunitoshi Iseki, MD
Dialysis Unit, University Hospital of The Ryukyus
207 Uehara, Nishihara
Okinawa 903–0215 (Japan)
Tel. +81 98 895 3331, Ext. 2360, Fax +81 98 895 1416
E-Mail chihokun@med.u-ryukyu.ac.jp

Obesity Paradox in Patients on Maintenance Dialysis

Kamyar Kalantar-Zadeh, Joel D. Kopple

Division of Nephrology and Hypertension, Los Angeles Biomedical Institute at Harbor-UCLA Medical Center, Torrance, Calif.; David Geffen School of Medicine at UCLA and UCLA School of Public Health, Los Angeles, Calif., USA

Abstract

Overweight (body mass index [BMI] = 25–30 kg/m^2) and obesity (BMI > 30 kg/m^2) have become mass phenomena with a pronounced upward trend in prevalence in most countries throughout the world and are associated with increased cardiovascular risk and poor survival. In patients with chronic kidney disease (CKD) undergoing maintenance hemodialysis an 'obesity paradox' has been consistently reported, i.e., a high BMI is incrementally associated with better survival. While this 'reverse epidemiology' of obesity is relatively consistent in maintenance hemodialysis patients, studies in peritoneal dialysis patients have yielded mixed results. A similar obesity paradox has been described in patients with chronic heart failure as well as in 20 million members of other distinct medically 'at risk' populations in the USA. Possible causes of the reverse epidemiology of obesity include: (1) time-discrepancies between the competing risks for the adverse events that are associated with overnutrition and undernutrition; (2) sequestration of uremic toxins in adipose tissue; (3) selection of a gene pool favorable to longer survival in dialysis patients during the course of CKD progression, which eliminates over 95% of the CKD population before they commence maintenance dialysis therapy; (4) a more stable hemodynamic status; (5) alterations in circulating cytokines; (6) unique neurohormonal constellations; (7) endotoxin–lipoprotein interactions; and (8) reverse causation. Examining the causes and consequences of the obesity paradox in dialysis patients can improve our understanding of similar paradoxes observed both for other conventional risk factors in chronic dialysis patients, such as blood pressure and serum cholesterol, and in other populations, such as patients with heart failure, cancer or AIDS or geriatric populations.

Copyright © 2006 S. Karger AG, Basel

Founding Source: Supported by a Young Investigator Award from the National Kidney Foundation, a research grant from DaVita, Inc., and the National Institute of Diabetes, Digestive and Kidney Disease grant # DK61162 (for KKZ).

Introduction

Individuals with chronic kidney disease (CKD) stage 5 who undergo maintenance dialysis treatment have a high mortality rate, currently 20% per year in the USA and 10–15% in Europe [1]. This high mortality has not changed substantially in recent years despite many advances in dialysis techniques and patient care [2]. Cardiovascular disease is the main cause of death in dialysis patients [3]. The currently estimated chronic dialysis population of 350,000 patients in the USA grows constantly and fast, and is projected to reach over one-half million by 2010 and over one million by 2020 [2].

It was once believed that the traditional cardiovascular risk factors and/or conditions related to dialysis treatment and technique are the main causes of poor clinical outcome; however, recent randomized controlled trials including the 4D Trial [4] and the HEMO and ADEMEX studies [5, 6] failed to show an improvement of mortality by lowering serum cholesterol levels or by increasing dialysis dose, respectively. Hence, it is not unlikely that conditions other than the traditional risk factors are related to the enormous cardiovascular epidemic and high death rate in this population.

An increasing number of epidemiologic studies, based on analyses of large samples of dialysis patients and national databases, have indicated paradoxical and inverse associations between classical cardiovascular risk factors and mortality in dialysis patients [7, 8]. Indeed, a worse survival among dialysis patients has been observed with a *low*, rather than a high, body mass index (BMI) [9] or weight-for-height [10], blood pressure [11], and serum concentrations of cholesterol [12], homocysteine [13] and creatinine [14]. Even more ironically are findings indicating that *high* values of these risk factors are paradoxically protective and associated with improved survival. This phenomenon has been referred to as 'reverse epidemiology' [7] or 'altered risk factor pattern' [8]. These epidemiologic findings have contributed to the growing confusion and have left physicians with the ongoing dilemma as to whether or not to treat obesity, hypercholesterolemia, hypertension, or hyperhomocysteinemia in chronic dialysis patients [8]. Among the above-mentioned cardiovascular risk factors that are inversely associated with mortality, the so-called 'obesity paradox' has been the most consistent and most extensively studied [15, 16]. In this manuscript, the paradoxical predictability of mortality of measures of body size and several hypotheses that have been advanced to explain these paradoxes are reviewed critically.

Obesity in the General Population

In recent years, overweight (BMI = $25-30\,kg/m^2$) and obesity (BMI > $30\,kg/m^2$) have become mass phenomena with a pronounced upward trend in prevalence in virtually all developed and developing countries [17, 18]. The prevalence of obesity has reached epidemic proportions in the USA: It ranged between 13 and 15% between 1960 and 1980 [19–21], but doubled to 23 and 31% during 1988–1994 and 1999–2000, respectively [17, 22]. Obesity is a strong risk factor for the development of diabetes mellitus, atherosclerotic cardiovascular disease, cancer and even CKD [23–26]. However, despite detrimental effects of being overweight, 'obese nations live paradoxically longer than ever' [18]. The increasing prevalence of obesity may be understood in the light of evolution, because energy metabolism is asymmetric with energy accumulation being the necessary condition of survival during 'hard times' [18]. According to this theory, this genetic characteristic, the so-called thrifty gene(s), was necessary for survival of humanity, because during the course of history there was never a long period of uninterrupted food abundance, whereas famines and other hardships occurred frequently. Therefore, fat accumulation, when food was available, meant survival at times of hardship. In contrast, the potential detrimental effects of overnutrition and overweight generally only became manifest at an older age to which most people did not live and therefore were not very relevant to survival [18]. The foregoing model may explain why in chronic disease states obesity confers survival advantages [27].

Obesity and Survival in Hemodialysis Patients

Obesity has recently been reported to be a risk factor of the CKD progression to stage 5 [26], although this epidemiologic observation may be severely confounded by a strong survival bias [28], especially because many CKD patients die before they reach stage 5 [29]. Maintenance hemodialysis (MHD) patients appear to have a lower BMI as compared with age- and sex-matched controls from the general population [30–32].

Most epidemiologic studies have shown an inverse association between larger body size and lower death risk in MHD patients, independent of other markers of nutritional status [15]. The Diaphane Collaborative Study [33] appears to be the first to report the association between low BMI and high death rate in a cohort (1972–1978) of 1,453 mostly nondiabetic French MHD patients, which was confirmed 15 years later by Leavey et al. [34] (3,607 MHD patients) using the United States Renal Data System (USRDS) database. Fleischmann et al. [35] identified for the first time the so-called 'obesity paradox', i.e., a significantly higher survival rate in overweight and obese MHD patients (BMI $\geq 27.5\,kg/m^2$) compared to those with a normal weight (BMI = 20–$27.5\,kg/m^2$) and underweight (BMI $< 20\,kg/m^2$) [35]. Wolf et al. [36] (9,165 MHD patients) and Port et al. [37] (45,967 incident MHD patients) reported similar paradoxical associations in the USRDS database. The *Dialysis Outcomes and Practice Patterns Study* (DOPPS) [38] (9,714 MHD patients in the USA and Western Europe from 1996 to 2000) confirmed an inverse BMI–mortality relationship in MHD subpopulations defined by continent, race, gender, tertiles of severity of illness and comorbid conditions. Glanton et al. [39] (151,027 incident dialysis patients from the USRDS) found that the obesity paradox was not uniform across different gender and race/ethnicity subgroups and was stronger in African-Americans. Johansen et al. [40] (418,055 maintenance dialysis patients in the USRDS data) found that even morbid obesity was associated with increased survival, except for in Asian-Americans. High BMI and Benn's index were also associated with a reduced risk of hospitalization. Survival rates based on estimates of adiposity and fat mass yielded similar results, and adjustments of body weights for differences in lean body mass did not substantially alter the paradoxical associations [40].

Some investigators have studied body size surrogates other than BMI. Kopple et al. [10] examined weight adjusted for height percentiles in 12,965 MHD patients from the Fresenius database and found that those patients with greater weight-for-height had lower mortality rates. Lowrie et al. [41] (43,334 MHD patients from the Fresenius database) examined body surface area and weight divided by height (wt/ht) in addition to the BMI. The log of risk decreased in rough linear fashion for weight, weight-for-height, and body surface area [41]. Beddhu et al. [42] (70,028 incident MHD patients in the USRDS database) used 24-hour urine creatinine excretion as a indicator of muscle mass and concluded that higher muscle mass was a stronger predictor of survival than was higher total body weight in heavy MHD patients. However, their data showed that obesity was associated with better survival within each fat/muscle category [42]. The inherent associations of urine creatinine with renal function, muscle mass and meat intake may restrict the generalizability of the foregoing conclusions [43].

Kalantar-Zadeh et al. [9] recently examined the effects of both absolute magnitude of BMI (using the time-dependent Cox model) and changes in BMI over time on all-cause and cardiovascular mortality in a 2-year nonconcurrent cohort of 54,535 MHD patients in the national database of the second largest dialysis care provider in the USA (DaVita, Inc.). They found that obesity, including morbid obesity (BMI $> 35\,kg/m^2$), was associated with survival advantages in virtually all subgroups of age, gender, race, dialysis vintage, serum albumin, and Kt/V (fig. 1). Moreover, they showed for the first time that independently of almost any BMI level, weight loss is associated with increased mortality, whereas weight gain confers survival advantages [9] (fig. 2). Finally, in another recent study, Kalantar-Zadeh et al. [44] measured total body fat directly in 535 MHD patients over 3.5 years and found that not only a lower

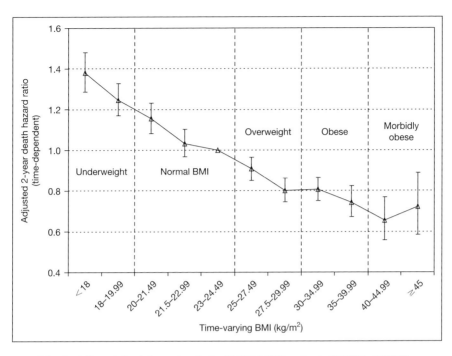

Fig. 1. Adjusted death hazard ratio in 54,535 MHD patients (7/2001–6/2003); recreated bases on data from reference [9].

total body fat was associated with higher mortality, but loss of body fat over time was associated with increased death risk.

Obesity and Survival in Peritoneal Dialysis Patients

Some [45–50], but not all [51–53], studies in chronic peritoneal dialysis (CPD) patients have reported inverse weight–mortality relationships [15]. In the CANUSA study, a 1% difference in the percent lean body mass was associated with a 3% change in the relative risk of death [45, 46]. McCusker et al. [47], Chung et al. [49], and Johnson et al. [48] found a significantly decreased survival rate in CPD patients with a lower body weight. The largest epidemiologic study of body weight and survival in CPD patients included nearly 46,000 CPD patients in 1990s [50], and showed that overweight and obese CPD patients had longer survival than those with normal BMI. These findings could not be adequately explained by lower rates of renal transplantation or lower technique survival rates. Abbott et al. [52] compared 1,675 MHD and 1,662 CPD patients and found that

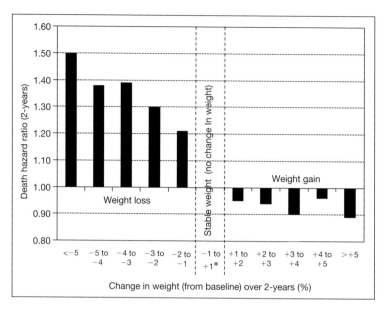

Fig. 2. Relative risk of death for changes in weight over time; recreated bases on data from reference [9].

5-year survival based on BMI cutoff of 30 kg/m^2 in CPD patients was not different than in MHD.

Several studies in CPD patients either have not found any survival advantage for obesity or have indicated a higher risk of death in obese CPD patients [53]. McDonald et al. [51] examined 9,679 CPD patients in Australia or New Zealand over an 11-year interval and found that obesity was independently associated with death and technique failure except among patients of New Zealand Maori/Pacific Islander origin. Stack et al. [54] examined CPD–MHD differences in a cohort of 134,728 new ESRD patients from the USRDS and concluded that the selection of HD over PD was associated with a survival advantage in patients with large body habits [54]. Beddhu et al. [42] hypothesized that the survival advantages of obesity is due to muscle mass both in MHD and CPD patients [55] using urinary creatinine as a indicator of muscle mass [56].

Other Populations with a Reverse Epidemiology

Patients with chronic heart failure (CHF) [57], geriatric populations [58], and patients with malignancy [59], AIDS [60], chronic obstructive pulmonary disease [61], or rheumatoid arthritis [62] also exhibit a risk factor reversal.

Table 1. Potential mechanisms that may result in the observed paradoxical associations between obesity and better survival in dialysis patients

---

Kidney Disease Wasting (Malnutrition–inflammation complex syndrome)
Time discrepancy between competitive risk factors: overnutrition vs. undernutrition
Unusual genetic constellation due to survival selection during CKD progression

Sequestration/storage of uremic toxins in fat tissue
Anti-inflammatory cytokines related to body mass, including adiponectins
Tumor necrosis factor alpha receptors
Endotoxin-lipoprotein hypothesis
Stability of hemodynamic status in obese patients
Neurohormonal alterations in obesity
Alteration of conventional risk factors in uremic milieu ('beyond Framingham')
Reverse causation
Survival bias

There appear to be at least 20 million Americans who may have a reverse epidemiology pattern [63]. There are striking similarities in the reported paradoxes between the patients with CHF, currently almost 5 million individuals in the USA, and MHD patients [64, 65]. Furthermore, millions of the fast-growing population of octogenarians and nonagenarians in the industrial countries appear to display a reverse epidemiology [63]. Studying the causal factors that engender the obesity paradox in chronic dialysis patients may confer better insight into understanding the pathophysiology and public health consequences of these phenomena in other populations.

Pathophysiology of the Obesity Paradox

Several hypotheses have been advanced to explain the obesity paradox in dialysis patients (table 1) [8, 15, 16]. Obesity and weight gain may be associated with a more stable short-term hemodynamic constellation and improved hemodynamic tolerance to afterload-reducing agents, especially because overweight and obese patients with heart failure tend to have higher systemic blood pressure values [66]. Thus, obese patients might better tolerate removal of large volumes of fluid during the hemodialysis procedure with less likelihood of hypotension. Obesity may mitigate stress responses and the heightened sympathetic and renin–angiotensin activity; the latter are associated with a poor prognosis in heart failure and fluid overload states such as in dialysis patients [67]. Hence, better outcome is expected if angiotensin axis can be blocked [68]. Altered cytokine and neuroendocrine profiles of obese patients may also play a role in conferring survival advantages to obese patients. Adipose tissue produces adiponectins [69], as well as soluble tumor necrosis factor alpha receptors which may neutralize the adverse biologic effects of tumor necrosis factor alpha [70]. It is also possible, although not yet proven, that the uremic milieu or volume overload modifies the cardiovascular constellations so that factors 'beyond Framingham paradigms' are more relevant for survival [71].

It has been postulated that higher concentrations of total cholesterol (lipoproteins) are beneficial for dialysis and CHF patients, since a richer pool of lipoproteins can actively bind to and remove circulating endotoxins; hence, the increase pool of these lipoproteins may attenuate the propensity of endotoxins which would otherwise cause inflammation and subsequent atherosclerosis if unbound [72]. This so-called 'endotoxin–lipoprotein' hypothesis was originally advanced to explain the hypercholesterolemia paradox in CHF patients [73].

It is also possible that uremic toxins are more effectively sequestered when abundant adipose tissue is present. It has been shown that weight loss and reduction in adipose tissue is associated with the imminent release of and significant increase in circulating lipophilic hexachlorobenzene and other chlorinated hydrocarbons [74]. Weight loss may also be associated with reduced skeletal muscle oxidative metabolism, leading to a mitigated anti-oxidant defense [75]. These findings may provide one explanation for why body fat loss has recently been found to be associated with increased death risk in dialysis patients [44].

It is, of course, possible that BMI is not a cause but a consequence of conditions that lead to poor outcome in dialysis patients or in similar populations with a paradoxical risk factor profile. 'Reverse causation' is a known possible source of bias in epidemiologic studies that examine associations without the direction of the causal pathway [76]. Comorbid states may lead to kidney disease wasting or cardiac cachexia and also to higher rate of mortality. However, even if the reverse causation is a cause of the reverse epidemiology, it does not explain why obesity including morbid obesity is associated with better outcome than the traditional normal or healthy weights in dialysis patients.

Of the currently estimated 20 million individuals with CKD in the USA [77], it is projected that over 90% will die before advancing to end-stage renal disease [29]. Hence, only less than 5% of the large CKD pool will be the 'unlucky lucky' individuals to reach the dialysis facility chair [7]. This may lead to a significant 'survival selection' [64] resulting in genetic constellations in dialysis patients that may be significantly different than their early CKD predecessors [78]. According to this theory, those few CKD patients who reach ESRD may have either a more accelerated rate of progression of CKD or special genes that protected them against the fatal ravages of cardiovascular disease which is inherent to CKD. Whether this is called 'survival bias' or 'survival selection' (similar to evolutionary natural selection), maintenance dialysis patients must be genetically or phenotypically dissimilar to their CKD predecessors who do not survive and may not have the survival characteristics and epidemiological features of their progenitors.

Survival advantages that exist in obese dialysis patients may, in the *short-term*, outweigh the harmful effects of these risk factors in causing cardiovascular disease and death in the *long-term* [7]. In other words, dialysis patients may not live long enough to die of the adverse effects of overnutrition, because they are more likely to die much faster of the consequences of undernutrition [64]. This so-called time-discrepancy between the two sets of competing risk factors, i.e., short-term killers (malnutrition–inflammation complex) vs. long-term killers (obesity and overnutrition), can explain why obesity treatment may be irrelevant or even harmful in many (but not all) dialysis patients if the issue at hand is the short-term survival. Currently 2/3 of all dialysis patients in the USA die within 5 years of commencing dialysis, a 5-year survival worse than many cancer patients [64]. Hence, treatment of malnutrition–inflammation complex, also known as kidney disease wasting, should be the target of efforts to improve survival in maintenance dialysis patients.

Conclusions and Future Steps

Studying the obesity paradox and other similar phenomena in dialysis patients leads to additional questions: Is the reverse epidemiology a true entity with clinical and public health implications in millions of patients with CKD, CHF, advanced age, malignancy, AIDS, etc., or is it a statistical fallacy that needs to be 'controlled away'? [79]. At which CKD stage does the reverse epidemiology start and in whom does it develop? Which groups of dialysis patients have a stronger, weaker or no obesity paradox? Can the so-called 'reversal of the reverse epidemiology' (or 'back-to normal) phenomenon upon successful renal transplantation of dialysis patients or with frequent (daily and/or nocturnal) dialysis treatment be confirmed? [64]. If this reversal of the altered risk factor relationships in these patients are real, what are the mechanisms for this phenomenon? What should be our therapeutic targets for body weight-for-height in our CKD and maintenance dialysis patients? Should we revise the current guidelines that recommend that obese dialysis patients on transplant waiting lists should lose weight as a prerequisite for renal transplantation? [40]. For that matter, what should be our therapeutic targets in maintenance dialysis patients for other clinical characteristics for which the usual risk factor relationships to mortality are altered, such as blood pressure or serum cholesterol or phosphorus, to mention a few? Is the evidence for these altered risk factor relationships sufficiently established to justify proposing to research granting agencies, the funding of randomized, prospective interventional trials to examine the appropriate therapeutic targets for BMI or some of these other clinical targets in chronic dialysis patients?

The field of altered risk factor relationships is in its infancy and appears to be evolving quickly. It is possible that, in the long run, overweight patients, if they survive sufficiently long, will suffer from more cardiovascular consequences. Until more information is available, it is prudent to avoid causal inferences for such observational data.

References

1 Goodkin DA, Mapes DL, Held PJ: The dialysis outcomes and practice patterns study (DOPPS): how can we improve the care of hemodialysis patients? Semin Dial 2001;14:157–159.
2 United States Renal Data System: Excerpts from the USRDS 2004 Annual Data Report. Am J Kid Dis 2005;45(suppl 1):S1–S280.
3 Foley RN, Parfrey PS, Sarnak MJ: Epidemiology of cardiovascular disease in chronic renal disease. J Am Soc Nephrol 1998;9:S16–S23.
4 Wanner C, Krane V, Marz W, Olschewski M, Mann JF, Ruf G, Ritz E: Atorvastatin in patients with type 2 diabetes mellitus undergoing hemodialysis. N Engl J Med 2005;353:238–248.
5 Eknoyan G, Beck GJ, Cheung AK, Daugirdas JT, Greene T, Kusek JW, Allon M, Bailey J, Delmez JA, Depner TA, Dwyer JT, Levey AS, Levin NW, Milford E, Ornt DB, Rocco MV, Schulman G, Schwab SJ, Teehan BP, Toto R: Effect of dialysis dose and membrane flux in maintenance hemodialysis. N Engl J Med 2002;347:2010–2019.
6 Paniagua R, Amato D, Vonesh E, Correa-Rotter R, Ramos A, Moran J, Mujais S: Effects of increased peritoneal clearances on mortality rates in peritoneal dialysis: ADEMEX, a prospective, randomized, controlled trial. J Am Soc Nephrol 2002;13:1307–1320.
7 Kalantar-Zadeh K, Block G, Humphreys MH, Kopple JD: Reverse epidemiology of cardiovascular risk factors in maintenance dialysis patients. Kidney Int 2003;63:793–808.
8 Kopple JD: The phenomenon of altered risk factor patterns or reverse epidemiology in persons with advanced chronic kidney failure. Am J Clin Nutr 2005;81:1257–1266.
9 Kalantar-Zadeh K, Kopple JD, Kilpatrick RD, McAllister CJ, Shinaberger CS, Gjertson DW, Greenland S: Association of morbid obesity and weight change over time with cardiovascular survival in hemodialysis population. Am J Kidney Dis 2005;46:489–500.
10 Kopple JD, Zhu X, Lew NL, Lowrie EG: Body weight-for-height relationships predict mortality in maintenance hemodialysis patients. Kidney Int 1999;56:1136–1148.
11 Kalantar-Zadeh K, Kilpatrick RD, McAllister CJ, Greenland S, Kopple JD: Reverse epidemiology of hypertension and cardiovascular death in the hemodialysis population: the 58th annual fall conference and scientific sessions. Hypertension 2005;45:811–817.
12 Nishizawa Y, Shoji T, Ishimura E, Inaba M, Morii H: Paradox of risk factors for cardiovascular mortality in uremia: is a higher cholesterol level better for atherosclerosis in uremia? Am J Kidney Dis 2001;38:S4–S7.
13 Kalantar-Zadeh K, Block G, Humphreys MH, McAllister CJ, Kopple JD: A low, rather than a high, total plasma homocysteine is an indicator of poor outcome in hemodialysis patients. J Am Soc Nephrol 2004;15:442–453.
14 Lowrie EG, Lew NL: Death risk in hemodialysis patients: the predictive value of commonly measured variables and an evaluation of death rate differences between facilities. Am J Kidney Dis 1990;15:458–482.
15 Kalantar-Zadeh K, Abbott KC, Salahudeen AK, Kilpatrick RD, Horwich TB: Survival advantages of obesity in dialysis patients. Am J Clin Nutr 2005;81:543–554.

16 Kalantar-Zadeh K: Causes and consequences of the reverse epidemiology of body mass index in dialysis patients. J Ren Nutr 2005;15:142–147.
17 Kuczmarski RJ, Flegal KM, Campbell SM, Johnson CL: Increasing prevalence of overweight among US adults. The National Health and Nutrition Examination Surveys, 1960 to 1991. JAMA 1994;272:205–211.
18 Lev-Ran A: Human obesity: an evolutionary approach to understanding our bulging waistline. Diabetes Metab Res Rev 2001;17:347–362.
19 Birkner R: Plan and Initial Program of the Health Examination Survey. Vital Health Stat 1 1965;125:1–43.
20 Miller HW: Plan and operation of the health and nutrition examination survey. United States – 1971–1973. Vital Health Stat 1 1973;1:1–46.
21 McDowell A, Engel A, Massey JT, Maurer K: Plan and operation of the Second National Health and Nutrition Examination Survey, 1976–1980. Vital Health Stat 1 1981;1:1–144.
22 Flegal KM, Carroll MD, Ogden CL, Johnson CL: Prevalence and trends in obesity among US adults, 1999–2000. JAMA 2002;288:1723–1727.
23 Byers T: Body weight and mortality. N Engl J Med 1995;333:723–724.
24 Manson JE, Willett WC, Stampfer MJ, Colditz GA, Hunter DJ, Hankinson SE, Hennekens CH, Speizer FE: Body weight and mortality among women. N Engl J Med 1995;333:677–685.
25 Lew EA, Garfinkel L: Variations in mortality by weight among 750,000 men and women. J Chronic Dis 1979;32:563–576.
26 Hsu C-y, McCulloch CE, Iribarren C, Darbinian J, Go AS: Body mass index and risk for end-stage renal disease. Ann Intern Med 2006;144:21–28.
27 Kalantar-Zadeh K, Abbott KC, Salahudeen AK, Kilpatrick RD, Horwich TB: Survival advantages of obesity in dialysis patients. Am J Clin Nutr 2005;81:543–554.
28 Kalantar-Zadeh K, Kopple JD: Body mass index and risk for end-stage renal disease. Ann Intern Med, 2006;144:21–28.
29 Keith DS, Nichols GA, Gullion CM, Brown JB, Smith DH: Longitudinal follow-up and outcomes among a population with chronic kidney disease in a large managed care organization. Arch Intern Med 2004;164:659–663.
30 United States Renal Data System: US Department of Public Health and Human Services. Public Health Service. Bethesda, National Institutes of Health, 2003.
31 Kopple JD: Nutritional status as a predictor of morbidity and mortality in maintenance dialysis patients. ASAIO J 1997;43:246–250.
32 Kalantar-Zadeh K, Kilpatrick RD, Kopple JD, Stringer WW: A matched comparison of serum lipids between hemodialysis patients and nondialysis morbid controls. Hemodial Int 2005;9: 314–324.
33 Degoulet P, Legrain M, Reach I, Aime F, Devries C, Rojas P, Jacobs C: Mortality risk factors in patients treated by chronic hemodialysis. Report of the Diaphane Collaborative Study. Nephron 1982;31:103–110.
34 Leavey SF, Strawderman RL, Jones CA, Port FK, Held PJ: Simple nutritional indicators as independent predictors of mortality in hemodialysis patients. Am J Kidney Dis 1998;31:997–1006.
35 Fleischmann E, Teal N, Dudley J, May W, Bower JD, Salahudeen AK: Influence of excess weight on mortality and hospital stay in 1346 hemodialysis patients. Kidney Int 1999;55:1560–1567.
36 Wolfe RA, Ashby VB, Daugirdas JT, Agodoa LY, Jones CA, Port FK: Body size, dose of hemodialysis, and mortality. Am J Kidney Dis 2000;35:80–88.
37 Port FK, Ashby VB, Dhingra RK, Roys EC, Wolfe RA: Dialysis dose and body mass index are strongly associated with survival in hemodialysis patients. J Am Soc Nephrol 2002;13: 1061–1066.
38 Leavey SF, McCullough K, Hecking E, Goodkin D, Port FK, Young EW: Body mass index and mortality in 'healthier' as compared with 'sicker' haemodialysis patients: results from the Dialysis Outcomes and Practice Patterns Study (DOPPS). Nephrol Dial Transplant 2001;16:2386–2394.
39 Glanton CW, Hypolite IO, Hshieh PB, Agodoa LY, Yuan CM, Abbott KC: Factors associated with improved short term survival in obese end stage renal disease patients. Ann Epidemiol 2003;13: 136–143.
40 Johansen KL, Young B, Kaysen GA, Chertow GM: Association of body size with outcomes among patients beginning dialysis. Am J Clin Nutr 2004;80:324–332.

41 Lowrie EG, Li Z, Ofsthun N, Lazarus JM: Body size, dialysis dose and death risk relationships among hemodialysis patients. Kidney Int 2002;62:1891–1897.
42 Beddhu S, Pappas LM, Ramkumar N, Samore M: Effects of body size and body composition on survival in hemodialysis patients. J Am Soc Nephrol 2003;14:2366–2372.
43 Kalantar-Zadeh K, Abbott KC, Salahudeen AK: Reverse epidemiology of obesity in dialysis patients: fat or muscle? Am J Clin Nutr 2005;82:910–911.
44 Kalantar-Zadeh K, Kuwae N, Wu DY, Shantouf RS, Fouque D, Anker SD, Block G, Kopple JD: Associations of body fat and its changes over time with quality of life and prospective mortality in hemodialysis patients. Am J Clin Nutr 2006;83:202–210.
45 Canada-USA Peritoneal Dialysis Study Group: Adequacy of dialysis and nutrition in continuous peritoneal dialysis: association with clinical outcomes. J Am Soc Nephrol 1996;7:198–207.
46 Hakim RM, Lowrie E: Obesity and mortality in ESRD: is it good to be fat? Kidney Int 1999;55: 1580–1581.
47 McCusker FX, Teehan BP, Thorpe KE, Keshaviah PR, Churchill DN, for the Canada-USA (CANUSA) Peritoneal Dialysis Study Group: How much peritoneal dialysis is necessary for maintaining a good nutritional status? Kidney Int 1996;56:(Suppl)S56–S61.
48 Johnson DW, Herzig KA, Purdie DM, Chang W, Brown AM, Rigby RJ, Campbell SB, Nicol DL, Hawley CM: Is obesity a favorable prognostic factor in peritoneal dialysis patients? Perit Dial Int 2000;20:715–721.
49 Chung SH, Lindholm B, Lee HB: Influence of initial nutritional status on continuous ambulatory peritoneal dialysis patient survival. Perit Dial Int 2000;20:19–26.
50 Snyder JJ, Foley RN, Gilbertson DT, Vonesh EF, Collins AJ: Body size and outcomes on peritoneal dialysis in the United States. Kidney Int 2003;64:1838–1844.
51 McDonald SP, Collins JF, Johnson DW: Obesity is associated with worse peritoneal dialysis outcomes in the Australia and New Zealand patient populations. J Am Soc Nephrol 2003;14: 2894–2901.
52 Abbott KC, Glanton CW, Trespalacios FC, Oliver DK, Ortiz MI, Agodoa LY, Cruess DF, Kimmel PL: Body mass index, dialysis modality, and survival: analysis of the United States Renal Data System Dialysis Morbidity and Mortality Wave II Study. Kidney Int 2004;65:597–605.
53 Aslam N, Bernardini J, Fried L, Piraino B: Large body mass index does not predict short-term survival in peritoneal dialysis patients. Perit Dial Int 2002;22:191–196.
54 Stack AG, Murthy BV, Molony DA: Survival differences between peritoneal dialysis and hemodialysis among 'large' ESRD patients in the United States. Kidney Int 2004;65:2398–2408.
55 Ramkumar N, Pappas LM, Beddhu S: Effect of body size and body composition on survival in peritoneal dialysis patients. Perit Dial Int 2005;25:461–469.
56 Foley RN: Body mass index and survival in peritoneal dialysis patients. Perit Dial Int 2005;25: 435–437.
57 Curtis JP, Selter JG, Wang Y, Rathore SS, Jovin IS, Jadbabaie F, Kosiborod M, Portnay EL, Sokol SI, Bader F, Krumholz HM: The obesity paradox: body mass index and outcomes in patients with heart failure. Arch Intern Med 2005;165:55–61.
58 Stevens J, Cai J, Pamuk ER, Williamson DF, Thun MJ, Wood JL: The effect of age on the association between body-mass index and mortality. N Engl J Med 1998;338:1–7.
59 Yeh S, Wu SY, Levine DM, Parker TS, Olson JS, Stevens MR, Schuster MW: Quality of life and stimulation of weight gain after treatment with megestrol acetate: correlation between cytokine levels and nutritional status, appetite in geriatric patients with wasting syndrome. J Nutr Health Aging 2000;4:246–251.
60 Chlebowski RT, Grosvenor M, Lillington L, Sayre J, Beall G: Dietary intake and counseling, weight maintenance, and the course of HIV infection. J Am Diet Assoc 1995;95:428–432; quiz 433–425.
61 Wilson DO, Rogers RM, Wright EC, Anthonisen NR: Body weight in chronic obstructive pulmonary disease. The National Institutes of Health Intermittent Positive-Pressure Breathing Trial. Am Rev Respir Dis 1989;139:1435–1438.
62 Escalante A, Haas RW, del Rincon I: Paradoxical effect of body mass index on survival in rheumatoid arthritis: role of comorbidity and systemic inflammation. Arch Intern Med 2005;165: 1624–1629.

63   Kalantar-Zadeh K, Kilpatrick RD, Kuwae N, Wu DY: Reverse epidemiology: a spurious hypothesis or a hardcore reality? Blood Purif 2005;23:57–63.
64   Kalantar-Zadeh K, Abbott KC, Kronenberg F, Anker SD, Horwich TB, Fonarow GC: Epidemiology of dialysis patients and heart failure patients; special review article for the 25th anniversary of the Seminars in Nephrology. Semin Nephrol 2006;26:118–133.
65   Kalantar-Zadeh K, Block G, Horwich T, Fonarow GC: Reverse epidemiology of conventional cardiovascular risk factors in patients with chronic heart failure. J Am Coll Cardiol 2004;43: 1439–1444.
66   Horwich TB, Fonarow GC, Hamilton MA, MacLellan WR, Woo MA, Tillisch JH: The relationship between obesity and mortality in patients with heart failure. J Am Coll Cardiol 2001;38:789–795.
67   Schrier RW, Abraham WT: Hormones and hemodynamics in heart failure. N Engl J Med 1999;341:577–585.
68   Yusuf S, Sleight P, Pogue J, Bosch J, Davies R, Dagenais G: Effects of an angiotensin-converting-enzyme inhibitor, ramipril, on cardiovascular events in high-risk patients. The Heart Outcomes Prevention Evaluation Study Investigators. N Engl J Med 2000;342:145–153.
69   Stenvinkel P, Marchlewska A, Pecoits-Filho R, Heimburger O, Zhang Z, Hoff C, Holmes C, Axelsson J, Arvidsson S, Schalling M, Barany P, Lindholm B, Nordfors L: Adiponectin in renal disease: relationship to phenotype and genetic variation in the gene encoding adiponectin. Kidney Int 2004;65:274–281.
70   Mohamed-Ali V, Goodrick S, Bulmer K, Holly JM, Yudkin JS, Coppack SW: Production of soluble tumor necrosis factor receptors by human subcutaneous adipose tissue in vivo. Am J Physiol 1999;277:E971–E975.
71   McClellan WM, Chertow GM: Beyond framingham: cardiovascular risk profiling in ESRD. J Am Soc Nephrol 2005;16:1539–1541.
72   Niebauer J, Volk HD, Kemp M, Dominguez M, Schumann RR, Rauchhaus M, Poole-Wilson PA, Coats AJ, Anker SD: Endotoxin and immune activation in chronic heart failure: a prospective cohort study. Lancet 1999;353:1838–1842.
73   Rauchhaus M, Coats AJ, Anker SD: The endotoxin-lipoprotein hypothesis. Lancet 2000;356: 930–933.
74   Jandacek RJ, Anderson N, Liu M, Zheng S, Yang Q, Tso P: Effects of yo-yo diet, caloric restriction, and olestra on tissue distribution of hexachlorobenzene. Am J Physiol Gastrointest Liver Physiol 2005;288:G292–G299.
75   Imbeault P, Tremblay A, Simoneau JA, Joanisse DR: Weight loss-induced rise in plasma pollutant is associated with reduced skeletal muscle oxidative capacity. Am J Physiol Endocrinol Metab 2002;282:E574–E579.
76   Macleod J, Davey Smith G: Psychosocial factors and public health: a suitable case for treatment? J Epidemiol Community Health 2003;57:565–570.
77   Jones CA, McQuillan GM, Kusek JW, Eberhardt MS, Herman WH, Coresh J, Salive M, Jones CP, Agodoa LY: Serum creatinine levels in the US population: third National Health and Nutrition Examination Survey. Am J Kidney Dis 1998;32:992–999.
78   Kalantar-Zadeh K, Balakrishnan VS: The kidney disease wasting: inflammation, oxidative stress and diet-gene interaction. Hemodial Int 2006, in press.
79   Liu Y, Coresh J, Eustace JA, Longenecker JC, Jaar B, Fink NE, Tracy RP, Powe NR, Klag MJ: Association between cholesterol level and mortality in dialysis patients: role of inflammation and malnutrition. JAMA 2004;291:451–459.

Kamyar Kalantar-Zadeh, MD, PhD, MPH
Associate Professor of Medicine and Pediatrics
Division of Nephrology and Hypertension
Los Angeles Biomedical Research Institute at Harbor-UCLA Medical Center
1124 West Carson Street
Torrance, CA 90502 (USA)
Tel. +1 310 222 3891, Fax +1 310 782 1837, E-mail kamkal@ucla.edu

The Adipose Tissue as an Endocrine Organ – A Nephrologists' Perspective

Jerzy Chudek, Marcin Adamczak, Teresa Nieszporek, Andrzej Więcek

Department of Nephrology, Endocrinology and Metabolic Diseases,
Medical University of Silesia, Katowice, Poland

Abstract

During the last decade white adipose tissue was recognized as an active endocrine organ and a source of many proinflammatory cytokines, chemokines, growth factors and complement proteins called 'adipokines' or 'adipocytokines'. The contribution of different cell types which compose the adipose tissue: adipocytes, preadipocytes, stromal/vascular cells and macrophages in secretion of above-mentioned adipokines varies remarkably. These adipokines seem to play an important role in the pathogenesis of obesity-related comorbidities. In this review, we have summarized the present knowledge on the most important adipokines in patients with obesity, arterial hypertension and chronic kidney diseases.

Copyright © 2006 S. Karger AG, Basel

Introduction

The epidemic of visceral obesity, insulin resistance including type 2 diabetes mellitus and obesity related arterial hypertension is a challenging health problem for modern societies. Until now no solution to increasing calamity of these well-known risk factors of cardiovascular morbidity and mortality is proposed. Promotion of an active life style and low-caloric diet as well as anorexigenic medications are still insufficient to counterbalance the easy access to high-caloric products in developed countries.

In the last decade, adipose tissue was recognized as an active endocrine organ that can affect the function of other organs and an important source of several proinflammatory cytokines, chemokines, growth factors and complement proteins called 'adipokines' or 'adipocytokines' [1, 2]. Many of them may influence function of the cardiovascular system and participate in the pathogenesis of atherosclerosis process directly. An incomplete list of such adipokines

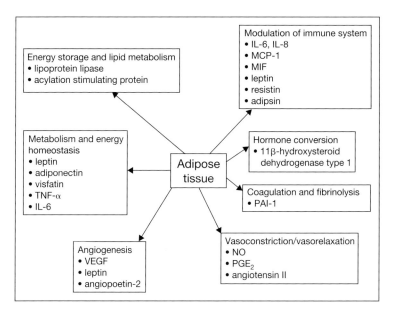

Fig. 1. The major physiological functions of adipose tissue. IL = Interleukin; MCP-1 = monocyte chemoattractant protein-1; MIF = migration inhibitory factor; NO = nitric oxide; PAI-1 = plasminogen activator inhibitor-1; PGE$_2$ = prostaglandin E$_2$; TNF-α = tumour necrosis factor-α; VEGF = vascular endothelial growth factor.

comprises: leptin, adiponectin, visfatin, apelin and resistin, Agouti signalling protein, acylation stimulating protein, nitric oxide, renin, angiotensin II, plasminogen activator inhibitor-1 (PAI-1), tumour necrosis factor-α (TNF-α), interleukin (IL)-6, IL-8, IL-10, IL-1β, monocyte chemoattractant protein-1, migration inhibitory factor, prostaglandin E$_2$, hepatocyte growth factor, vascular endothelial growth factor, nerve growth factor, heparin binding–epidermal growth factor, insulin growth factor-1, tissue factor, complement factor D (adipsin) [3]. In addition in adipose tissue conversion of steroid hormone take place like conversion from cortisone to cortisol by 11β-hydroxysteroid dehydrogenase type 1 (fig. 1).

Adipose tissue is not a homogenous organ. It consists of a variety of different cell types: adipocytes, preadipocytes, stromal/vascular cells and macrophages [3] (table 1). Each of these cells present their own secretion profile and specific regulation. Obesity is related not only to increase the number and size of adipocytes [4] but also infiltration of adipose tissue by macrophages [5].

It is already well-known that mature adipocytes are the main source of leptin and adiponectin, macrophages produce almost all TNF-α, resistin and visfatin, while prostaglandin E$_2$, interleukins, vascular endothelial growth factor, hepatocyte growth factor are synthesized by stromal and vascular cells [3, 6, 7].

Table 1. The most important adipokines release by adipocytes and matrix of adipose tissue

	Adipocytes	Tissue matrix
Prostaglandin E_2	+/–	+++
Prostacycline	+/–	+++
Adiponectin	+++	+
Leptin	+++	–
Resistin	+/–	+++
Interleukin (IL)-8	+	+++
IL-6	+/–	+++
IL-10	+/–	+++
IL-1β	+/–	+++
Tumour necrosis factor α (TNF-α)	+/–	+++
Platelet activator inhibitor 1 (PAI-1)	++	+++
Hepatocyte growth factor (HGF)	+/–	+++
Vascular endothelial growth factor (VEFG)	+/–	+++
Angiotensin II	+	+++
Visfatin	+/–	+++

There are also some differences in the adipokines production between visceral and peripheral fat tissue (table 2). Adipokines released from the fat tissue may exert their action on the endocrine, paracrine or autocrine way. In this review mainly the endocrine action of adipokines will be discussed. However, the paracrine action cannot be neglected in cardiovascular diseases. As an example, secretion of nitric oxide by the periadventitial fat tissue can be mentioned [8]. Its action on the neighbouring smooth muscles cells may actively influence the vascular tonus and arterial blood pressure [8].

The knowledge of adiposity related mechanisms influencing the cardiovascular system and the kidneys may contribute in the developing of a new drugs for prevention and treatment of the metabolic syndrome. The role of some adipokines in the pathogenesis of cardiovascular and renal diseases is discussed in this review.

Leptin

Leptin is a protein predominantly produced by adipocytes [9]. It is encoded by the ob gene [9]. Leptin gene expression is increased about 100-fold during the differentiation of preadipocytes. Initially, leptin was implicated in the regulation of appetite as a satiety hormone [9]. It was found that rats with homozygous non-sense mutation of the ob gene were suffering from marked obesity,

Table 2. Comparison of gene expression of adipokines by adipocytes in human visceral or subcutaneous adipose tissue

	Visceral	Subcutaneous
Leptin	+	++
Adiponectin	+++	+
IL-6	+++	+
TNF-α	+	+
PAI-1	++	+
Angiotensinogen	++	+
Estrogens	+	+
IGF-1	+	+
Visfatin	++	+

while parenteral leptin substitution was decreasing both their appetite and body mass [10]. A few years later, leptin appeared mostly as a marker of nutrition unable to decrease food intake even in obese humans opposite to the earlier expectations. Obese individuals, especially females, are characterized by high plasma leptin concentrations [11]. Thus even highly elevated plasma leptin concentration does not suppress the appetite in these individuals [12].

According to Flier [13], leptin should not be regarded as an antiobesity hormone but as a signal of energy deficiency and integrator of neuroendocrine function. Low plasma leptin concentration – a 'starvation signal' – modulate the hypothalamic–pituitary–adrenal axis, suppresses thyroid and gonadal axes [13]. Such endocrine changes resemble clinical status of patients with anorexia nervosa [14].

Low plasma leptin concentration decreases energy expenditure and stimulates the search for food in deprived of food rodents [15, 16]. Only in animals, with low body fat stores, it was demonstrated that leptin is involved in regulation of food intake and energy expenditure. Similarly, only in obese children with inactivating mutations of both leptin alleles, parenteral substitution of this peptide decreases appetite with a subsequent reduction of body mass [17].

It is well known that sexual maturation and fertility are linked to nutritional status and adiposity. Injection of leptin in prepubertal female mouse causes earlier maturation of the reproductive tract [18], suggesting that leptin acts as a signal for puberty. Moreover, leptin may promote angiogenesis and stimulate both proliferation and differentiation of haemopoietic cells, including

T cells [19, 20]. In acute infections, the stimulation by leptin shift of the immune system toward the predominance of the proinflammatory Th_1 T cells population seems to be beneficial [21]. Chronic stimulation of the immune system by leptin may lead to acceleration of atherosclerosis, as leptin-deficient mice are protected from atherosclerosis despite all the other metabolic factors that contribute to such vascular disease [22].

In obese humans, leptin exerts a deteriorative impact on cardiovascular system and kidneys by significant contribution to the pathogenesis of obesity-related arterial hypertension. It was shown that high plasma leptin concentration stimulates the activity of the sympathetic nervous system via the receptors located in the brain trunk [23]. Moreover, leptin exaggerates insulin resistance in obese patients [24], stimulate activity of the renin–angiotensin system [25] and modulate function of endothelial cells (vascular remodelling) [19]. Many of these above-mentioned mechanisms interfere with the tubular sodium reabsorption. The contra-regulatory mechanisms responsible for the lack of arterial hypertension in 10–20% of obese humans were not identified.

Leptin is also involved in the pathogenesis of obesity-related nephropathy and the progression of chronic glomerulopathies [26]. Within the glomerulus, leptin directly stimulates endocapillary cell proliferation and mesangial collagen types I and IV deposition [27, 28]. In cultured rat endothelial cells, mouse recombinant leptin stimulates proliferation and increases TGF-β mRNA and TGF-β secretion [27]. It also stimulates expression of TGF receptors [27]. Leptin infusion in rats enhances urinary protein excretion [27]. In addition, leptin activates sympathetic nervous system, that may contribute to renal damage directly or indirectly via elevated blood pressure.

Patients with advanced chronic kidney disease (CKD) are characterized by markedly elevated plasma leptin concentration compared to body mass index (BMI) and sex matched healthy individuals [29–31]. CAPD patients have even higher plasma leptin concentration than haemodialysis ones [29, 30]. However, increased plasma leptin concentration in CKD is not accounted by oversecretion of this hormone. Leptin gene expression is even lower in adipocytes from CKD patients than of healthy individuals [32]. In renal failure decreased renal clearance of leptin contributes to elevated plasma leptin concentrations [33]. Thus it is not surprising that plasma leptin concentration is normalized by successful kidney transplantation [34].

Elimination of leptin is partly independent from glomerular filtration rate [35, 36]. Kinetic studies suggest that the active uptake of this peptide hormone by the renal tissue contribute to leptin catabolism [37]. Cumin et al. [35] studied plasma leptin in Zucker obese rats subjected to bilateral nephrectomy or bilateral ureteral ligation. Following the bilateral nephrectomy plasma leptin concentrations increased by 300%, more than the 50% increase was observed after bilateral ligation of the ureters. Similarly in patients with non-inflammatory, reversible acute renal failure requiring haemodialysis treatment, plasma leptin concentration although significantly higher than in healthy subjects was much lower than in CKD patients on haemodialysis [36].

In chronic uremia, elevated plasma leptin concentration seems to participate in the pathogenesis of uremia-associated cachexia via signalling through the central melanocortin system [38]. On the other hand, it should be stressed that studies addressing the influence of plasma leptin concentration on future changes of the body mass in haemodialysis patients failed to find such a relationship [39, 40]. Therefore, plasma leptin concentration may serve in these patients also as a marker of well-being [39, 41]. Elucidation of the mechanisms regulating appetite will be useful in our understanding of cachexia in uremic patients. From the other side it is well-known that low body mass (predominantly muscle mass) is combined with the increased morbidity and mortality in these patients. Therefore, there is no doubt that these new findings related to the role of leptin in the pathogenesis of cachexia in uremia may have important clinical and therapeutic implications [38].

Adiponectin

Adiponectin is a protein hormone secreted almost exclusively by adipocytes with antiatherogenic and insulin-sensitizing properties. Adiponectin gene expression is increased 100 folds during differentiation of human cultured preadipocytes [42]. It shows structural homology with collagen VIII, X and complement factor C1q [43]. Plasma concentration of adiponectin is relatively high (almost 0.01% of total plasma protein) [44]. Adiponectin is present as full-length molecules (almost all) or smaller globular C-terminal domain fragments, see for review [45, 46]. Within the circulation, adiponectin forms a wide range of multimers: from trimers (low-molecular weight), hexamers (medium-molecular weight) to dodecamres or 18-mers (high-molecular weight) [46]. It was shown that high-molecular weight is an active form of adiponectin improving insulin-sensitivity [47, 48]. In contrast to leptin its concentration is lower in obese than in non-obese subjects [44]. Lower adiponectinaemia was also found in males than in females, and in patients with coronary artery disease, diabetes mellitus type II and essential hypertension than in healthy subjects [49–51]. Some of the common polymorphism in the promoter region, exon and intron 2 an the rare mutations in exon 3 of the human adiponectin gene are associated with phenotypes related to body weight, glucose metabolism, insulin sensitivity and the risk of diabetes mellitus and coronary artery disease, see for review [52].

Adiponectin receptors were found in the skeletal muscle, liver and endothelial cells [53]. Ouchi et al. [54] found that endothelium-dependent vasorelaxation is impaired in subjects with low plasma adiponectin concentration, suggesting that it might be one of the mechanism responsible for hypertension in obesity. It has been suggested, that adiponectin inhibits formation of initial atherosclerotic lesions by decreasing expression of adhesion molecules (VCAM-1; ICAM-1, E-selectin) in endothelial cells in response to inflammatory stimuli such as TNF-α [55]. Adiponectin also suppresses production of cytokines, such as TNF-α by macrophages [56]. Moreover, adiponectin suppresses accumulation of lipids in human monocyte-derived macrophages and inhibits transformation of macrophages into foam cells [57]. It also inhibits cell proliferation stimulated by oxidised LDL and suppresses superoxide generation [58]. Finally, adiponectin interferes with atherogenesis in ApoE-deficient mice, a well-known model of spontaneous atherosclerosis [59].

Adiponectin participates also in the stabilization of atherosclerotic plaques by increasing expression of tissue inhibitor of metalloproteinase-1 in infiltrating macrophages [60].

Adiponectin improves insulin sensitivity [61] by stimulating glucose utilization and fatty acid oxidation in skeletal muscles and in liver and by suppressing enhanced glucose production in the liver [61]. It was already shown that low plasma adiponectin concentration is an independent predictor of the risk for development of insulin resistance and type 2 diabetes in the general population [62]. The impaired multimerization of adiponectin seems to participate in the pathogenesis of type 2 diabetes mellitus [63]. Adiponectin also modulates inflammatory processes by modifying the expression of inflammatory cytokines as shown in the model of anti-GBM glomerulonephritis [64]. Adiponectin deficient mice were more prone to renal injury after administration of anti-GBM serum [64].

Plasma adiponectin concentration is almost 3-times higher in haemodialysis patients with CKD then in healthy subjects [65, 66]. However, increased plasma adiponectin concentration in these patients could not be explained by its oversecretion by adipose tissue, because the expression of the adiponectin gene (ApM1) is decreased by adipocytes in patients with advanced CKD [67]. Kidneys are the main organ participating in the biodegradation and elimination of adiponectin from the circulation. Thus, as expected, successful kidney transplantation is accompanied by prompt reduction, but not normalization of plasma adiponectin concentration [68]. In addition, an inverse relationship between plasma adiponectin concentrations and GFR was found in patients with essential hypertension and in apparently healthy individuals [58, 69]. Measurements of plasma adiponectin concentrations in the renal veins and aorta of patients with haemodynamically important renal artery stenosis confirmed the role of the kidneys in the elimination of adiponectin [70].

Lower adiponectin gene expression in CKD patients may be partially caused by the microinflammation. Also in the general population, an inverse relationship was found between concentrations of adiponectin and C-reactive protein (CRP) [71, 72]. The same is true for haemodialysis patients [66, 73, 74]. This clinical observation was already confirmed by the experimental studies indicating that TNF-α and IL-6 inhibit adiponectin gene expression in cultured adipocytes [75, 76].

Low plasma adiponectin concentration is now recognized as a new potential risk factor of cardiovascular morbidity and mortality. In a large cohort of haemodialysis patients lower plasma adiponectin concentration was recognized as an independent predictor of fatal and non-fatal cardiovascular complications in these patients [65, 66].

Taking into consideration that lower plasma adiponectin concentration is the new cardiovascular risk factor, an introduction of the drugs increasing plasma adiponectin concentration to the therapy may be a promising strategy for cardiovascular diseases prevention. This might provide scientific rationale for more often use of drugs increasing to some extend plasma adiponectin concentration like: thiazolidinediones [77, 78], temocapril and ramipril (angiotensin-converting enzyme inhibitors) [79, 80], losartan and candesartan (angiotensin II receptor 1 blockers) [79–82], rilmenidine (clonidine-like sympathoinhibitory antihypertensive agent) [83], fenofibrate [84] and rimonabant (receptor 1 blocker) [85]. It was shown that both treatment with thiazolidinediones or weight reduction increases particularly insulin-sensitizing high-molecular weight adiponectin form [47, 86, 87].

Resistin

Resistin is a large polypeptide which is produced exclusively by adipose tissue matrix and by macrophages [88, 89]. Plasma resistin concentration is markedly elevated in obese overfed insulin-resistant mice and also in ob/ob and db/db mice with obesity and diabetes mellitus [90]. In mice, neutralization of resistin by anti-resistin antibodies lowers glucose concentration and improves insulin sensitivity. Conversely, intraperitoneal administration of resistin causes in these animals glucose intolerance and insulin resistance [90]. Moreover, resistin impairs glucose uptake by adipocytes in vitro. It is important to stress that rosiglitazone, an agonist of peroxisome proliferator-activated receptor-γ reduces gene expression and secretion of resistin [90]. Despite of initial finding in mice, the true role of resistin in humans is unknown. In humans, no association between plasma resistin concentration and adiposity or insulin resistance was found [91, 92]. In addition, in humans little or no relationship was observed between resistin gene expression and insulin resistance or BMI [93–95]. One of the reasons of such a difference between the physiological role of resistin in humans and mice may be incomplete homology of resistin (59%) between these two species [96]. Although the resistin mRNA is detectable in human adipocytes [97] the main source of circulating resistin in humans in contrast to mice is not adipose tissue but mononuclear cells and macrophages [88]. It was shown that resistin expression in human monocytes is markedly increased after treatment with endotoxin or proinflammatory cytokines [98, 99]. In consequence, endotoxaemia is related with the markedly increases resistin plasma concentration in humans [88]. Resistin stimulates expression of adhesion molecules and cytokines by endothelial cells [100, 101]. The role of resistin in the pathogenesis of atherosclerosis was further confirmed by clinical observation of Reilly et al. [102], who showed that plasma resistin concentration is correlated with levels of the inflammatory markers (IL-6, soluble TNF-R2) and adhesive molecules (ICAM-1), as well as plasma resistin concentration predicts to coronary atherosclerosis, measured by calcification index.

Similarly to other adipokines, plasma resistin concentration is elevated in subjects with impairment of kidney function [103]. However, no association was found between plasma resistin concentration and insulin sensitivity in these patients [103].

Visfatin

Recently, Fukuhara et al. [104] described adipose-tissue derived protein previously identified as growth factor for early B-lymphocytes termed pre-B colony enhancing factor. This adipokine is predominantly secreted by stromal cells identified as CD 14+ macrophages [105]. Visfatin locally regulates adipocyte maturation. It was shown that overexpression of visfatin in preadipocytes facilitates both differentiation of these cells into mature adipocytes as well as promotes fat deposition in these cells [104]. In humans visfatin gene is more extensively expressed in visceral than subcutaneous tissue [104]. In obese and type 2 diabetic patients plasma visfatin concentration is higher than in healthy subjects [106, 107] and correlates with visceral fat content [104]. Moreover, significant weight loss achieved by gastric banding was followed by reduction of plasma visfatin concentration [106].

The relationship between visfatin, metabolic syndrome and type 2 diabetes mellitus is complex. Visfatin an insulin-mimicking hormone, stimulates glucose uptake by adipocytes and myocytes and inhibits glucose release from the liver [104]. Visfatin also stimulates phosphorylation of proteins involved in postreceptor insulin signal transduction [104] and binds to insulin receptor.

Despite recent great progress concerning the visfatin its physiological and pathophysiological role still remains to be elucidated.

Apelin

Apelin (identified previously as endogenous ligand of orphan receptor APJ) [108] is involved in the regulation of cardiovascular function and fluid homeostasis, see for review [109]. Apelin exerts a potent vasodilator and positive inotropic actions in rats. Buocher et al. [110] showed that apelin is expressed in adipose tissue. Apelin expression and its plasma concentration are increased in mice models of obesity as well as in human obese subjects [110, 111]. Expression of this peptide is increasing with the adipocyte maturation [110]. Apelin receptors are localized in human cardiomyocytes, vascular smooth muscle cells and endothelial cells [112].

In obese patients, both plasma apelin and insulin levels were significantly elevated, suggesting that insulin could influence apelin secretion [110]. Plasma apelin concentration correlates with BMI in human obese subjects [111].

A direct regulation of apelin expression by insulin was observed in both human and mouse adipocytes and was clearly associated with the stimulation of phosphatidylinositol 3-kinase, protein kinase C, and MAPK [110, 113]. These data provide evidence that insulin exerts a direct control of apelin gene expression in adipocytes. Interestingly in vitro it was shown that glucocorticoids suppress adipocyte expression of apelin [113]. This leads to the speculation that hypertension induced by glucocorticoids might be partially mediated by reduction of apelin plasma concentration.

Interleukin-6

IL-6 is one of the main proinflammatory mediator primarily secreted by the immune system cells [6]. However, 20–30% of these cytokine in the circulation is produced by the adipose tissue [114]. In obese subjects with high waist–hip ratio the participation is even greater [114]. It was also shown that omental adipose tissue is releasing 2–3 times more IL-6 than subcutaneous one [115].

The expression of IL-6 gene is related to the adipose cell size [116] and increases postprandially [117]. Moreover, several studies documented the positive relationship between BMI and plasma IL-6 concentration [118, 119].

In the liver, IL-6 stimulates synthesis of acute-phase response proteins including CRP, fibrinogen, serum amyloid-A and α-1 antichymotrypsine [120]. It is already well-known that CRP is both a marker and important risk factor of cardiac events and atherosclerosis in both general population and CKD patients [121]. It is also worth to stress that IL-6 stimulates fibrinogen production and platelets activity which increase the risk of clot formation [122]. In animal model Il-6 increases hepatic component of insulin resistance without effecting (or enhancing) insulin sensitivity in skeletal muscles [123]. However, in humans low-dose Il-6 infusion did increase insulin resistance [124].

In CKD patients plasma concentration of IL-6 is higher than in the general population, mainly as a consequence of prolonged plasma half-life of this cytokine and presence of chronic infections like Chlamydia pneumonia or others [125]. Both elevated concentration of IL-6 and CRP are strong predictors of cardiovascular morbidity and mortality in haemodialysis patients [126, 127].

Tumour Necrosis Factor-Alpha

TNF-α is predominantly synthesized by macrophages infiltrating adipose tissue [128, 129]. Subcutaneous adipose tissue is claimed to produce more TNF-α than visceral one [130]. This cytokine is involved in the pathogenesis of inflammation and/or insulin resistance [131].

The excess of TNF-α production by adipose tissue is one of the many factors influencing insulin resistance, but certainly not fundamental one. Mice lacking TNF-α or TNF-α receptor function demonstrate only a modest protection against hyperglycemia or insulin resistance in experimental obesity [132, 133].

There are three mechanisms linking TNF-α with insulin resistance. The first one is related to inactivation of insulin receptor signalling pathway (involving NFκB and/or INK) by phosphorylation [134, 135]. The second mechanism engages adiponectin, which secretion from adipocytes is markedly reduced by TNF-α [56]. The third mechanism compromises induction of adipocytes lipolysis and stimulation of hepatic lipogenesis [136].

In humans with type 2 diabetes mellitus increase plasma TNF-α concentration was found only in some studies and was not consistently associated with insulin resistance [137–139]. Systemic infusion of TNF-α inhibits peripheral insulin mediated, glucose uptake in healthy humans [140]. Additionally, administration of antibodies against TNF-α improves or has no effect on insulin sensitivity [141, 142]. The consequence of elevated plasma TNF-α concentration in CKD patients [143] is now only a matter of speculation.

Plasminogen Activator Inhibitor-1

Adipose tissue is a site of abundant PAI-1 synthesis [144, 145]. Cigolini et al. [146] showed that adipose visceral tissue mass in humans assesses by CT correlates with plasma PAI-1 concentration. Furthermore, weight loss is associated with reduction of PAI-1 activity [147–149].

PAI-1 is a procoagulative agent and fibrinolysis inhibitor. Increased PAI-1 concentration promotes release of platelet-derived growth factor which is known to play a role in vascular injury. Therefore, high circulating plasma PAI-1 concentration is a strong risk factor for thrombotic complications. It was shown that plasma level of PAI-1 is an independent predictor of coronary artery disease [150]. These characteristics may explain more frequent coronary artery disease in obese patients.

IL-6 within adipose tissue induces PAI-1 synthesis, which is the possible link between inflammation, obesity and cardiovascular diseases in obesity [151]. Recently, it was also shown in cultured adipocytes that absence or inhibition of PAI-1 protects against insulin resistance by promoting glucose uptake and adipocyte differentiation via increased peroxisome proliferator activated receptor-γ expression [152]. Therefore, PAI-1 is probably also involved in pathogenesis of obesity related insulin resistance.

Interestingly, in atherosclerotic rabbits atorvastatin reduces plasma PAI-1 concentration and PAI-1 expression in adipose tissue and adipocytes itself [153].

The Renin–Angiotensin System

All components of the renin–angiotensin system are present in adipose tissue [154, 155]. However, it is still unclear whether angiotensin II secreted by adipose tissue has an important systemic haemodynamic and/or non-haemodynamic effects [156].

Conclusion

Visceral adiposity is the key feature of metabolic syndrome (fig. 2). Adipokines oversecreted by the excessive amount of visceral adipose tissue are involved in the pathogenesis of metabolic abnormalities related to obesity, like insulin resistance, dyslipidaemia and chronic microinflammation. Accelerated atherosclerosis, diabetes type 2 and obesity related hypertension are the most frequent causes of morbidity and mortality of the general population. Chronic kidney diseases are also common consequence of diabetic, hypertensive or ischaemic nephropathies. Only in markedly (morbidly) obese humans obesity per se may directly impair the kidney function as isolated entity–obesity related glomerulopathy.

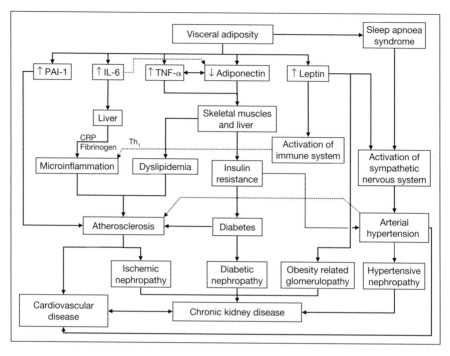

Fig. 2. Visceral adiposity and diseases of cardiovascular system and the kidneys.

However, even a moderate impairment of kidney function, further increases the risk of cardiovascular morbidity and mortality (vicious circle) [157].

Therefore, it is expected and brings a new hope that pharmacological intervention in paracrine/endocrine or immune activity of the adipose tissue may improve the outcome of metabolic syndrome and its cardiovascular and renal complications in the forthcoming future.

References

1 Havel PJ: Update on adipocyte hormones: regulation of energy balance and carbohydrate/lipid metabolism. Diabetes 2004;53:S143–S151.
2 Więcek A, Kokot F, Chudek J, Adamczak M: The adipose tissue – a novel endocrine organ of interest to the nephrologist. Nephrol Dial Transplant 2002;17:191–195.
3 Wisse BE: The inflammatory syndrome: the role of adipose tissue cytokines in metabolic disorders linked to obesity. J Am Soc Nephrol 2004;15:2792–2800.
4 Cinti S: Adipocyte differentiation and transdifferentiation: plasticity of the adipose organ. J Endocrinol Invest 2002;25:823–835.
5 Wellen KE, Hotamisligil GS: Obesity-induced inflammatory changes in adipose tissue. J Clin Invest 2003;112:1785–1788.

6   Fain JN, Madan AK, Hiler ML, Cheema P, Bahouth SW: Comparison of the release of adipokines by adipose tissue, adipose tissue matrix, and adipocytes from visceral and subcutaneous abdominal adipose tissues of obese humans. Endocrinology 2004;145:2273–2282.

7   Curat CA, Wegner V, Sengenes C, Miranville A, Tonus C, Busse R, Bouloumie A: Macrophages in human visceral adipose tissue: increased accumulation in obesity and a source of resistin and visfatin. Diabetologia 2006;49:744–747.

8   Löhn M, Dubrovska G, Lauterbach B, Luft FC, Gollasch M, Sharma AM: Periadventitial fat releases a vascular relaxing factor. FASEB J 2002;16:1057–1063.

9   Zhang Y, Proenca R, Maffei M, Barone M, Leopold L, Friedman JM: Positional cloning of the mouse obese gene and its human homologue. Nature 1994;372:425–432.

10 Muzzin P, Eisensmith RC, Copeland KC, Woo SL: Correction of obesity and diabetes in genetically obese mice by leptin gene therapy. Proc Natl Acad Sci USA 1996;93:14804–14808.

11 Horn R, Geldszus R, Potter E, von zur Muhlen A, Brabant G: Radioimmunoassay for the detection of leptin in human serum. Exp Clin Endocrinol Diabetes 1996;104:454–458.

12 Caro JF, Kolaczynski JW, Nyce MR, et al: Decreased cerebrospinal-fluid/serum leptin ratio in obesity: a possible mechanism for leptin resistance. Lancet 1996;348:159–161.

13 Flier JS: Physiology: is brain sympathetic to bone? Nature 2002;420:621–622.

14 Grinspoon S, Gulick T, Askari H, Landt M, Lee K, Anderson E, Ma Z, Vignati L, Bowsher R, Herzog D, Klibanski A: Serum leptin levels in women with anorexia nervosa. J Clin Endocrinol Metab 1996;81:3861–3863.

15 Ormseth OA, Nicolson M, Pelleymounter MA, Boyer BB: Leptin inhibits prehibernation hyperphagia and reduces body weight in arctic ground squirrels. Am J Physiol 1996;271:R1775–R1779.

16 Hwa JJ, Fawzi AB, Graziano MP, Ghibaudi L, Williams P, Van Heek M, Davis H, Rudinski M, Sybertz E, Strader CD: Leptin increases energy expenditure and selectively promotes fat metabolism in ob/ob mice. Am J Physiol 1997;272:R1204–R1209.

17 Gibson WT, Farooqi IS, Moreau M, DePaoli AM, Lawrence E, O'Rahilly S, Trussell RA: Congenital leptin deficiency due to homozygosity for the Delta133G mutation: report of another case and evaluation of response to four years of leptin therapy. J Clin Endocrinol Metab 2004;89:4821–4826.

18 Chehab FF, Mounzih K, Lu R, Lim ME: Early onset of reproductive function in normal female mice treated with leptin. Science 1997;275:88–90.

19 Sierra-Honigmann MR, Nath AK, Murakami C, Garcia-Cardena G, Papapetropoulos A, Sessa WC, Madge LA, Schechner JS, Schwabb MB, Polverini PJ, Flores-Riveros JR: Biological action of leptin as an angiogenic factor. Science 1998;281:1683–1686.

20 Lord GM, Matarese G, Howard JK, Baker RJ, Bloom SR, Lechler RI: Leptin modulates the T-cell immune response and reverses starvation-induced immunosuppression. Nature 1998;394: 897–901.

21 Bornstein SR, Licinio J, Tauchnitz R, Engelmann L, Negrao AB, Gold P, Chrousos GP: Plasma leptin levels are increased in survivors of acute sepsis: associated loss of diurnal rhythm, in cortisol and leptin secretion. J Clin Endocrinol Metab 1998;83:280–283.

22 Hasty AH, Shimano H, Osuga J, Namatame I, Takahashi A, Yahagi N, Perrey S, Iizuka Y, Tamura Y, Amemiya-Kudo M, Yoshikawa T, Okazaki H, Ohashi K, Harada K, Matsuzaka T, Sone H, Gotoda T, Nagai R, Ishibashi S, Yamada N: Severe hypercholesterolemia, hypertriglyceridemia, and atherosclerosis in mice lacking both leptin and the low density lipoprotein receptor. J Biol Chem 2001;276:37402–37408.

23 Rahmouni K, Haynes WG, Mark AL: Cardiovascular and sympathetic effects of leptin. Curr Hypertens Rep 2002;4:119–125.

24 Szanto I, Kahn CR: Selective interaction between leptin and insulin signaling pathways in a hepatic cell line. Proc Natl Acad Sci USA 2000;97:2355–2360.

25 Adamczak M, Kokot F, Więcek A: Relationship between plasma renin profile and leptinaemia in patients with essential hypertension. J Hum Hypertens 2000;14:503–950.

26 Kambham N, Markowitz GS, Valeri AM, Lin JK: Obesity-related glomerulopathy: an emerging epidemic. Kidney Int 2001;59:1498–1509.

27 Wolf G, Hamann A, Han DC, Helmchen U, Thaiss F, Ziyadeh FN, Stahl RA: Leptin stimulates proliferation and TGF-beta expression in renal glomerular endothelial cells: potential role in glomerulosclerosis. Kidney Int 1999;56:860–872.

28 Han DC, Isono M, Chen S, Casaretto A, Hong SW, Wolf G, Ziyadeh FN: Leptin stimulates type I collagen production in db/db mesangial cells. Glucose uptake and TGF-β type II receptor expression. Kidney Int 2001;59:1315–1323.
29 Howard JK, Lord GM, Clutterbucck EJ, Ghatei MA, Pusey CD, Bloom SR: Plasma immunoreactive leptin concentration in endstage renal disease. Clin Sci 1997;93:119–126.
30 Heimbürger O, Lönnqvist F, Danielsson A, Nordenström J, Stenvinkel P: Serum immunoreactive leptin concentration and its relation to the body fat content in chronic renal failure. J Am Soc Nephrol 1997;8:1423–1430.
31 Kokot F, Chudek J, Karkoszka H, Adamczak M, Więcek A, Klimek D: Does PTH influence leptin concentration in haemodialysed uraemic patients? Nephron 1999;82:372–373.
32 Nordfors L, Lönnqvist F, Heimbürger O, Danielsson A, Schalling M, Stenvinkel P: Low leptin gene expression and hyperleptinemia in chronic renal failure. Kidney Int 1998;54:1267–1275.
33 Cumin F, Baum HP, Levens N: Mechanism of leptin removal from the circulation by the kidney. J Endocrinol 1997;155:577–585.
34 Kokot F, Adamczak M, Więcek A: Plasma leptin concentration in kidney transplant patients during the early post-transplant period. Nephrol Dial Transplant 1998;13:2276–2280.
35 Cumin F, Baum HP, de Gasparo M, Levens N: Removal of endogenous leptin from the circulation by the kidney. Int J Obes Relat Metab Disord 1997;21:495–504.
36 Ficek R, Kokot F, Chudek J, Adamczak M, Ficek J, Więcek A: Plasma leptin concentration in patients with acute renal failure. Clin Nephrol 2004;62:84–91.
37 Zeng J, Patterson BW, Klein S, Martin DR, Dagogo-Jack S, Kohrt WM, Miller SB, Landt M: Whole body leptin kinetics and renal metabolism in vivo. Am J Physiol 1997;273:E1102–E1106.
38 Cheung W, Yu PX, Little BM, Cone RD, Marks DL, Mak RH: Role of leptin and melanocortin signaling in uremia-associated cachexia. J Clin Invest 2005;115:1659–1665.
39 Chudek J, Adamczak M, Kania M, Holowiecka A, Rozmus W, Kokot F, Więcek A: Does plasma leptin concentration predict the nutritional status of hemodialyzed patients with chronic renal failure? Med Sci Monit 2003;9:CR377–CR382.
40 Merabet E, Dagogo-Jack S, Coyne DW, Klein S, Santiago JV, Hmiel SP, Landt M: Increased plasma leptin concentration in end-stage renal disease. J Clin Endocrinol Metab 1997;82:847–850.
41 Bossola M, Muscaritoli M, Valenza V, Panocchia N, Tazza L, Cascino A, Laviano A, Liberatori M, Lodovica Moussier M, Rossi Fanelli F, Luciani G: Anorexia and serum leptin levels in hemodialysis patients. Nephron Clin Pract 2004;97:c76–c82.
42 Lihn AS, Pedersen SB, Richelsen B: Adiponectin: action, regulation and association to insulin sensitivity. Obes Rev 2005;6:13–21.
43 Yokota T, Oritani K, Ishikawa JTI, Ishikawa J, Matsuyama A, Ouchi N, Kihara S, Funahashi T, Tenner AJ, Tomiyama Y, Matsuzawa Y: Adiponectin, a new member of the family of soluble defense collagens, negatively regulates the growth of myelomonocytic progenitors and the function of macrophages. Blood 2000;96:1723–1732.
44 Arita Y, Kihara S, Ouchi N, Takahashi M, Maeda K, Miyagawa J, Hotta K, Shimomura I, Nakamura T, Miyaoka K, Kuriyama H, Nishida M, Yamashita S, Okubo K, Matsubara K, Muraguchi M, Ohmoto Y, Funahashi T, Matsuzawa Y: Paradoxical decrease of an adipose-specific protein, adiponectin, in obesity. Biochem Biophys Res Commun 1999;257:79–83.
45 Peake, PW, Kriketos AD, Denyer GS, Campbell LV, Charlesworth JA: The postprandial response of adiponectin to a high-fat meal in normal and insulin-resistant subjects. Int J Obes Relat Metab Disord 2003;27:657–662.
46 Kadowaki T, Yamauchi T: Adiponectin and adiponectin receptors. Endocr Rev 2005;26:439–451.
47 Pajvani UB, Hawkins M, Combs TP, Rajala MW, Doebber T, Berger JP, Wagner JA, Wu M, Knopps A, Xiang AH, Utzschneider KM, Kahn SE, Olefsky JM, Buchanan TA, Scherer PE: Complex distribution, not absolute amount of adiponectin, correlates with thiazolidinedione-mediated improvement in insulin sensitivity. J Biol Chem 2004;279:12152–12162.
48 Fisher M, Trujillo ME, Hanif W, Barnett AH, McTernan PG, Scherer PE, Kumar S: Serum high molecular weight complex of adiponectin correlates better with glucose tolerance than total serum adiponectin in Indo-Asian males. Diabetologia 2005;48:1084–1087.
49 Hotto K, Funahashi T, Arita Y, Takahashi M, Matsuda M, Okamoto Y, Iwahashi H, Kuriyama H, Ouchi N, Maeda K, Nishida M, Kihara S, Sakai N, Nakajima T, Hasegawa K, Muraguchi M,

Ohmoto Y, Nakamura T, Yamashita S, Hanafusa T, Matsuzawa Y: Plasma concentration of a novel, adipose-specific protein, adiponectin, in type 2 diabetic patients. Arterioscler Thromb Vasc Biol 2000;20:1595–1599.
50 Dzielińska Z, Januszewicz A, Więcek A, Demkow M, Makowiecka-Cieśla M, Prejbisz A, Kadziela J, Mielniczuk R, Florczak E, Janas J, Januszewicz M, Rużyllo W: Decreased plasma concentration of a novel anti-inflammatory protein-adiponectin-in hypertensive men with coronary artery disease. Thromb Res 2003;110:365–369.
51 Adamczak M, Więcek A, Funahashi T, Chudek J, Kokot F, Matsuzawa Y: Decreased plasma adiponectin concentration in patients with essential hypertension. Am J Hypertens 2003;16:72–75.
52 Yang W, Chuang L: Human genetics of adiponectin in the metabolic syndrome. J Mol Med 2006;84:112–121.
53 Yamauchi T, Kamon J, Ito Y, et al: Cloning of adiponectin receptors that mediate antidiabetic metabolic effects. Nature 2003;423:762–769.
54 Ouchi N, Ohishi M, Kihara S, Funahashi T, Nakamura T, Nagaretani H, Kumada M, Ohashi K, Okamoto Y, Nishizawa H, Kishida K, Maeda N, Nagasawa A, Kobayashi H, Hiraoka H, Komai N, Kaibe M, Rakugi H, Ogihara T, Matsuzawa Y: Association of hypoadiponectinemia with impaired vasoreactivity. Hypertension 2003;42:231–234.
55 Ouchi N, Kihara S, Arita Y, Maeda K, Kuriyama H, Okamoto Y, Hotta K, Nishida M, Takahashi M, Nakamura T, Yamashita S, Funahashi T, Matsuzawa Y: Novel modulator for endothelial adhesion molecules adipocyte-derived plasma protein adiponectin. Circulation 1999;100:2473–2476.
56 Ouchi N, Kihara S, Arita Y, Okamoto Y, Maeda K, Kuriyama H, Hotta K, Nishida M, Takahashi M, Muraguchi M, Ohmoto Y, Nakamura T, Yamashita S, Funahashi T, Matsuzawa Y: Adiponectin and adipocyte-derived plasma protein, inhibits endothelial NF-kappaB signaling through a cAMP-dependent pathway. Circulation 2000;102:1296–1301.
57 Ouchi N, Kihara S, Arita Y, Nishida M, Matsuyama A, Okamoto Y, Ishigami M, Kuriyama H, Kishida K, Nishizawa H, Hotta K, Muraguchi M, Ohmoto Y, Yamashita S, Funahashi T, Matsuzawa Y: Adipocyte – derived plasma protein, adiponectin, suppress lipid accumulation and class A receptor expression in human monocyte-derived macrophages. Circulation 2001;103: 1057–1063.
58 Motoshima H, Wu X, Mahadev K, Goldstein BJ: Adiponectin suppresses proliferation and superoxide generation and enhances eNOS activity in endothelial cells treated with oxidized LDL. Biochem Biophys Res Commun 2004;315:264–271.
59 Okamoto Y, Kihara S, Ouchi N, Nishida M, Arita Y, Kumada M, Ohashi K, Sakai N, Shimomura I, Kobayashi H, Terasaka N, Inaba T, Funahashi T, Matsuzawa Y. Adiponectin reduces atherosclerosis in apolipoprotein E-deficient mice. Circulation 2002;106:2767–2770.
60 Kumada M, Kihara S, Ouchi N, Kobayashi H, Okamoto Y, Ohashi K, Maeda K, Nagaretani H, Kishida K, Maeda N, Nagasawa A, Funahashi T, Matsuzawa Y: Adiponectin specifically increased tissue inhibitor of metalloproteinase-1 through interleukin-10 expression in human macrophages. Circulation 2004;109:2046–2049.
61 Yamauchi T, Kamon J, Waki H, Terauchi Y, Kubota N, Hara K, Mori Y, Ide T, Murakami K, Tsuboyama-Kasaoka N, Ezaki O, Akanuma Y, Gavrilova O, Vinson C, Reitman ML, Kagechika H, Shudo K, Yoda M, Nakano Y, Tobe K, Nagai R, Kimura S, Tomita M, Froguel P, Kadowaki T: The fat-derived hormone adiponectin reverses insulin resistance associated with both lipoatrophy and obesity. Nat Med 2001;7:941–946.
62 Spranger J, Kroke A, Mohlig M, Bergmann MM, Ristow M, Boeing H, Pfeiffer AF: Adiponectin and protection against type 2 diabetes mellitus. Lancet. 2003;361:226–228.
63 Waki H, Yamauchi T, Kamon J Ito Y, Uchida S, Kita S, Hara K, Hada Y, Vasseur F, Froguel P, Kimura S, Nagai R, Kadowaki T: Impaired multimerization of human adiponectin mutants associated with diabetes. Molecular structure and multimer formation of adiponectin. J Biol Chem 2003;278:40352–40363.
64 Li P, Wu W, Gersch CM, Feng L: Anti-inflammatory effects of adiponectin on anti-GBM glomerulonephritis. J Am Soc Nephrol 2004;15:9A.
65 Zoccali C, Mallamaci F, Tripepi G, Benedetto FA, Cutrupi S, Parlongo S, Malatino LS, Bonanno G, Seminara G, Rapisarda F, Fatuzzo P, Buemi M, Nicocia G, Tanaka S, Ouchi N, Kihara S, Funahashi T, Matsuzawa Y: Adiponectin, metabolic risk factors, and cardiovascular events among patients with end-stage renal disease. J Am Soc Nephrol 2002;13:134–141.

66 Ignacy W, Chudek J, Adamczak M, Funahashi T, Matsuzawa Y, Kokot F, Więcek F: Reciprocal association of plasma adiponectin and serum C-reactive protein concentration in haemodialysis patients with end-stage kidney disease – a follow-up study. Nephron Clin Pract 2005;101:c18–c24.

67 Marchlewska A, Stenvinkel P, Lindholm B, Danielsson A, Pecoits-Filho R, Lonnqvist F, Schalling M, Heimburger O, Nordfors L: Reduced gene expression of adiponectin in fat tissue from patients with end-stage renal disease. Kidney Int 2004;66:46–50.

68 Chudek J, Adamczak M, Karkoszka H, Budziński G, Ignacy W, Funahashi T, Matsuzawa Y, Cierpka L, Kokot F, Więcek A: Plasma adiponectin concentration before and after successful kidney transplantation. Transplant Proc 2003;35:2186–2189.

69 Mallamaci F, Zoccali C, Cuzzola F, Tripepi G, Cutrupi S, Parlongo S, Tanaka S, Ouchi N, Kihara S, Funahashi T, Matsuzawa Y: Adiponectin in essential hypertension. J Nephrol 2002;15:507–511.

70 Adamczak M, Czerwieńska B, Chudek J, Więcek A: Role of kidneys in metabolism of adiponectin. studies in patients with unilateral renal artery stenosis. Nephrol Dial Transplant 2005;20(suppl 5):v55.

71 Ouchi N, Kihara S, Funahashi T, Nakamura T, Nishida M, Kumada M, Okamoto Y, Ohashi K, Nagaretani H, Kishida K, Nishizawa H, Maeda N, Kobayashi H, Hiraoka H, Matsuzawa Y: Reciprocal association of C-reactive protein with adiponectin in blood stream and adipose tissue. Circulation 2003;107:671–674.

72 Engeli S, Feldpausch M, Gorzelniak K, Hartwig F, Heintze U, Janke J, Mohlig M, Pfeiffer AF, Luft FC, Sharma AM: Association between adiponectin and mediators of inflammation in obese women. Diabetes 2003;52:942–947.

73 Huang JW, Yen CJ, Chiang HW, Hung KY, Tsai TJ, Wu KD: Adiponectin in peritoneal dialysis patients: a comparison with hemodialysis patients and subjects with normal renal function. Am J Kidney Dis 2004;43:1047–1055.

74 Lee CT, Lee CH, Su Y, Chuang YC, Tsai TL, Cheni JB: The relationship between inflammatory markers, leptin and adiponectin in chronic hemodialysis patients. Int J Artif Organs 2004;27:835–841.

75 Fasshauer M, Kralisch S, Klier M, Lossner U, Bluher M, Klein J, Paschke R: Adiponectin gene expression and secretion is inhibited by interleukin-6 in 3T3-L1 adipocytes. Biochem Biophys Res Commun 2003;301:1045–1050.

76 Kappes A, Loffler G: Influences of ionomycin, dibutyryl-cycloAMP and tumour necrosis factor-alpha on intracellular amount and secretion of apM1 in differentiating primary human preadipocytes. Horm Metab Res 2000;32:548–554.

77 Maeda N, Takahashi M, Funahashi T, Kihara S, Nishizawa H, Kishida K, Nagaretani H, Matsuda M, Komuro R, Ouchi N, Kuriyama H, Hotta K, Nakamura T, Shimomura I, Matsuzawa Y: PPARgamma ligands increase expression and plasma concentrations of adiponectin. An adipose-derived protein. Diabetes 2001;50:1094–1099.

78 Hirose H, Kaqwai T, Yamamoto Y, Taniyama M, Tomita M, Matsubara K, Okazaki Y, Ishii T, Oguma Y, Takei I, Saruta T: Effects of pioglitasone on metabolic parameters, body fat distribution and serum adiponectin levels in Japanese male patients with type 2 diabetes. Metab Clin Exp 2002;51:314–317.

79 Furuhashi M, Ura N, Higashiura K, Murakami H, Tanaka M, Moniwa N, Yoshida D, Shimamoto K: Blockade of the renin-angiotensin system increases adiponectin concentrations in patients with essential hypertension. Hypertension 2003;42:76–82.

80 Koh K, Quon M, Han S, et al: Vascular and metabolic effects of combined therapy with ramipril and simvastatin in patients with type 2 diabetes. Hypertension 2005;45:1088–1093.

81 Koh K, Quon M, Han S, Chung WJ, Ahn JY, Seo YH, Kang MH, Ahn TH, Choi IS, Shin EK: Additive beneficial effects of losartan combined with simvastatin in the treatment of hypercholesterolemic, hypertensive patients. Circulation 2004;110:3687–3692.

82 Koh K, Quon M, Han S, Chung WJ, Lee Y, Shin EK: Anti-inflammatory and metabolic effects of candesartan in hypertensive patients. Int J Cardiol 2005;108:96–100.

83 Nowak L, Adamczak M, Więcek A: Blockade of sympathetic nervous system activity by rilmenidine increases plasma adiponectin concentration in patients with essential hypertension. Am J Hypertens 2005;18:1470–1475.

84 Koh K, Quon M, Han S, Chung WJ, Ahn JY, Seo YH, Choi IS, Shin EK: Additive beneficial effects of fenofibrate combined with atorvastatin in the treatment of combined hyperlipidemia. J Am Coll Cardiol 2005;45:1649–1653.

85 Despres JP, Golay A, Sjostrom L: Effects of rimonabant on metabolic risk factors in overweight patients with dyslipidemia. N Engl J Med 2005;353:2121–2134.

86 Tiikkainen M, Hakkinen AM, Korsheninnikova E, Nyman T, Makimattila S, Yki-Jarvinen H: Effects of rosiglitazone and metformin on liver fat content, hepatic insulin resistance, insulin clearance and gene expression in adipose tissue in patients with type 2 diabetes. Diabetes 2004;53:2169–2176.

87 Bobbert T, Rochlitz H, Wegewitz U, Akpulat S, Mai K, Weickert MO, Mohlig M, Pfeiffer AF, Spranger J: Changes of adiponectin oligomer composition by moderate weight reduction. Diabetes 2005;54:2712–2719.

88 Lehrke M, Reilly MP, Millington SC, Iqbal N, Rader DJ, Lazar MA: An inflammatory cascade leading to hyperresistinemia in humans. Plos Med 2004;1:e45.

89 Fain JN, Cheema PS, Bahouth SW, Lloyd Hiler M: Resistin release by human adipose tissue explants in primary culture. Biochem Biophys Res Commun 2003;300:674–678.

90 Steppan CM, Bailey ST, Bhat S, Brown EJ, Banerjee RR, Wright CM, Patel HR, Ahima RS, Lazar MA: The hormone resistin links obesity to diabetes. Nature 2001;18:307–312.

91 Lee JH, Chan JL, Yiannakouris N, Kontogianni M, Estrada E, Seip R, Orlova C, Mantzoros CS: Circulating resistin levels are not associated with obesity or insulin resistance in humans and are not regulated by fasting or leptin administration: cross-sectional and interventional studies in normal, insulin-resistant, and diabetic subjects. J Clin Endocrinol Metab 2003;88:4848–4856.

92 Heilbronn LK, Rood J, Janderova L, Albu JB, Kelley DE, Ravussin E, Smith SR: Relationship between serum resistin concentration and insulin resistance in nonobese, obese and obese diabetic subjects. J Clin Endocrinol Metab 2004;89:1844–1848.

93 Janke J, Engeli S, Gorzelniak K, Luft FC, Sharma AM: Resistin gene expression in human adipocytes is not related to insulin resistance. Obes Res 2002;10:1–5.

94 Nagaev I, Smith U: Insulin resistance and type 2 diabetes are not related to resistin expression in human fat cells or skeletal muscle. Biochem Biophys Res Commun 2001;285:561–564.

95 Savage DB, Sewter CP, Klenk ES: Resistin/Fizz3 expression in relation to obesity and peroxisome proliferator-activated receptor-gamma action in humans. Diabetes 2001;50:2199–2202.

96 Ghosh S, Singh AK, Aruna B, Mukhopadhyay S, Ehtesham NZ: The genomic organization of mouse resistin reveals major differences from the human resistin: functional implications. Gene 2003;305:27–34.

97 McTernan CL, McTernan PG, Harte AL, Levick PL, Barnett AH, Kumar S: Resistin, central obesity, and type 2 diabetes. Lancet 2002;359:46–47.

98 Kaser S, Kaser A, Sandhofer A, Ebenbichler CF, Tilg H, Patsch JR: Resistin messenger-RNA expression is increased by proinflammatory cytokines in vitro. Biochem Biophys Res Commun 2003;309:286–290.

99 Lu SC, Shieh WY, Chen CY, Hsu SC, Chen HL: Lipopolysaccharide increases resistin gene expression in vivo and in vitro. FEBS Lett 2002;530:158–162.

100 Verma S, Li SH, Wang CH, Fedak PW, Li RK, Weisel RD, Mickle DA: Resistin promotes endothelial cell activation: further evidence of adipokine-endothelial interaction. Circulation 2003;108:736–740.

101 Kawanami D, Maemura K, Takeda N, Harada A, Nojiri Y, Imai Y, Manabe I, Utsunomiya K, Nagai R: Direct reciprocal effects of resistin and adiponectin on vascular endothelial cells: a new insight into adipocytokine-endothelial cell interactions. Biochem Biophys Res Commun 2004;314: 415–419.

102 Reilly M, Lehrke M, Wolfe ML, Rohatgi A, Lazar MA, Rader DJ: Resistin is an inflammatory marker of atherosclerosis in humans. Circulation 2005;111:932–939.

103 Kielstein JT, Becker B, Graf S, Brabant G, Haller H, Fliser D: Increased resistin blood levels are not associated with insulin resistance in patients with renal disease. Am J Kidney Dis 2003;42:62–66.

104 Fukuhara A, Matsuda M, Nishizawa M, Segawa K, Tanaka M, Kishimoto K, Matsuki Y, Murakami M, Ichisaka T, Murakami H, Watanabe E, Takagi T, Akiyoshi M, Ohtsubo T, Kihara S, Yamashita S, Makishima M, Funahashi T, Yamanaka S, Hiramatsu R, Matsuzawa Y, Shimomura I: Visfatin: a protein secreted by visceral fat that mimics the effects of insulin. Science 2005;307: 426–430.

105 Curat CA, Wegner V, Sengenes C, Miranville A, Tonus C, Busse R, Bouloumie A: Macrophages in human visceral adipose tissue: increased accumulation in obesity and a source of resistin and visfatin. Diabetologia 2006;49:744–747.

106 Haider DG, Schindler K, Schaller G, Prager G, Wolzt M, Ludvik B: Increased plasma visfatin concentrations in morbidly obese subjects are reduced after gastric banding. J Clin Endocrinol Metab 2006;91:41578–41581.

107 Chen MP, Chung FM, Chang DM, Tsai JC, Huang HF, Shin SJ, Lee YJ: Elevated plasma level of visfatin/pre-B cell colony-enhancing factor in patients with type 2 diabetes mellitus. J Clin Endocrinol Metab 2006;91:295–299.

108 Tatemoto K, Hosoya M, Habata Y, Fujii R, Kakegawa T, Zou MX, Kawamata Y, Fukusumi S, Hinuma S, Kitada C, Kurokawa T, Onda H, Fujino M: Isolation and characterization of a novel endogenous peptide ligand for the human APJ receptor. Biochem Biophys Res Commun 1998;251:471–476.

109 Kleinz MJ, Davenport AP: Emerging roles of apelin in biology and medicine. Pharmacol Ther 2005;107:198–211.

110 Boucher J, Masri B, Daviaud D, Gesta S, Guigne C, Mazzucotelli A, Castan-Laurell I, Tack I, Knibiehler B, Carpene C, Audigier Y, Saulnier-Blache JS, Valet P: Apelin, a newly identified adipokine up-regulated by insulin and obesity. Endocrinology 2005;146:1764–1771.

111 Heinonen MV, Purhonen AK, Miettinen P, Paakkonen M, Pirinen E, Alhava E, Akerman K, Herzig KH: Apelin, orexin-A and leptin plasma levels in morbid obesity and effect of gastric banding. Regul Pept 2005;130:7–13.

112 Kleinz MJ, Skepper JN, Davenport AP: Immunocytochemical localisation of the apelin receptor, APJ, to human cardiomyocytes, vascular smooth muscle and endothelial cells. Regul Pept 2005;126:233–240.

113 Wei L, Hou X, Tatemoto K: Regulation of apelin mRNA expression by insulin and glucocorticoids in mouse 3T3-L1 adipocytes. Regul Pept 2005;132:27–32.

114 Mohamed-Ali V, Goodrick S, Rawesh A, Jauffred S, Fischer CP, Steensberg A, Pedersen BK: Subcutaneous adipose tissue releases interleukin-6, but not tumor necrosis factor-alpha, in vivo. J Clin Endocrinol Metab 1997;82:4196–4200.

115 Fried SK, Bunkin DA, Greenberg AS: Omental and subcutaneous adipose tissues of obese subjects release interleukin-6: depot difference and regulation by glucocorticoid. J Clin Endocrinol Metab 1998;83:847–850.

116 Sopasakis VR, Sandqvist M, Gustafson B, Hammarstedt A, Schmelz M, Yang X, Jansson PA, Smith U: High local concentrations and effects on differentiation implicate interleukin-6 as a paracrine regulator. Obes Res 2004;12:454–460.

117 Orban Z, Remaley AT, Sampson M, Trajanoski Z, Chrousos GP: The differential effect of food intake and beta-adrenergic stimulation on adipose-derived hormones and cytokines in man. J Clin Endocrinol Metab 1999;84:2126–2133.

118 Park HS, Park JY, Yu R: Relationship of obesity and visceral adiposity with serum concentrations of CRP, TNF-alpha and IL-6. Diabetes Res Clin Pract 2005;69:29–35.

119 Maachi M, Pieroni L, Bruckert E, Jardel C, Fellahi S, Hainque B, Capeau J, Bastard JP: Systemic low-grade inflammation is related to both circulating and adipose tissue TNFalpha, leptin and IL-6 levels in obese women. Int J Obes Relat Metab Disord 2004;28:993–997.

120 Zoccali C, Mallamaci F, Tripepi G: Adipose tissue as a source of inflammatory cytokines in health and disease: focus on end-stage renal disease. Kidney Int 2003;84:S65–S68.

121 Burke AP, Tracy RP, Kolodgie F, Malcom GT, Zieske A, Kutys R, Pestaner J, Smialek J, Virmani R: Elevated C-reactive protein values and atherosclerosis in sudden coronary death: association with different pathologies. Circulation 2002;105:2019–2023.

122 Burstein SA, Peng J, Friese P, Wolf RF, Harrison P, Downs T, Hamilton K, Comp P, Dale GL: Cytokine-induced alteration of platelet and hemostatic function. Stem Cells 1996;14(suppl 1):154–162.

123 Carey AL, Febbraio MA: Interleukin-6 and insulin sensitivity: friend or foe? Diabetologia 2004;47:1135–1142.

124 Steensberg A, Fischer CP, Sacchetti M, Keller C, Osada T, Schjerling P, van Hall G, Febbraio MA, Pedersen BK: Acute interleukin-6 administration does not impair muscle glucose uptake or whole-body glucose disposal in healthy humans. J Physiol 2003;548:631–638.

125 Jacobs P, Glorieux G, Vanholder R: Interleukin/cytokine profiles in haemodialysis and in continuous peritoneal dialysis. Nephrol Dial Transplant 2004;19(suppl 5):v41–v45.

126 Kimmel PL, Phillips TM, Simmens SJ, Peterson RA, Weihs KL, Alleyne S, Cruz I, Yanovski JA, Veis JH: Immunologic function and survival in hemodialysis patients. Kidney Int 1998;54:236–244.

127 Stenvinkel P, Wanner C, Metzger T, Heimburger O, Mallamaci F, Tripepi G, Malatino L, Zoccali C: Inflammation and outcome in end-stage renal failure: does female gender constitute a survival advantage? Kidney Int 2002;62:1791–1798.

128 Weisberg SP, McCann D, Desai M, Rosenbaum M, Leibel RL, Ferrante AW Jr: Obesity is associated with macrophage accumulation in adipose tissue. J Clin Invest 2003;112:1796–1808.

129 Fain JN, Bahouth SW, Madan AK: TNF alpha release by the nonfat cells of human adipose tissue. Int J Obes Relat Metab Disord 2004;28:616–622.

130 Winkler G, Kiss S, Keszthelyi L, Sapi Z, Ory I, Salamon F, Kovacs M, Vargha P, Szekeres O, Speer G, Karadi I, Sikter M, Kaszas E, Dworak O, Gero G, Cseh K: Expression of tumor necrosis factor (TNF)-alpha protein in the subcutaneous and visceral adipose tissue in correlation with adipocyte cell volume, serum TNF-alpha, soluble serum TNF-receptor-2 concentrations and C-peptide level. Eur J Endocrinol 2003;149:129–135.

131 Sethi J, Hotamisligil GS: The role of TNF-alfa in adipocyte metabolism. Semin Cell Dev Biol 1999;10:19–29.

132 Uysal KT, Wiesbrock SM, Marino MW, Hotamisligil GS: Protection from obesity-induced insulin resistance in mice lacking TNF-alpha function. Nature 1997;389:610–614.

133 Ventre J, Doebber T, Wu M, MacNaul K, Stevens K, Pasparakis M, Kollias G, Moller DE: Targeted disruption of the tumor necrosis factor-alpha gene: metabolic consequences in obese and nonobese mice. Diabetes 1997;46:1526–1531.

134 Hirosumi J, Tuncman G, Chang L, Gorgun CZ, Uysal KT, Maeda K, Karin M, Hotamisligil GS: A central role for JNK in obesity and insulin resistance. Nature 2002;420:333–336.

135 Hotamisligil GS: Inflammatory pathways and insulin action. Int J Obes Relat Metab Disord 2003;27(suppl 3):S53–S55.

136 Feingold KR, Grunfeld C: Role of cytokines in inducing hyperlipidemia. Diabetes 1992;41(suppl 2):97–101.

137 Carey AL, Bruce CR, Sacchetti M, Anderson MJ, Olsen DB, Saltin B, Hawley JA, Febbraio MA: Interleukin-6 and tumor necrosis factor-α are not increased in patients with type 2 diabetes: evidence that plasma interleukin-6 is related to fat mass and not insulin responsiveness. Diabetologia 2004;47:1029–1037.

138 Katsuki A, Sumida Y, Murashima S, Murata K, Takarada Y, Ito K, Fujii M, Tsuchihashi K, Goto H, Nakatani K, Yano Y: Serum levels of tumor necrosis factor-α are increased in obese patients with noninsulin-dependent diabetes mellitus. J Clin Endocrinol Metab 1998;83:859–862.

139 Ohya M, Taniguchi A, Fukushima M, Nakai Y, Kawasaki Y, Nagasaka S, Kuroe A, Taki Y, Yoshii S, Hosokawa M, Inagaki N, Seino Y: Three measures of tumor necrosis factor alpha activity and insulin resistance in nonobese Japanese type 2 diabetic patients. Metabolism 2005;54:1297–1301.

140 Plomgaard P, Bouzakri K, Krogh-Madsen R, Mittendorfer B, Zierzth JR, Pedersen BK: Tumor necrosis factor-α induces skeletal muscle insulin resistance in healthy human subjects via inhibition of Akt substrate 160 phosphorylation. Diabetes 2005;54:2939–2945.

141 Ofei F, Hurel S, Newkirk J, Sopwith M, Taylor R: Effects of an engineered human anti-TNF-α antibody (CDP571) on insulin sensitivity and glycemic control in patients with NIDDM. Diabetes 1996;45:881–885.

142 Yazdani-Biuki B, Stelzl H, Brezinschek HP, Hermann J, Mueller T, Krippl P, Graninger W, Wascher TC: Improvement of insulin sensitivity in insulin resistant subjects during prolonged treatment with the anti-TNF-α antibody infliximab. Eur J Clin Invest 2004;34:641–642.

143 Nishimura M, Hashimoto T, Kobayashi H, Fukukda T, Okino K, Yamamoto N, Nakamura N, Yoshikawa T, Takahashi H, Ono T: Possible involvement of TNF-alpha in left ventricular remodeling in hemodialysis patients. J Nephrol 2003;16:641–649.

144 Allessi MC, Peiretti F, Morange P, Henry M, Nalbone G, Juhan-Vague I: Production of plasminogen activator inhibitor-1 by human adipose tissue: possible link between visceral fat accumulation and vascular disease. Diabetes 1997;46:860–867.

145 Sawdey MS, Loskutoff DJ: Regulation of murine type 1 plasminogen activator inhibitor gene expression in vivo: tissue specificity and induction by lipopolysaccharide, tumor necrosis factor-alpha, and transforming growth factor-beta. J Clin Invest 1991;88:1346–1353.
146 Cigolini M, Targher G, Bergamo Andreis IA, Tonoli M, Agostino G, De Sandre G: Visceral fat accumulation and its relation to plasma hemostatic factors in healthy men. Arterioscler Thromb Vasc Biol 1996;16:368–374.
147 Primrose JN, Davies JA, Prentice CR, Hughes R, Johnston D: Reduction in factor VII, fibrinogen and plasminogen activator inhibitor-1 activity after surgical treatment of morbid obesity. Thromb Haemost 1992;68:396–399.
148 Svendsen OL, Hassager C, Christiansen C, Nielsen JD, Winther K: Plasminogen activator inhibitor-1, tissue-type plasminogen activator, and fibrinogen: effect of dieting with or without exercise in overweight postmenopausal women. Arterioscler Thromb Vasc Biol 1996;16:381–385.
149 Murakami T, Horigome H, Tanaka K, Nakata Y, Ohkawara K, Katayama Y, Matsui A: Impact of weight reduction on production of platelet-derived microparticles and fibrinolytic parameters in obesity. Thromb Res 2006; [Epub ahead of print].
150 Segarra A, Chacon P, Martinez-Eyarre C, Argelaguer X, Vila J, Ruiz P, Fort J, Bartolome J, Camps J, Moliner E, Pelegri A, Marco F, Olmos A, Piera L: Circulating levels of plasminogen activator inhibitor type-1, tissue plasminohen activator, and thrombomodulin in hemodialysis patients: biochemical correlations and role as independent predictors of coronary artery stenosis. J Am Soc Nephrol 2001;12:1255–1263.
151 Rega G, Kaun C, Weiss TW, Demyanets S, Zorn G, Kastl SP, Steiner S, Seidinger D, Kopp CW, Frey M, Roehle R, Maurer G, Huber K, Wojta J: Inflammatory cytokines interleukin-6 and oncostatin m induce plasminogen activator inhibitor-1 in human adipose tissue. Circulation 2005;111:1938–1945.
152 Liang X, Kanjanabuch T, Mao SL, Hao CM, Tang YW, Declerck PJ, Hasty AH, Wasserman DH, Fogo AB, Ma LJ: Plasminogen activator inhibitor-1 modulates adipocyte differentiation. Am J Physiol Endocrinol Metab 2006;290:E103–E113.
153 Li JQ, Zhao SP, Li QZ, Cai YC, Wu LR, Fang Y, Li P: Atorvastatin reduces plasminogen activator inhibitor-1 expression in adipose tissue of atherosclerotic rabbits. Clin Chim Acta 2006; [Epub ahead of print].
154 Gorzelniak K, Engeli S, Janke J, Luft FC, Sharma AM: Hormonal regulation of the human adipose-tissue renin-angiotensin system: relationship to obesity and hypertension. J Hypertens 2002;20:965–973.
155 Karlsson C, Lindell K, Ottosson M, Sjostrom L, Carlsson B, Carlsson LM: Human adipose tissue expresses angiotensinogen and enzymes required for its conversion to angiotensin II. J Clin Endocrinol Metab 1998;83:3925–3929.
156 Klahr S, Morrissey J: Angiotensin II and gene expression in the kidney. Am J Kidney Dis 1998;31:171–176.
157 Go AS, Chertow GM, Fan D, McCulloh CE, Hsu C: Chronic kidney disease and the risk of death, cardiovascular events, and hospitalization. N Engl J Med 2004;351:1296–1305.

Prof. Dr. hab. med. Andrzej Więcek, FRCP (Edin)
Department of Nephrology, Endocrinology and Metabolic Diseases
Medical University of Silesian, Katowice
ul. Francuska 20/24
PL–40–027 Katowice (Poland)
Tel. +48 322552695, Fax +48 322553726, E-Mail awiecek@spskm.katowice.pl

Renal Handling of Adipokines

Hitomi Kataoka[a], Kumar Sharma[b]

[a]Dorrance Hamilton Research Laboratories, Division of Nephrology, Thomas Jefferson University, [b]Center for Diabetic Kidney Disease and Cell and Molecular Biology of Kidney Disease, Philadelphia, Pa., USA

Abstract

Chronic kidney disease (CKD) is now considered as one of the strongest risk factors for all cause mortality and cardiovascular events. However, the link between CKD and systemic events is unclear. The role of the kidney is primarily considered a target organ during the development of obesity as altered production of adipokines from visceral adipocytes, however, it should also be recognized that the kidney itself could alter the clearance and production of adipokines. In this chapter, we provide a discussion of renal handling of a variety of adipokines. Specifically, there is a growing body of data supporting a major role for the kidney in clearance of insulin, leptin, and TGF-β. In addition, plasminogen activator inhibitor-1, vascular endothelial growth factor, angiotensin II, and resistin may also be altered by the kidney. The mechanistic regulation of renal handling by the kidney of a variety of circulating adipokines, however is poorly defined. We conclude that the kidney has pivotal roles in the regulation of adipokines and that altered renal handling of adipokines may contribute to the imbalance of factors that ultimately lead to progressive cardiovascular and systemic disease.

Copyright © 2006 S. Karger AG, Basel

Introduction

Chronic kidney disease (CKD) is now considered as one of the strongest risk factors for all cause mortality and cardiovascular (CV) events. However, the link between CKD and systemic events is unclear. The role of the kidney is primarily considered a target organ during the development of obesity as altered production of adipokines from visceral adipocytes may cause functional and structural damage. It should also be recognized that the kidney itself could alter the clearance and production of adipokines. Thus, it is quite likely that altered renal handling of adipokines may actually contribute to the imbalance of factors that ultimately lead to progressive CV and systemic disease. In this chapter, we will focus on the important role of the kidney in the regulation of adipokines.

Insulin

Although insulin is not an adipokine, insulin obviously has a role in the development, maturation of adipocytes and insulin resistance is sine qua non of type 2 diabetes. Insulin is perhaps the most widely known example of renal handling of a circulating hormone. Experimental and human studies have revealed that the kidney is an important organ of glucose homeostasis and is also responsive to insulin [1–3].

Insulin is produced in islet cells of the pancreas in response to elevation in blood glucose and a signaling pathway characterized by calcium signaling. Upon release of insulin by the pancreas, there is an immediate effect in hepatic tissue based on the venous drainage system. However, insulin levels are dramatically elevated in the systemic circulation as well after a glucose load. The kidney is a major site of removal of insulin from the circulation. In addition, insulin has direct effects on renal cells to promote glucose uptake. Direct measurements to document insulin clearance were performed in the 70s by Rabkin et al. [4]. Renal clearance of insulin appears to be contributed by filtration as well as basolateral uptake, primarily at the proximal tubular level [5]. Both of these pathways involve receptor and non-recept mediated uptake. Insulin delivered to proximal tubular cells is degraded to oligopeptides and amino-acids by enzymatic pathways [6]. Fawcett et al. demonstrated the mechanism of insulin degradation in cultured proximal-tubular like cells. After insulin is internalized, a portion of the insulin undergoes partial cleavage in endosomes and these products are then delivered to lysosomes where they are degraded to completion [7]. The importance of renal clearance of insulin is evident in that the degree of hypoglycemia following oral insulin releasing agents is markedly increased with decline in renal function. Furthermore, many patients with type 1 and type 2 diabetes require much less insulin, once they have end-stage renal disease (ESRD). In fact, patients who are insulin dependent before ESRD may not require any insulin administration with dialysis. It is interesting to note that diabetes risk is increased after renal transplantation, however, the role of insulin clearance in this context has not been adequately investigated.

Apart from altered insulin clearance, insulin resistance has been described in CKD. Insulin resistance in renal patients is accompanied by hyperinsulinemia and glucose intolerance as well as by complex derangements of insulin secretion [8, 9]. Results of several clinical studies have indicated that insulin resistance is present already in patients even with mild degrees of renal dysfunction [10, 11].

Thus, kidney disease itself may promote insulin resistance. Of note, although insulin resistance is observed with renal insufficiency, the kidney itself appears to maintain insulin sensitivity in states of systemic insulin resistance [12].

Leptin

Function of Leptin

Leptin, a 16-kDA (167 aa) protein encoded by the *ob* gene, is expressed almost exclusively by adipose tissue. It is produced in adipocytes and its primary site of action is on the brain to limit the satiety center [13]. The leptin receptor belongs to the class I cytokine receptor family. Five isoforms have been identified. Ob-Ra, the short leptin receptor, is present in almost all tissues examined and it is unknown if it retains functional signaling activity. Ob-Rb, the long-leptin receptor, has a 302-amino-acid intracellular domain which contains interaction sites for Jak kinase and STAT proteins, that are characteristic second messengers for cytokine receptor signaling. It is interesting to note that the kidney is one of the few extra-neural tissues that express the long-form of the leptin receptor. Although the functional role of leptin binding sites and leptin receptors in the kidney are unknown, the action of leptin may be important in inflammatory disorders of the kidney [14]. Circulating leptin levels are proportional to the size of adipose tissue depots, and weight loss results in reduced circulating leptin concentration in both rodents and humans [15]. Lack of leptin or impaired receptor function leads to uncontrolled food intake and consequent obesity and diabetes. Thermogenesis and the sympathetic system also seem to be regulated by leptin. Leptin circulates both free of and bound to soluble receptors. Interestingly, both the structure of leptin and that of its receptors suggest that leptin should be classified as a cytokine.

Leptin Clearance in the Kidney

Recent studies have found an important role for the kidney to be the major site of removal of leptin from the circulation. Several groups demonstrated renal clearance of leptin in 1996–1997 [16, 17]. Our study demonstrated clearance of leptin across the renal vascular bed in humans. In patients with intact renal function there was net renal uptake of 12% of circulating leptin, whereas in patients with renal insufficiency there was no renal uptake of leptin [18]. Impaired renal clearance of leptin was likely a major reason why leptin levels were elevated with renal failure. Urinary leptin levels were below detection, suggesting that leptin was being degraded in the kidney. In a separate study, arterio-venous measurement of leptin across the rat kidney was performed

[17]. In rats, there was a 25% decrease of circulating endogenous leptin across the renal bed [17]. In addition, leptin was not found in the urine and no leptin metabolites were found in renal vein, suggesting that leptin was filtered at the glomerulus and likely taken up by proximal tubular cells. This study was performed in the *fa/fa* rats, which have a mutation in the long-form of leptin receptor, thus excluding this receptor as playing an important role in renal clearance of leptin. In support of these findings that the kidney plays a major role in clearing leptin, a report of 57 type 2 diabetic patients evaluating determinants of plasma leptin levels using multiple regression analysis found a significant positive correlation between plasma circulating leptin levels and the serum creatinine [19]. Also, Meyer et al. reported renal leptin uptake could account for over 80% of all leptin removal from plasma. They indicated that the human kidney played a considerable role in leptin removal from plasma [20]. Thus, in both rats and humans, it seems that the kidney is the major site of leptin clearance.

Leptin and Kidney Disease

We also demonstrated that in a cohort of 36 patients with ESRD on hemodialysis (HD), peripheral leptin levels factored for body mass index was increased by >4-fold as compared to a group of healthy controls [18]. A separate study found that both HD patients and peritoneal dialysis (PD) patients had elevated leptin levels [21]. It has been speculated that hyperleptinemia may contribute to various uremic complications, such as anorexia and malnutrition [22]. In a recent longitudinal study, it was found that increases in leptin levels during PD were associated with both inflammation and a loss of lean body mass [23]. The elevated leptin levels may have deleterious effects as leptin receptors have been demonstrated in the kidney as well as in many peripheral tissues. The exact physiologic function of leptin peripherally is unclear. Interestingly, leptin stimulated matrix production in wild type mesangial cells [24]. Wolf et al. reported that leptin stimulated cellular proliferation, transforming growth factor (TGF)-β1 synthesis, and type IV collagen production in glomerular endothelial cells. Conversely, in mesangial cells, leptin upregulated synthesis of the TGF-β type II receptor, but not TGF-β1, and stimulates glucose transport and type I collagen production through signal transduction pathways involving PI-3-K. Both cell types increased their expression of extracellular matrix expression in response to leptin. Other direct and indirect effects of leptin on the kidney included natriuresis, increased sympathetic nervous activity, and stimulation of reactive oxygen species. These findings collectively implicated that the kidney was not only a site of leptin metabolism, but also a target organ for leptin action in pathophysiological states [25].

Transforming Growth Factor-β

Function of TGF-β

TGF-β represents a group of 25-kDa proteins that are actively involved in the development and differentiation of various tissues [26]. TGF-β was originally isolated from platelets, however, it has also been to be secreted by most cell types and in high concentrations by monocytes/macrophages and mesangial cells [27]. Recently, it has been demonstrated to be produced by adipocytes as well [28]. TGF-β is known to be a multi-functional cytokine. Three isoforms of TGF-β have been identified in mammalian species: TGF-β1, TGF-β2 and TGF-β3. TGF-β1, the most important isoform in humans, is secreted from cells in the form of a high-molecular-weight latent complex. Normally, TGF-β1 release ceases by feedback mechanisms when the healing process has been completed [29]. It is a key regulatory molecule in the control of the activity of fibroblasts and has been implicated in several disease states characterized by excessive fibrosis. In the kidney, TGF-β promotes tubuloepithelial cell hypertrophy and regulates the glomerular production of almost every known molecule of the extracellular matrix, including collagens, fibronectin, tenascin, and proteoglycans, as well as the integrins that are the receptors for these molecules. Furthermore, TGF-β blocks the destruction of newly synthesized extracellular matrix by upregulating the synthesis of protease inhibitors and downregulating the synthesis of matrix-degrading proteases, such as stromelysin and collagenase. Strong evidence supports the concept that persistent overproduction of TGF-β1 in glomeruli after the acute inflammatory stage of glomerulonephritis is a major pathway leading to glomerulosclerosis. TGF-β is clearly important in a variety of other chronic renal disorders characterized by hypertrophy and sclerosis, such as diabetic nephropathy [30]. Studies in mouse and rat models using neutralizing antibodies to TGF-β have demonstrated that TGF-β is perhaps the predominant cytokine involved in mesangial matrix accumulation in diabetes [31–34].

Clearance of TGF-β

Our prior studies were the first to demonstrate that the human kidney is a major site of clearance of circulating TGF-β [35]. Interestingly, there is an opposite pattern in diabetics, even with similar levels of plasma creatinine. In this study, aortic, renal vein, and urinary levels of TGF-β was measured in 14 type 2 diabetic patients and 11 nondiabetics. Renal blood flow was measured in all patients to calculate net mass balance across the kidney. Diabetic patients demonstrated net renal production of immunoreactive TGF-β1 (830 ± 429 ng/min [mean $\pm$ SE]), whereas nondiabetic patients demonstrated net renal extraction of circulating TGF-β1 ($-3{,}479 \pm 1{,}010$ ng/min, $p < 0.001$). Urinary levels of bioassayable TGF-β were also significantly increased in diabetic patients compared with nondiabetic patients (2.435 ± 0.385 vs. 0.569 ± 0.190 ng/mg creatinine, respectively; $p < 0.001$). Therefore, type 2 diabetes is associated with enhanced net renal production of TGF-β1 [35]. Urinary concentration of TGF-β is considered to be an efficient marker in other nondiabetic various kidney diseases, such as membranous glomerulonephritis, crescentic nephritis, IgA nephropathy [36–38].

TGF-β and Kidney Disease

As a direct correlation with proteinuria and urinary levels of TGF-β exists, there is likely a significant portion of TGF-β that is filtered and lost in the urine. There is also likely production of TGF-β from renal cells including glomerular mesangial and podocytes as well as proximal tubular cells [39]. The production of renal TGF-β must also be reflected in injury levels. It may be expected that decreased renal function would lead to accumulation of circulating TGF-β. Although this is potentially true the degree of production of TGF-β may also be altered during states of renal failure. Furthermore, the local production of TGF-β may be much more significant than circulating levels in the blood. Nevertheless, circulating levels should be examined and correlated with markers of vascular injury during renal failure. Although the role of circulating TGF-β is unclear, it has been suggested that active levels of TGF-β in the circulation are correlated with coronary artery disease (CAD). Wang et al. demonstrated that an increase in active TGF-β1 levels with both the occurrence and severity of CAD, which is independent of standard CAD risk factors. They suggested that it might reflect a 'double-edged sword' effect of TGF-β1 in which it may reduce atherogenesis by inhibiting smooth muscle cell proliferation but, when there is ongoing vessel wall injury, enhance it by promoting excessive extracellular matrix accumulation [40]. A window of renal production of TGF-β may be examined clinically by measuring urine levels of TGF-β. Our group as well as many others have found that urine TGF-β1 levels are increased in diabetic patients with nephropathy and are sensitive to inhibitors of the renin–angiotensin system (RAS) [41, 42]. It has also been suggested that reno-protection offered by reduction in systolic pressure, reduction of proteinuria, and use of RAS inhibitors may be largely due to inhibition of renal TGF-β production [41, 43].

Adiponectin

Function of Adiponectin

Adiponectin, one of the most abundant adipokines, is a 30-kDa (244 aa) protein produced by the *apM1* gene, highly expressed in human adipose cells [44]. Adiponectin plays an important pathophysiologic role in patients with impaired glucose homeostasis [45, 46]. It is synthesized and secreted by adipose tissue. In contrast to other adipokines, adiponectin demonstrates potent anti-inflammatory and anti-atherosclerotic effects. They involve inhibition of monocytic cell adhesion to endothelial cells [47], suppression of vascular smooth muscle cell proliferation [48], and inhibition of foam cell formation from macrophages [49]. It is suggested to be a component of a novel signaling network among adipocytes, insulin-sensitive tissues, and vasculature [45]. In contrast to leptin, adiponectin levels are reduced in conditions of increased visceral obesity [50].

Adiponectin and Kidney Disease

Plasma concentration of adiponectin was decreased in patients with CAD [46] and type 2 diabetes [51, 52]. Prospective studies in a large male population with normal renal function and in patients with ESRD revealed a significant relationship between adiponectin plasma levels and CV events [53, 54]. In a recent study by Becker et al. demonstrated that patients with CKD who experienced CV events during the follow-up period (median follow-up was 54 months) had markedly lower adiponectin plasma concentration [55]. These results supported the hypothesis that adiponectin was a vasoprotective factor and that in patients with mild to moderate renal dysfunction, hypoadiponectinemia is a CV risk factor. The renal handling of adiponectin has not been studied to date.

Plasminogen Activator Inhibitor-1

Function and Regulation of PAI-1

Plasminogen activator inhibitor-1 (PAI-1) is a member of the family of serine protease inhibitors (serpins) and is the main regulator of the endogenous fibrinolytic system [56]. In vitro studies demonstrated that a variety of cell types, including endothelial and mesothelial cells, were able to produce PAI-1 [57]. In 1991, the expression of PAI-1 in mouse adipose tissue was reported for the first time [57]. Subsequently, expression and secretion of PAI-1 was also demonstrated in adipose tissue from humans [58–60]. Birgel et al. [61] reported that TGF-β1 increased PAI-1 secretion in a dose- and time-dependent manner in newly differentiated human adipocytes. In addition, Alessi et al. [58] reported that the adipose tissue content of PAI-1 antigen was found to be strongly correlated with that of TGF-β1 antigen in humans. Elevated plasma levels of PAI-1 have been found in obese subjects in many clinical and epidemiological studies, and PAI-1 levels were closely correlated with visceral adiposity [62]. PAI-1 levels were also positively associated with obesity, insulin resistance and CV disease [63]. The mechanisms responsible for elevated PAI-1 plasma levels in obesity/insulin resistance syndrome are not yet completely understood.

PAI-1 in HD Patients

It has been shown clearly that circulating levels of PAI-1 and other endothelial cell glycoproteins were increased in HD patients [64–67]. Although this increase has been considered as a subclinical sign of endothelial cell injury, this association is merely a hypothesis that remains to be proved. Seggara et al. revealed that dialysis patients showed increased levels of PAI-1, CRP, tissue plasminogen activator, thrombomodulin and evidence of increased thrombin-dependent fibrin formation. They reported increased levels of active PAI-1 that were associated to a great extent with other classical vascular risk factors; age, male gender, smoking, hypertension, hyperlipidemia etc. Their data indicated a pathogenic link among activated inflammatory response, endothelial injury, and CAD in HD patients. Assessment of circulating PAI-1 levels could be a promising tool to identify high-risk HD patients for developing atheromatous CV disease [68].

Vascular Endothelial Growth Factor

Function of VEGF

Vascular endothelial growth factor (VEGF), stimulates endothelial cell proliferation and differentiation, increases vascular permeability, mediates endothelium-dependent vasodilatation, and supports vascular survival by preventing endothelial apoptosis [69, 70]. Moreover, VEGF enhances plasminogen activator, PAI-1 and interstitial collagenase expression; important factors for matrix remodeling. VEGF also promotes monocyte chemotaxis and adhesion molecule expression [69, 70]. Hypoxia is the main stimulus for VEGF expression and/or production. Several growth factors and cytokines such as epidermal growth factor, TGF-β, platelet-derived growth factor, insulin-like growth factor I, angiotensin II (Ang II), interleukin-1 (IL-1), and IL-6 also have the potential to up-regulate VEGF. Angiogenesis is physiologically important for wound healing, embryonic development, normal growth and development and reproduction. Regarding adipose tissue, the role of VEGF as a stimulus for new vessel growth in brown fat has been described [71]. White adipose tissue from rats was also examined for insulin-responsive VEGF secretion and mRNA expression [72]. Physiological insulin concentrations stimulated VEGF formation and expression in cultured rodent white adipocytes. In addition, increased serum VEGF concentrations were suggested to associate with visceral fat accumulation and to influence vascular endothelial function [73].

VEGF and Kidney Disease

Plasma VEGF levels have been described to be higher in type 2 diabetics than in controls, and VEGF tended to rise with increasing urinary albumin excretion [74]. Urinary VEGF excretion also increased with the progression of diabetic nephropathy and correlated weakly with the levels of serum creatinine, creatinine clearance, microalbuminuria, and proteinuria [75]. In biopsies with early stage of diabetic nephropathy, VEGF was upregulated in glomerular podocytes and distal tubular cells. In biopsies with advanced stage, VEGF staining was decreased or negative in sclerotic glomeruli but remained intense in tubules [75]. Plasma VEGF levels were higher in uremic patients than in control subjects [74, 76]. The mechanisms for higher VEGF levels in uremia are not clarified, but excess production, tissue hypoxia, or reduced clearance of VEGF have been suggested [74]. The negative correlation between plasma VEGF levels and residual renal function observed in PD patients, and they suggested that renal clearance might contribute to the elimination of VEGF [77]. In addition, a recent report revealed plasma VEGF was elevated in HD patients and correlated with oxidative stress marker [78].

Angiotensin II

Function of Angiotensin II
Ang II is an important regulator of blood pressure and of salt and water balance. Angiotensinogen, the precursor of angiotensin, is synthesized primarily by the liver, and released into the circulation. Then it is cleaved by renin, produced in the kidney, to Ang I. Ang I is subsequently converted to Ang II by angiotensin-converting enzyme. In addition to the classical pathway of Ang II synthesis, alternative tissue RAS have been identified in a number of organs. It suggests that various tissues have the ability to synthesize Ang II, independent of the circulating RAS [79, 80]. Under normal conditions, the glomerular level of Ang II was elevated about 1,000-fold above the plasma concentration of [81]. An intriguing concept has emerged that Ang II not only mediates intraglomerular hypertension but also behaves as a growth factor that contributes to diabetic hypertrophy and sclerosis [82]. Adipose tissue has also been described to possess a local RAS. Human adipose tissue expresses all components required for the local production of Ang II [83], which has numerous functions in adipose tissue, ranging from regulation of local blood flow to influences on tissue homeostasis. Overexpression of angiotensinogen in adipocytes leads to hypertension [84] suggesting that adipose-derived RAS has direct effect on the kidney.

Ang II and Adipokine Interaction
The hemodynamic mechanisms of Ang II-induced renal injury have been well documented. However, the nonhemodynamic actions of Ang II are equally

important and becoming more widely recognized. Ang II has been suggested to be relevant to other adipokines and mediate their expression. We postulated that diabetes-induced TGF-β production contributes to impaired Ang II response of vascular smooth muscle cells in macrovessels and microvessels [33]. Cassis et al. [85] revealed inhibition of Ang II synthesis and release from adipocytes decreased leptin release from rat epididymal adipocytes. In addition, exogenously applied Ang II stimulated leptin release from isolated adipocytes and increased leptin mRNA expression. However, when Ang II was elevated systemically, negative regulation of leptin through sympathetic stimulation overwhelmed the direct stimulatory effect of Ang II [85]. A separate study revealed Ang II stimulated the podocyte to produce α 3 type IV collagen protein via mechanisms involving TGF-β and VEGF signaling. Alterations in α 3 type IV collagen production could possibly contribute to glomerular basement membrance thickening and proteinuria in diabetes [86]. The regulation of circulating Ang II by the kidney is complex and still not yet fully elucidated.

Resistin

Function of Resistin
Resistin is a recently identified as a novel adipose-specific cysteine-rich protein secreted by adipocytes and inhibits insulin action in vitro. It is a 12.5-kDa polypeptide, whose functions are not clearly established, although it has been suggested as a link between obesity and insulin resistance [87]. Resistin has been found to be expressed in preadipocytes in addition to adipocytes, which may contribute to the elevation of resistin content in adipose tissue of obese humans [88, 89]. Resistin mRNA has also been found in human monocytes [90, 91]. Differentiation of monocytes into macrophages in vitro increases resistin mRNA [92]. However, the role of resistin in the pathophysiology of obesity and insulin resistance in humans is controversial. The study in the population of Pima Indians revealed that higher serum resistin levels predicted future increases in percent body fat, but did not affect in insulin resistance [93].

Resistin and Kidney Disease
In a study in male patients with IgA glomerulonephritis with various stages of renal disease, glomerular filtration rate (GFR) was the only independent predictor of plasma resistin concentrations [94]. The authors proposed that insufficient filtration by the kidney may account for the observed increase in resistin concentration with declining renal function. Diez et al. suggested that serum resistin levels in patients with ESRD undergoing PD or HD were significantly higher, compared with controls. Although logistic regression analysis did not reveal a

relationship between serum resistin levels and the presence of vascular disease of any type, subgroup analysis demonstrated a significant relationship between resistin levels and previous heart disease [95]. A recent report from Axelsson demonstrated that circulating resistin levels were strongly associated with both GFR and inflammatory biomarkers. As the significant relationship between plasma resistin levels and insulin resistance was lost following the correction for GFR, resistin may not be a major mediator of insulin resistance in patients with renal disease [96]. In another report, resistin was not associated with coronary heart disease but was elevated in survivors of myocardial infarction. Moreover, an association of raised resistin with diabetic nephropathy was suggested [97]. In summary, renal dysfunction seems to affect the clearance of serum resistin level, and resistin levels may impact CVD. However, these mechanisms are still unclear and requires further clarification.

Interaction between CKD and CV Complications Related to Adipokines

Insulin resistance is one of the characteristic features of CKD patients, as stated above. In addition, various adipokines are handled by the kidney in their clearance and excretion. CKD patients are known to be at high risk of CV complications. Although many local vascular pro-athelogenic factors have been implicated in CKD patients, altered plasma concentration of adipokines due to renal handling may be an important contributor. Wang et al. [40] identified a strong positive association between increased active TGF-β1 levels and the occurrence and severity of CAD as assessed by the number of significantly diseased vessels. Becker et al. [55] reported a relationship among insulin resistance, adiponectin, and CV morbidity in patients with mild and moderate kidney disease. They suggested that in patients with CKD, insulin resistance was present even in the earliest stage of renal dysfunction, and several components of this syndrome were associated with CV events. Moreover, hypoadiponectinemia was a novel putative CV risk factor in patients with mild and moderate renal failure. In addition, PAI-1 and resistin also have relationship between CV complication in renal patients stated above [68, 95]. Further studies are needed to characterize the regulatory mechanisms of adipokine production, clearance, regulation in CKD, and the relationship of adipokines to the CV system.

References

1 Wirthensohn G, Guder W: Renal substrate metabolism. Physiol Rev 1986;66:469–497.
2 Stumvoll M, et al: Uptake and release of glucose by the human kidney. J Clin Invest 1995;96:2528–2533.

3 Cersosimo E, Garlick P, Ferretti J: Insulin regulation of renal glucose metabolism in humans. Am J Physiol 1999;276:E78–E84.
4 Rabkin R, et al: Effect of renal disease on renal uptake and excretion of insulin in man. N Engl J Med 1970;282:182–187.
5 Rabkin R, Kitabchi AE: Factors influencing the handling of insulin by the isolated rat kidney. J Clin Invest 1978;62:169–175.
6 Rabkin R, Ryan MP, Duckworth WC: The renal metabolism of insulin. Diabetologia 1984;27:351–357.
7 Fawcett J, Rabkin R: Sequential processing of insulin by cultured kidney cells. Endocrinology 1995;136:39–45.
8 Alvestrand A: Carbohydrate and insulin metabolism in renal failure. Kidney Int Suppl 1997;62:S48–S52.
9 Feneberg R, Sparber M, Veldhuis JD, Mehls O, Ritz E, Schaefer F: Altered temporal organization of plasma insulin oscillations in chronic renal failure. J Clin Endocrinol Metab 2002;87:1965–1973.
10 Eidemak I, et al: Insulin resistance and hyperinsulinemia in mild to moderate progressive chronic renal failure and its association with aerobic work capacity. Diabetologia 1995;38:565–572.
11 Fliser D, et al: Insulin resistance and hyperinsulinaemia are present already in patients with incipient renal disease. Kidney Int 1998;53:1243–1247.
12 Feliers D, et al: Activation of renal signaling pathways in db/db mice with type 2 diabetes. Kidney Int 2001;60:495–504.
13 Cohen SL, et al: Human leptin characterization. Nature 1996;382:589.
14 Sharma K, Considine RV: The Ob protein (leptin) and the kidney. Kidney Int 1998;53:1483–1487.
15 Howlett R: Fat regulation. Prime time for neuropeptide Y. Nature 1996;382:113.
16 Cumin F, Baum HP, Levens N: Leptin is cleared from the circulation primarily by the kidney. Int J Obes Relat Metab Disord 1996;20:1120–1126.
17 Cumin F, et al: Removal of endogenous leptin from the circulation by the kidney. Int J Obes Relat Metab Disord 1997;21:495–504.
18 Sharma K, et al: Plasma leptin is partly cleared by the kidney and is elevated in hemodialysis patients. Kidney Int 1997;51:1980–1985.
19 Shoji T, et al: Renal function and insulin resistance as determinants of plasma leptin levels in patients with NIDDM. Diabetologia 1997;40:676–679.
20 Meyer C, et al: Role of the kidney in human leptin metabolism. Am J Physiol 1997;273 (pt 1):E903–E907.
21 Howard JK, et al: Plasma immunoreactive leptin concentration in end-stage renal disease. Clin Sci (Lond) 1997;93:119–126.
22 Young GA, et al: Increased plasma leptin/fat ratio in patients with chronic renal failure: a cause of malnutrition? Nephrol Dial Transplant 1997;12:2318–2323.
23 Stenvinkel P, et al: Increases in serum leptin levels during peritoneal dialysis are associated with inflammation and a decrease in lean body mass. J Am Soc Nephrol 2000;11:1303–1309.
24 Lee MP, Orlov D, Sweeney G: Leptin induces rat glomerular mesangial cell hypertrophy, but does not regulate hyperplasia or apoptosis. Int J Obes (Lond) 2005;29:1395–1401.
25 Wolf G, et al: Leptin and renal disease. Am J Kidney Dis 2002;39:1–11.
26 Roberts AB, Sporn MB: Physiological actions and clinical applications of transforming growth factor-beta (TGF-β). Growth Factors 1993;8:1–9.
27 Border WA, Noble NA: Transforming growth factor beta in tissue fibrosis. N Engl J Med 1994;331:1286–1292.
28 Rahimi N, et al: Autocrine secretion of TGF-β1 and TGF-β2 by pre-adipocytes and adipocytes: a potent negative regulator of adipocyte differentiation and proliferation of mammary carcinoma cells. In Vitro Cell Dev Biol Anim 1998;34:412–420.
29 Border WA, Ruoslahti E: Transforming growth factor-beta in disease: the dark side of tissue repair. J Clin Invest 1992;90:1–7.
30 Sharma K, Ziyadeh FN: The emerging role of transforming growth factor-beta in kidney diseases. Am J Physiol 1994;266(pt 2):F829–F842.

31 Sharma K, et al: Neutralization of TGF-β by anti-TGF-β antibody attenuates kidney hypertrophy and the enhanced extracellular matrix gene expression in STZ-induced diabetic mice. Diabetes 1996;45:522–530.
32 Ziyadeh FN, et al: Long-term prevention of renal insufficiency, excess matrix gene expression, and glomerular mesangial matrix expansion by treatment with monoclonal antitransforming growth factor-beta antibody in db/db diabetic mice. Proc Natl Acad Sci USA 2000;97: 8015–8020.
33 Sharma K, et al: Involvement of transforming growth factor-beta in regulation of calcium transients in diabetic vascular smooth muscle cells. Am J Physiol Renal Physiol 2003;285: F1258–F1270.
34 Kim YS, et al: Novel interactions between TGF-β1 actions and the 12/15-lipoxygenase pathway in mesangial cells. J Am Soc Nephrol 2005;16:352–362.
35 Sharma K, et al: Increased renal production of transforming growth factor-beta1 in patients with type II diabetes. Diabetes 1997;46:854–859.
36 Honkanen E, et al: Urinary transforming growth factor-beta 1 in membranous glomerulonephritis. Nephrol Dial Transplant 1997;12:2562–2568.
37 Goumenos DS, et al: Urinary transforming growth factor-beta 1 as a marker of response to immunosuppressive treatment, in patients with crescentic nephritis. BMC Nephrol 2005;6:16.
38 Haramaki R, et al: Steroid therapy and urinary transforming growth factor-beta1 in IgA nephropathy. Am J Kidney Dis 2001;38:1191–1198.
39 Abbate M, et al.: Transforming growth factor-$β_1$ is up-regulated by podocytes in response to excess intraglomerular passage of proteins: a central pathway in progressive glomerulosclerosis. Am J Pathol 2002;161:2179–2193.
40 Wang XL, Liu SX, Wilcken DE: Circulating transforming growth factor beta 1 and coronary artery disease. Cardiovasc Res 1997;34:404–410.
41 Agarwal R, et al: Add-on angiotensin II receptor blockade lowers urinary transforming growth factor-beta levels. Am J Kidney Dis 2002;39:486–492.
42 Houlihan CA, et al: Urinary transforming growth factor-beta excretion in patients with hypertension, type 2 diabetes, and elevated albumin excretion rate: effects of angiotensin receptor blockade and sodium restriction. Diabetes Care 2002;25:1072–1077.
43 Sharma K, et al: Captopril-induced reduction of serum levels of transforming growth factor-beta1 correlates with long-term renoprotection in insulin-dependent diabetic patients. Am J Kidney Dis 1999;34:818–823.
44 Maeda K, et al: cDNA cloning and expression of a novel adipose specific collagen-like factor, apM1 (adipose most abundant gene transcript 1). Biochem Biophys Res Commun 1996;221: 286–289.
45 Goldstein B, Scalia R: Adiponectin: a novel adipokine linking adipocytes and vascular function. J Clin Endocrinol Metab 2004;89:2563–2568.
46 Hotta K, et al. Plasma concentrations of a novel, adipose-specific protein, adiponectin, in type 2 diabetic patients. Arterioscler Thromb Vasc Biol 2000;20:1595–1599.
47 Ouchi N, et al: Novel modulator for endothelial adhesion molecules: adipocyte-derived plasma protein adiponectin. Circulation 1999;100:2473–2476.
48 Matsuzawa Y, Funahashi T, Nakamura T: Molecular mechanism of metabolic syndrome X: contribution of adipocytokines adipocyte-derived bioactive substances. Ann NY Acad Sci 1999;892: 146–154.
49 Ouchi N, et al: Adipocyte-derived plasma protein, adiponectin, suppresses lipid accumulation and class A scavenger receptor expression in human monocyte-derived macrophages. Circulation 2001;103:1057–1063.
50 Kwon K, et al: Reciprocal association between visceral obesity and adiponectin: in healthy premenopausal women. Int J Cardiol 2005;101:385–390.
51 Hotta K, et al: Circulating concentrations of the adipocyte protein adiponectin are decreased in parallel with reduced insulin sensitivity during the progression to type 2 diabetes in rhesus monkeys. Diabetes 2001;50:1126–1133.
52 Weyer C, et al: Hypoadiponectinemia in obesity and type 2 diabetes: close association with insulin resistance and hyperinsulinemia. J Clin Endocrinol Metab 2001;86:1930–1935.

53   Pischon T, et al: Plasma adiponectin levels and risk of myocardial infarction in men. JAMA 2004;291:1730–1737.
54   Zoccali C, et al: Adiponectin, metabolic risk factors, and cardiovascular events among patients with end-stage renal disease. J Am Soc Nephrol 2002;13:134–141.
55   Becker B, et al: Renal insulin resistance syndrome, adiponectin and cardiovascular events in patients with kidney disease: the mild and moderate kidney disease study. J Am Soc Nephrol 2005;16:1091–1098.
56   Wiman B, Chmielewska J, Ranby M: Inactivation of tissue plasminogen activator in plasma. Demonstration of a complex with a new rapid inhibitor. J Biol Chem 1984;259:3644–3647.
57   Sawdey MS, Loskutoff DJ: Regulation of murine type 1 plasminogen activator inhibitor gene expression in vivo. Tissue specificity and induction by lipopolysaccharide, tumor necrosis factor-alpha, and transforming growth factor-beta. J Clin Invest 1991;88:1346–1353.
58   Alessi MC, et al: Production of plasminogen activator inhibitor 1 by human adipose tissue: possible link between visceral fat accumulation and vascular disease. Diabetes 1997;46:860–867.
59   Eriksson P, et al: Adipose tissue secretion of plasminogen activator inhibitor-1 in non-obese and obese individuals. Diabetologia 1998;41:65–71.
60   Cigolini M, et al: Expression of plasminogen activator inhibitor-1 in human adipose tissue: a role for TNF-alpha? Atherosclerosis 1999;143:81–90.
61   Birgel M, et al: Role of cytokines in the regulation of plasminogen activator inhibitor-1 expression and secretion in newly differentiated subcutaneous human adipocytes. Arterioscler Thromb Vasc Biol 2000;20:1682–1687.
62   Janand-Delenne B, et al: Visceral fat as a main determinant of plasminogen activator inhibitor 1 level in women. Int J Obes Relat Metab Disord 1998;22:312–317.
63   Juhan-Vague I, Alessi MC: PAI-1, obesity, insulin resistance and risk of cardiovascular events. Thromb Haemost 1997;78:656–660.
64   Gris JC, et al: Increased cardiovascular risk factors and features of endothelial activation and dysfunction in dialyzed uremic patients. Kidney Int 1994;46:807–813.
65   Haaber AB, et al: Vascular endothelial cell function and cardiovascular risk factors in patients with chronic renal failure. J Am Soc Nephrol 1995;5:1581–1584.
66   Ishii Y, et al: Evaluation of blood coagulation-fibrinolysis system in patients receiving chronic hemodialysis. Nephron 1996;73:407–412.
67   Tomura S, et al: Fibrinogen, coagulation factor VII, tissue plasminogen activator, plasminogen activator inhibitor-1, and lipid as cardiovascular risk factors in chronic hemodialysis and continuous ambulatory peritoneal dialysis patients. Am J Kidney Dis 1996;27:848–854.
68   Segarra A, et al: Circulating levels of plasminogen activator inhibitor type-1, tissue plasminogen activator, and thrombomodulin in hemodialysis patients: biochemical correlations and role as independent predictors of coronary artery stenosis. J Am Soc Nephrol 2001;12:1255–1263.
69   Ferrara N, Gerber HP: The role of vascular endothelial growth factor in angiogenesis. Acta Haematol 2001;106:148–156.
70   Neufeld G, et al: Vascular endothelial growth factor (VEGF) and its receptors. FASEB J 1999;13:9–22.
71   Fredriksson JM, et al: Norepinephrine induces vascular endothelial growth factor gene expression in brown adipocytes through a beta -adrenoreceptor/cAMP/protein kinase A pathway involving Src but independently of Erk1/2. J Biol Chem 2000;275:13802–13811.
72   Mick GJ, Wang X, McCormick K: White adipocyte vascular endothelial growth factor: regulation by insulin. Endocrinology 2002;143:948–953.
73   Miyazawa-Hoshimoto S, et al: Elevated serum vascular endothelial growth factor is associated with visceral fat accumulation in human obese subjects. Diabetologia 2003;46:1483–1488.
74   Wasada T, et al: Plasma concentration of immunoreactive vascular endothelial growth factor and its relation to smoking. Metabolism 1998;47:27–30.
75   Cha DR, et al: Role of vascular endothelial growth factor in diabetic nephropathy. Kidney Int Suppl 2000;77:S104–S112.
76   Harper S, Downs L, Tomson C: Elevated plasma vascular endothelial growth factor levels in non-diabetic predialysis uraemin. Nephron 2002;90:341–343.

77   Stompor T, et al: Selected growth factors in peritoneal dialysis: their relationship to markers of inflammation, dialysis adequacy, residual renal function, and peritoneal membrane transport. Perit Dial Int 2002;22:670–676.
78   Pawlak K, et al: Possible association between circulating vascular endothelial growth factor and oxidative stress markers in hemodialysis patients. Med Sci Monit 2006;12:CR181–CR185.
79   Phillips M, Speakman E, Kimura B: Levels of angiotensin and molecular biology of the tissue renin angiotensin systems. Regul Pept 1993;43:1–20.
80   Paul M, Wangner J, Dzau V: Gene expression of the renin-angiotensin system in human tissues. Quantitative analysis by the polymerase chain reaction. J Clin Invest 1993;91:2058–2064.
81   Seikaly MG, Arant BS Jr, Seney FD Jr: Endogenous angiotensin concentrations in specific intrarenal fluid compartments of the rat. J Clin Invest 1990;86:1352–1357.
82   Wolf G, Ziyadeh FN: The role of angiotensin II in diabetic nephropathy: emphasis on nonhemodynamic mechanisms. Am J Kidney Dis 1997;29:153–163.
83   Karlsson S, et al: Human adipose tissue expresses angiotensinogen and enzymes required for its conversion to angiotensin II. J Clin Endocrinol Metab 1998;83:3925–3929.
84   Massiera F, et al: Adipose angiotensinogen is involved in adipose tissue growth and blood pressure regulation. FASEB J 2001;15:2727–2729.
85   Cassis LA, et al: Differential effects of local versus systemic angiotensin II in the regulation of leptin release from adipocytes. Endocrinology 2004;145:169–174.
86   Chen S, et al: Angiotensin II stimulates alpha3(IV) collagen production in mouse podocytes via TGF-β and VEGF signalling: implications for diabetic glomerulopathy. Nephrol Dial Transplant 2005;20:1320–1328.
87   Steppan CM, et al: The hormone resistin links obesity to diabetes. Nature 2001;409:307–312.
88   McTernan PG, et al: Increased resistin gene and protein expression in human abdominal adipose tissue. J Clin Endocrinol Metab 2002;87:2407.
89   Janke J, et al: Resistin gene expression in human adipocytes is not related to insulin resistance. Obes Res 2002;10:1–5.
90   Roche HM, et al: Isomer-dependent metabolic effects of conjugated linoleic acid: insights from molecular markers sterol regulatory element-binding protein-1c and LXRalpha. Diabetes 2002;51:2037–2044.
91   Nagaev I, Smith U: Insulin resistance and type 2 diabetes are not related to resistin expression in human fat cells or skeletal muscle. Biochem Biophys Res Commun 2001;285:561–564.
92   Patel L, et al: Resistin is expressed in human macrophages and directly regulated by PPAR gamma activators. Biochem Biophys Res Commun 2003;300:472–476.
93   Vozarova de Courten B, et al: High serum resistin is associated with an increase in adiposity but not a worsening of insulin resistance in Pima Indians. Diabetes 2004;53:1279–1284.
94   Kielstein JT, et al: Increased resistin blood levels are not associated with insulin resistance in patients with renal disease. Am J Kidney Dis 2003;42:62–66.
95   Diez JJ, et al: Serum concentrations of leptin, adiponectin and resistin, and their relationship with cardiovascular disease in patients with end-stage renal disease. Clin Endocrinol (Oxf) 2005;62:242–249.
96   Axelsson J, et al: Elevated resistin levels in chronic kidney disease are associated with decreased glomerular filtration rate and inflammation, but not with insulin resistance. Kidney Int 2006;69:596–604.
97   Burnett MS, et al: Cross-sectional associations of resistin, coronary heart disease, and insulin resistance. J Clin Endocrinol Metab 2006;91:64–68.

Kumar Sharma, MD
Suite 353, Jefferson Alumni Hall, Thomas Jefferson University
1020 Locust Street
Philadelphia, PA 19107 (USA)
Tel. +1 215 503 6950, Fax +1 215 923 7212, E-Mail Kumar.Sharma@jefferson.edu

Lipid Metabolism and Renal Disease

Christine K. Abrass

Department of Medicine, University of Washington School of Medicine, Seattle, Wash., USA

Abstract

Metabolic syndrome is associated with dyslipidemia, which is thought to contribute in part to the development of chronic kidney disease (CKD). This review discusses the factors that regulate intracellular handling of lipids and their relationship to disordered mesangial cell function. Specific attention is paid to those factors such as fatty acid translocase/scavenger receptor BII, proliferator-activated receptor δ, insulin-like growth factor-1, inflammation and hypertriglyceridemia that are altered in the metabolic syndrome. CKD also causes an increase in triglycerides and a decrease in high-density lipoprotein that mimic the lipid abnormalities of metabolic syndrome, which accelerate the progression of CKD and increase the risk for cardiovascular mortality. There is a special emphasis on foam cells in the kidney and lipid-mediated changes in intrinsic kidney cells that lead to glomerulosclerosis and interstitial fibrosis. Correlates to whole animal and humans studies are included.

Copyright © 2006 S. Karger AG, Basel

Introduction

Epidemiologic, as well as experimental evidence, establish that abnormalities in lipid metabolism contribute to progressive chronic kidney disease (CKD). Moreover, CKD patients develop a secondary form of dyslipidemia that heightens the risk for development of cardiovascular disease (CVD) [1, 2]. This secondary dyslipidemia mimics metabolic syndrome as it is characterized by an increase in serum triglyceride (TG) (elevated VLDL-remnants/IDL), and a reduction in high-density lipoprotein (HDL) cholesterol [2, 3]. The growing epidemic of metabolic syndrome contributes metabolic (abnormal glucose metabolism, hyperinsulinemia, and hypertension), dyslipidemic (elevated TG and low HDL), and cytokine abnormalities that conspire to increase the risk for progression of CKD and mortality from CVD (fig. 1). Reversal of metabolic syndrome is essential to slow the progression of CKD.

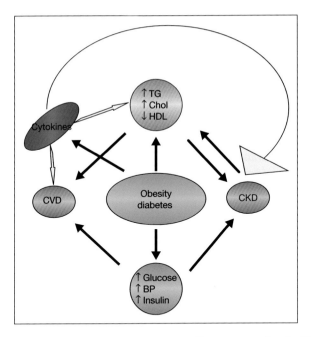

Fig. 1. Obesity and diabetes mellitus are associated with abnormalities in lipids and glucose, insulin and BP which conspire to induce CKD and CVD. Obesity and diabetes lead to increased levels of inflammatory cytokines that lead to disordered lipid metabolism in kidney and blood vessel cells. Furthermore, CKD per se leads to abnormalities in lipid metabolism that are characterized by elevated triglycerides and reduced HDL, which secondarily accelerates this process. Defining methods to interrupt this cycle of disordered lipid metabolism is crucial to reduce the rate of progression of CKD and reduce the mortality from CVD. BP = Blood pressure; Chol = cholesterol; CKD = chronic kidney disease; CVD = cardiovascular disease; HDL = high-density lipoprotein; TG = triglycerides.

In vivo Evidence for the Role of Lipids in Chronic Kidney Disease

A link between lipid abnormalities and CVD has been established for a long time and strategies that improve lipid abnormalities reduce cardiovascular risk and mortality [4, 5]. Although initial studies suggested that elevated cholesterol was the primary determinant of vascular disease, the important role of HDL in reverse cholesterol transport and the participation of elevated TG in the pathogenesis of foam cell formation has broadened our understanding of the role that lipids play [6, 7]. Many of the principles derived from extensive studies in CVD are relevant to CKD; yet, this has only been examined recently. Studies of a variety of populations of individuals with CKD show that lipids contribute to the rate of progression, and interventions that lower lipids slow progression of CKD [3].

Patients with lecithin-cholesterol acyltransferase deficiency have glomerular lipid accumulation and eventually develop renal failure from glomerulosclerosis [8]. Samuelsson et al. [9] demonstrated a strong correlation between TG-rich, Apo-B-containing lipoproteins and the rate of progression in patients with CKD. Foam cells are prominent in biopsies from patients with focal and segmental glomerulosclerosis [10]. Lipidpharesis reduces proteinuria [11] and improves the response to corticosteroids [12], which argues for a pathogenic relationship in focal and segmental glomerulosclerosis. The presence of oxidized lipids [13] and foam cells [14] in some renal biopsies further supports a role for lipids in mediating tissue injury in the kidney. As lipid abnormalities (e.g. elevated TG and low HDL) are a prominent feature of the metabolic syndrome, there is growing evidence for a link between metabolic syndrome and CKD [15–21].

Animal studies have begun to address the mechanisms of lipid-mediated kidney injury. Hypercholesterolemia leads to macrophage infiltration and foam cell formation in rats [22] and is associated with accelerated loss of kidney function in a variety of models including polycystic kidney disease [23], obesity [24], diabetes [25], and hypertension [24]. Dominquez et al. [24] established the role of lipids in crosses of obesity-prone and spontaneously hypertensive rats. Glomerulosclerosis and interstitial fibrosis were dependent on lipoxidation; as without hypercholesterolemia, glycoxidation per se was not nephrotoxic. In a study of food restriction in Zucker rats, changes in protein excretion and progression best correlated with reductions in circulating lipids [25]. Reduction of hypertriglyceridemia in obese Zucker rats is associated with marked reduction in the degree of glomerulosclerosis [26]. By independently manipulating cholesterol and TG levels in ovariectomized analbuminemic female rats, Joles et al. [27] showed that both contribute to podocyte injury, proteinuria and interstitial fibrosis. In aggregate these data indicate that abnormalities in lipids can contribute to renal injury independent of the initiating insult.

Work in atherosclerosis has focused on the adverse effects of oxidized low density lipoprotein (LDL) and its accumulation in foam cells that is primarily dependent upon fatty acid translocase (FAT)/scavenger receptor BII/fatty acid translocase (CD36) activity [28]. Recent studies indicate that intracellular accumulation of TG might also account for some of the abnormalities in function of lipid-laden cells. In models of type 1 diabetes, sterol regulatory element binding protein (SREBP)-1 is increased in kidney cortex, resulting in up-regulation of enzymes responsible for fatty acid synthesis, and as a consequence, high renal TG content. When SREBP-1 activity was inhibited, there was less mesangial matrix expansion and less glomerulosclerosis, which implies that lipids contribute to the development of diabetic nephropathy. In animals that have normal glucose and serum lipid levels, over-expression of SREBP-1 leads to elevated renal TG content, increased expression of transforming growth factor-β, mesangial matrix expansion with collagen IV and fibronectin, proteinuria, and glomerulosclerosis [29]. These findings indicate that increased intracellular accumulation of TG may play a central role in mediating the development of diabetic nephropathy. In nephrotic syndrome, additional evidence implicates TGs in renal injury [30]. Elevated very low density lipoprotein (VLDL) is associated with a reduction in lipoprotein lipase and the VLDL particle shows defective binding to cells, which leads to decreased clearance. The abnormal VLDL particle also mediates inflammatory responses that lead to mesangial sclerosis [31]. In nephrotic syndrome, VLDL-rich particles predominate because of reduced catabolism [30]. Increased filtration of these lipid-laden particles has been postulated to mediate interstitial fibrosis. As TGs play a key role in signal transduction cascades, the metabolism and pathogenic significance of intra-cellular TGs deserves further study.

Systemic versus Cellular Lipid Metabolism

Lipid metabolism is a complex process whereby the liver regulates circulating levels through uptake, degradation, and excretion in bile. In turn, plasma lipids provide the substrate for energy in skeletal muscle and for membrane synthesis and generation of signal transduction molecules in other tissues. Details of systemic lipid metabolism have been reviewed extensively elsewhere [1, 32–35]. Systemic effects that enhance lipid uptake in the liver and lead to catabolism or excretion in the bile protect other organs from lipid excess by reducing plasma levels. These same mechanisms may act to enhance lipid uptake in other tissues, but the lowered plasma concentration abrogates the effect on those tissues. For example metabolic syndrome is associated with low plasma levels of insulin-like growth factor-1 (IGF-1) [36], which is associated with hypertriglyceridemia and increased risk for the development of CVD [37]. IGF-1 directly modulates lipid uptake by cells [38–41]; thus, one would expect lipid uptake in the kidney to be reduced in metabolic syndrome; however, this benefit would be abrogated by elevated lipid levels in the circulation. Furthermore, in some cases systemic and local effects might be opposite. Inflammation, as a characteristic of metabolic syndrome, increases cytokines that influence local synthesis of growth factors including IGF-1 [42]. Elevated IGF-1 becomes particularly apparent in the kidney once hyperglycemia develops [40, 43] and in the tunica media of atherosclerotic vessels [44]. Thus, local increases in IGF-1 may lead to enhanced lipid uptake in kidney cells, which would be aggravated by elevated circulating levels [42]. Such dichotomies between systemic effects, local changes, and cellular consequences make the role of IGF-1 in metabolic syndrome, diabetes, vascular disease and kidney disease controversial [39, 44–46].

Cholesterol Synthesis and Uptake

Cellular lipids play important roles in the synthesis of plasma membranes, membranes of organelles, and in signal transduction cascades that direct cell function [47]. Yet, abnormal accumulation of free cholesterol leads to cellular toxicity; thus, regulation of esterification of cholesterol is a key requirement for controlling the toxicity of increased intracellular cholesterol. In order to meet critical cellular needs, uptake, synthesis, metabolism and disposal of lipids by cells are tightly regulated. Lipid synthesis is stimulated by a fall in intracellular cholesterol [48] through activation of SREBPs, which are leucine zipper transcription factors that reside in the plasma membrane. SREBPs are cleaved, translocate to the nucleus and regulate cholesterol (primarily SREBP-2) and fatty acid synthesis (primarily SREBP-1) through induction of hydroxymethylglutaryl (HMG) CoA reductase and fatty acid synthase, respectively [29]. Increased lipid content suppresses SREBP transcription and decreases activity of HMG CoA reductase [49], which restores lipid levels to normal and prevents toxicity from progressive accumulation of cholesterol.

Intracellular lipid content is also influenced by the rate of uptake of lipids from the circulation. Cholesteryl ester-rich LDL is taken up by the LDL receptor whose expression is regulated by SREBP, and both are normally down-regulated when intracellular lipid levels rise [48]. This negative feedback regulation serves to maintain intracellular cholesterol content relatively constant. Modified LDL, including oxidized LDL, is taken up by scavenger receptors including SR-A1 and FAT/CD36 [5, 50]. Acetylated and oxidized LDL are endocytosed following binding to SR-A1. The cholesteryl ester is hydrolyzed in lysosomes to produce free cholesterol, which can be toxic to cells when levels become elevated. Unlike LDL receptors, scavenger A receptor expression is not regulated by intracellular cholesterol content [51]. FAT/CD36 has a high affinity for LDL that has been minimally oxidized by myeloperoxidase, which is enhanced in inflammatory states and diabetes [5]. FAT/CD36 expression is upregulated by oxidized LDL and cellular cholesterol, which can lead to progressive lipid accumulation [28]. It is increased by ligands for peroxisome proliferator-activated receptor (PPAR)γ, including oxidized LDL, prostaglandin E and synthetic agonists (e.g. thioglitazones). Transforming growth factor-β down-regulates FAT/CD36 by phosphorylation and inactivation of PPARγ [52]. Modifications in the function of receptors that mediate cholesterol uptake occur in local inflammatory sites that may lead to lipid accumulation in the absence of elevated circulating levels.

Triglycerides

Initially it was thought that TGs had little atherogenic potential, yet, recent data have shown that they have significant toxicity [53]. Lipoprotein lipase hydrolyzes TGs in chylomicrons and VLDL, which controls plasma levels and reduces cellular uptake. When TG levels are high, VLDL can contain large amounts of apoE, which induces a conformational change that is associated with increased uptake by scavenger receptors [7]. Elevated TG levels can also be associated with formation of small dense LDL particles that are cleared slowly, which renders them more susceptible to oxidation and subsequent uptake by scavenger receptors. Unsaturated fatty acids (oleate, linolate and palmitate) are transported into the cell by fatty acid transfer proteins and converted to acyl-CoA derivatives [54]. Acyl-CoA activates phospholipase D2 leading to diacylglycerol, which activates a specific kinase that serine phosphorylates ATP-binding cassette (ABC)A1 and targets it for degradation. Loss of ABCA1 reduces the ability to enhance cholesterol efflux. In this way, TG accumulation can lead to increased cholesterol content.

VLDL receptors bind apoE-VLDL and VLDL remnants, thereby influencing uptake and metabolism in fatty-acid active tissues, but they do not bind LDL. VLDL receptors are abundant in heart, muscle, adipose tissue, brain, macrophages, cardiomyocytes, and endothelium of capillaries and small arteries, but they are not present in liver [55]. This implies that VLDL receptors contribute to fatty acid uptake for utilization by individual cells, but they do not significantly influence plasma levels. The VLDL receptor is not regulated by intracellular lipid levels, but thyroxine and estrogen increase its expression. The VLDL receptor has been implicated in the development of TG-rich foam cells, as its expression is increased in endothelium and vascular smooth muscle cell (VSMC) in atherosclerotic sites, and interferon-γ inhibits its expression and prevents the development of macrophage foam cells. VLDL receptor null mice are lean and do not develop insulin-resistance in the face of a high fat diet, which further suggests an important role for this receptor in the metabolic syndrome. Few studies have elucidated its role in specific tissues.

Reverse Cholesterol Transport and Lipid Removal from Cells

Toxicity of intracellular cholesterol is controlled by esterification and then removal from the cell by reverse cholesterol transport; thus, cholesterol efflux is an important contributor to net cholesterol accumulation. HDL is the primary molecule that accepts excess cellular lipids that are excreted from the cell by various transporters and then delivers them to the liver for clearance through bile. As a lipid-poor lipoprotein, HDL serves as an efficient acceptor for cholesterol and phospholipids that are transported passively across cells by SR-B1 or actively by the ABC transporters. Cholesterol and TG efflux are regulated by members of the PPAR including α, γ, and δ, through their effects on ABC activity [56, 57].

ABC transporters are a family of 49 proteins that use ATP to generate the energy needed to transport metabolites across the cell [34]. ABCA1 is a liver X receptor (LXR)-dependent protein regulated by expression of PPARγ that increases cholesterol export, but has no effect on TG or phospholipid clearance. ABCG1 is LXR-independent, regulated by PPARδ, and is critically important for efflux of TG and phospholipids, in addition to influencing export of cholesterol. HDL stimulates increased expression of PPARγ mRNA and protein and leads to PPARγ phosphorylation, which reduces intracellular lipid accumulation by increasing ABCA1 [58]. HDL also blocks the normal increase in FAT/CD36 that accompanies ligation and phosphorylation of PPARγ, thereby decreasing uptake of oxidized LDL and fatty acids via FAT/CD36 [58]. HDL reduces oxidation of LDL and inhibits cytokine-induced expression of adhesion molecules on endothelial cells, which further reduce the adverse consequences of lipids.

Role of PPARs

There is growing evidence that the PPAR family of ligand activated nuclear transcription factors are major controllers of lipid metabolism [59, 60]. PPARα is primarily expressed on liver where it influences hepatic fatty acid oxidation during fasting by increasing lipoprotein lipase. On other cells its effects are predominantly anti-inflammatory [61]. PPARγ is expressed on adipocytes where it has an important role in lipid synthesis and storage. This receptor is selectively expressed on other cells where it exhibits a wide variety of effects under different physiological conditions. PPARδ is ubiquitously expressed, and it increases glycolysis/lipogenesis in liver while activating fat burning in muscle. It is the major PPAR expressed in macrophages where it influences metabolism of VLDL and foam cell formation [62]. Unligated PPARδ suppresses fatty acid utilization through active repression, which is reversed upon ligand binding. Activation of both PPARγ and PPARδ improve insulin sensitivity through different mechanisms, which has made them potential therapeutic targets in managing metabolic syndrome.

Under normal circumstances, cholesterol synthesis, uptake, and efflux are balanced to maintain stable intracellular lipid levels. In inflammatory states, lipid deposits in tissues by binding to extracellular matrix proteins and then becomes oxidized. Inflammatory cytokines and growth factors enhance expression of influx pathways, particularly SR-A1 and FAT/CD36, inhibit efflux pathways and may enhance synthesis, leading to significant intracellular lipid accumulation and foam cell formation. As discussed below, intracellular accumulation of lipids can cause abnormal cell function.

Abnormal Cellular Lipid Metabolism in Disease

Foam cells are the hallmark of abnormal lipid metabolism. In order for foam cells to form, homeostatic mechanisms must be dysregulated leading to net lipid accumulation. Inflammatory cytokines mediate this dysregulation by blocking down-regulation of LDL receptors and HMG CoA reductase that normally would occur in response to increases in intracellular cholesterol [49]. Scavenger receptor uptake of oxidized lipids is a particularly potent stimulus for release of inflammatory cytokines from macrophages. In turn, monocyte chemoattractant protein-1 (MCP-1) increases expression of the scavenger receptor, FAT/CD36, which further enhances uptake of OxLDL [4, 63–65]. In this way, accumulation of oxidized lipids within tissues can accelerate injury. The critical role of FAT/CD36 was demonstrated by mutation studies in which absence of functional FAT/CD36 prevented LDL accumulation and formation of foam cells [66]. Furthermore, atherosclerosis-prone animals were protected from atherosclerosis when FAT/CD36 was absent [67]. Further support for the role of inflammatory cytokines in mediating disordered cellular lipid metabolism comes from studies of interleukin (IL)-10. As an anti-inflammatory cytokine, IL-10 redirects cholesterol handling by macrophages towards reverse cholesterol transport, which may in part contribute to its anti-atherosclerotic action. IL-10 suppresses PPARγ-mediated expression of FAT/CD36 [68], and it increases expression of ABCA1 and ABCG1. All of these effects act to limit lipid accumulation. Although oxidized LDL clearly has effects on cellular function, the lipid content of foam cells was recently found to be predominantly TG, indicating abnormal fatty acid metabolism [69, 70]. These findings enhance the significance of elevated TGs and abnormal PPAR function in metabolic syndrome. Along with the VLDL receptor, FAT/CD36 is critical for cellular uptake of long-chain fatty acids. FAT/CD36 is expressed on endothelial and VSMC; thus, disordered function of this scavenger receptor may influence foam cell formation in intrinsic vascular cells. FAT/CD36 is defective in SHR rats, where it contributes to glucose intolerance and defective adipocyte transport of long-chain fatty acids [71, 72].

Lipids and the Kidney

Tubular Cells

The significance of foam cells in the kidney and their relationship to progressive disease is a subject of intense interest [73]. Zager et al. [74–76] elucidated the details of alterations in intracellular lipids that occur in proximal tubular cells following ischemic injury. Both free cholesterol and cholesteryl ester levels rise, which in part is due to increased HMG CoA reductase driven cholesterol synthesis and increased cholesterol traffic to rafts and caveolar structures [76]. When cholesterol accumulation is prevented by treatment with HMG CoA reductase inhibitors, the cytoprotection from recurrent ischemia that normally develops in cells that survive is lost. These data argue that in the short-term lipid accumulation in response to injury has a protective effect. This phenomenon was not restricted to ischemic injury as Johnson et al. [77] showed that lipid accumulation occurs in the glomerulus and proximal tubules in two animal models of glomerulonephritis, (nephrotoxic serum nephritis and passive Heymann nephritis). In these models, lipid accumulation was associated with a decreased SR-B1, increased ABCA1 and increased HMG CoA reductase indicating that the normal adaptive response in SR-B1-mediated influx and ABCA1-mediated efflux occurred with cellular cholesterol loading; however, increased synthesis contributed to lipid accumulation. These data show that the intracellular lipid accumulation that follows injury can be crucial to cell survival; yet, when it persists in the face of chronic inflammatory stimuli, it becomes maladaptive and contributes to on-going injury.

Glomerular Cells

In addition to studies of animal models of glomerular injury [77], abnormal responses in specific kidney cells have been reported. Hypercholesterolemia and hypertriglyceridemia are associated with severe podocyte injury, which secondarily leads to mesangial sclerosis [27]. Although these findings are striking, little is known about the specific effects of lipids on podocytes. The majority of work in the kidney has shown that mesangial cells (MC) metabolize lipids and respond to oxidized LDL in ways that are similar to VSMC. MC express LDL receptors, and scavenger receptors, SR-A, SR-B1, and FAT/CD36 [49]. Ruan et al. [49] demonstrated dysregulation of the MC LDL receptor by a variety of inflammatory cytokines. Tumor necrosis factor-α and IL-1β increased SREBP-1, which in turn increased expression of the LDL receptor. This increase was no longer suppressed in the normal fashion by intracellular accumulation of cholesterol, which has important implications for disease progression. Inflammatory cytokines, including IL-1β, induce intracellular accumulation of lipid by up-regulating expression of SR-A1 that take up oxidized LDL. Oxidized LDL induces

MC to increase synthesis of collagens I, III and IV [78]. LDL simulates MC to increase synthesis of aldosterone [79], which has been shown to increase renal fibrosis. Angiotensin II leads to up-regulation of MC scavenger receptors SR-A1, which indicates a link between hypertension and lipid-mediated injury [80]. This may in part explain why angiotensin-converting enzyme inhibitors slow the rate of fibrosis in kidney disease. LDL and minimally modified LDL are potent stimulants of MC proliferation [81]. Thus, abnormalities in lipid metabolism occur in MC that are similar to macrophages and VSMC.

Fatty acid activation of FAT/CD36 induces expression of IL-6, which is a potent stimulus for mesangial sclerosis [82]. Similar to oxidized LDL, VLDL stimulates expression of MCP-1 in MC [31]. In turn, MCP-1 is chemotactic for macrophages, which can infiltrate the glomerulus and become foam cells [22] that release cytokines and further modulate MC proliferation and matrix accumulation. In nephrotic syndrome, TG-rich lipids may aggravate glomerular and tubular injury, particularly when uptake is increased and efflux pathways are reduced by inflammatory states. Cholesterol efflux by the PPAR–LXRα–ABCA1 pathway is impaired in IL-1β-treated MC [57]. This can be reversed by PPARγ agonists, which protect MC from IL-1β-induced intracellular lipid accumulation by activating the ABCA1 efflux pathway. Other transporters may also be important in the MC, as ABCA1 expression is quite low [38].

Based on data described above, IGF-1 may play a particularly important role in disordered lipid metabolism in metabolic syndrome. IGF-1 induces MC to become elongated and grow in a swirling pattern, as compared to the parallel, strap-like morphology associated with insulin [83–85]. After 7 days of exposure to IGF-1, but not insulin, MC become lipid-laden foam cells [86]. IGF-1 increases both cholesterol and TG uptake through enhanced endocytosis [83, 86]. Although the lipid moiety that accumulates can be manipulated by adjusting the composition and concentration of lipids in the medium, IGF-1-treated MC predominantly accumulate TGs [38, 87]. Triglyceride accumulation was associated with a marked decrease in expression of FAT/CD36 and PPARδ, two systems that influence TG uptake and disposal (fig. 2). Absence of PPARδ in cardiac muscle is associated with lipid toxicity and decreased contractility [59, 61, 62]. PPARδ agonists increase FAT/CD36, which enhances uptake for oxidation; thus, a decrease in PPARδ in IGF-1-treated MC may account of the fall in FAT/CD36 [61]. Additional studies are needed to identify factors that control PPARδ and FAT/CD36 in MC, and to understand the consequences to fatty acid metabolism and cell function in disease.

The majority of work on the significance of foam cells in atherosclerosis has focused on the macrophage and its release of cytokines that alter the behavior of cells intrinsic to the vessel wall. Like MC, VSMC can become foam cells; yet, few studies have determined the functional consequence of this change.

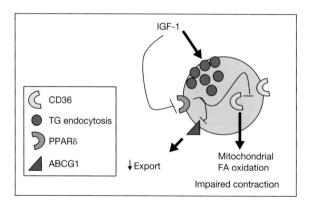

Fig. 2. IGF-1 increases TG uptake by enhanced endocytosis. IGF-1 also reduces expression of PPARδ, which in turn is associated with a reduction in expression of the transport protein, ABCG1, which participates in TG removal from the cell. Normally, PPARδ increases the concentration of FAT/CD36, which allows for enhanced fatty acid uptake to support muscle contraction. FAT/CD36 is also present on mitochondrial membranes, where it transports fatty acids into mitochondria for oxidation and energy production. Both of these effects of FAT/CD36 are reduced in cells that lack activation of PPARδ. These mechanisms may account for the reduction in angiotensin-II-mediated contraction that is observed in mesangial foam cells that develop with IGF-1 treatment. ABCG1 = ATP-binding Cassette G1; CD36 = scavenger receptor BII/fatty acid translocase; FAT = fatty acid translocase; IGF-1 = insulin-like growth factor-1; PPAR = peroxisome proliferator-activated receptor; TG = triglyceride.

Studies of lipid-laden MC demonstrate a variety of abnormalities. Shortly after exposure to IGF-1, the MC cytoskeleton reorganizes leading to a phenotype characteristic of a migrating cell [85]; however, later, after lipid has accumulated, the cytoskeleton becomes abnormally arrayed, and no longer reorganizes in response to stimuli such as PDGF or IGF binding protein-5 [84]. To establish the role of lipid accumulation in these changes, MC were loaded with either cholesterol or TGs. In cholesterol-loaded MC, the membrane was insufficiently mobile to completely encircle particles for phagocytosis [83], or to allow MC migration [88, 89]. These defects were reversed when cholesterol was removed from the cell [84]. In addition to the abnormal migratory response, IGF-1-induced MC foam cells do not contract in response to angiotensin II [84]. This defect was mimicked by TG accumulation, and abrogated by its removal, an effect similar to cardiac lipotoxicity from impaired function of PPARδ [62].

MC contraction may also have been affected by reductions in PPARδ and FAT/CD36 [38] that were associated with impaired oxidation of long-chain fatty acids and limit energy generation [41]. PPARδ activates fatty acid metabolism by increasing oxidative enzymes. FAT/CD36 on mitochondrial membranes is essential for uptake and oxidation of fatty acids [90]. The abnormal contractile response observed in lipid-laden MC may contribute to increased intracapillary pressures that influence glomerular injury in diabetic nephropathy [91]. Studies showing that IGF-1 drives foam cell formation are relevant to disease, as rats transgenic for growth hormone or IGF-1 develop glomerulosclerosis [92], IGF-1 is increased in diabetic nephropathy [93], and can contribute to VSMC proliferation in atherosclerotic plaques [44]. Additional studies are needed to understand the intra-renal changes in these systems in metabolic syndrome.

Conclusion

Under normal conditions, the intracellular content of cholesterol, fatty acids and phospholipids is tightly regulated, as these lipids are essential to the structural integrity of the cell and its signal transduction cascades. Following acute cellular injury, cholesterol and TGs accumulate, which may protect cells from additional injury and prevent apoptosis by allowing repair of injured cell membranes. When inflammatory stimuli persist, lipid accumulation worsens and foam cells develop, which release cytokines that perpetuate lipid accumulation. This process is accelerated when circulating lipid levels are elevated. In response to foam cell formation, MC fail to contract properly, and secrete extracellular matrix that contributes to glomerulosclerosis. The relevance of in vitro observations to disease is supported by animal and human studies, which demonstrate the accumulation of foam cells in diseased kidneys [10]. Also, studies show that elevated levels of lipids accelerate disease progression, which is improved by a variety of manipulations that lower circulating lipids or prevent intra-cellular lipid accumulation [94]. Considerable additional information is needed to understand abnormalities in intra-renal lipid metabolism that accompany metabolic syndrome and the role that is plays in CKD progression.

Acknowledgements

These studies were supported by the Department of Veterans Affairs Medical Research Service and the National Institutes of Health (CKA R01DK49971–08).

References

1 Vaziri ND: Dyslipidemia of chronic renal failure: the nature, mechanisms, and potential consequences. Am J Physiol Renal Physiol 2006;290:F262–F272.
2 Wanner C, Krane V: Uremia-specific alterations in lipid metabolism. Blood Purif 2002;20:451–453.
3 Cases A, Coll E: Dyslipidemia and the progression of renal disease in chronic renal failure patients. Kidney Int 2005;99:S87–S93.

4   Brewer HB Jr: The lipid-laden foam cell: an elusive target for therapeutic intervention. J Clin Invest 2000;105:703–705.
5   de Winther MPJ, Hofker MH: Scavenging new insights into atherogenesis. J Clin Invest 2000;105: 1039–1041.
6   Maxfield FR, Tabas I: Role of cholesterol and lipid organization in disease. Nature 2005;438: 612–621.
7   Rothblat GH, de la Llera-Moya M, Atger V, Kellner-Weibel G, Williams DL, Phillips MC: Cell cholesterol efflux: integration of old and new observations provides new insights. J Lipid Res 1999;40:781–796.
8   Lambert G, Sakai N, Vaisman BL, Neufeld EB, Marteyn B, Chan CC, Paigen B, Lupia E, Thomas A, Striker LJ, Blanchette-Mackie J, Csako G, Brady JN, Costello R, Striker GE, Remaley AT, Brewer HB Jr, Santamarina-Fojo S: Analysis of glomerulosclerosis in lecithin cholesterol acyltransferase-deficient mice. J Biol Chem 2001;276:15090–15098.
9   Samuelsson O, Mulec H, Knight-Gibson C, Attman P-O, Kron B, Larsson R, Weiss L, Wedel H, Alaupovic P: Lipoprotein abnormalities are associated with increased rate of progression of human chronic renal insufficiency. Nephrol Dial Transplant 1997;12:1908–1915.
10 Noel LH: Morphological features of primary focal and segmental glomerulosclerosis. Nephrol Dial Transplant 1999;14:53–57.
11 Yorioka N, Taniguchi Y, Nishida Y, Okushin S, Amimoto D, Yamakido A: Low-density lipoprotein apheresis for focal glomerular sclerosis. Ther Apher 1997;1:370–371.
12 Yokoyama K, Sakai S, Sigematsu T, Takemoto F, Hara S, Yamada A, Kawaguchi Y, Hosoya T: LDL adsorption improves the response of focal glomerulosclerosis to corticosteroid therapy. Clin Nephrol 1998;50:1–7.
13 Lee HS, Kim YS: Identification of oxidized low density lipoproteins in human renal biopsies. Kidney Int 1998;54:848–856.
14 Magil AB: Interstitial foam cells and oxidized lipoprotein in human glomerular disease. Mod Pathol 1999;12:33–40.
15 Adelman RD: Obesity and renal disease. Curr Opin Nephrol Hypertens 2002;11:331–335.
16 Bagby SP: Obesity-initiated metabolic syndrome and the kidney: a recipe for chronic kidney disease? J Am Soc Nephrol 2004;15:2775–2791.
17 Hayden MR, Whaley-Connell A, Sowers JR: Renal redox stress and remodeling in metabolic syndrome, type 2 diabetes mellitus, and diabetic nephropathy: paying homage to the podocyte. Am J Nephrol 2005;25:553–569.
18 Keane WF: The role of lipids in renal disease: Future challenges. Kidney Int 2003;57:S27–S31.
19 Kurella M, Lo JC, Chertow GM: Metabolic syndrome and the risk for chronic kidney disease among nondiabetic adults. J Am Soc Nephrol 2005;16:2134–2140.
20 Sedor JR, Schelling JR: Association of metabolic syndrome in nondiabetic patients with increased risk for chronic kidney disease: The fat lady sings. J Am Soc Nephrol 2005;16:1880–1882.
21 Tanaka H, Shiohira Y, Uezu Y, Higa A, Iseki K: Metabolic syndrome and chronic kidney disease in Okinawa, Japan. Kidney Int 2006;69:369–374.
22 Hattori M, Nikolic-Paterson DJ, Miyazaki K, Isbel NM, Lan HY, Atkin C, Kawaguchi H, Ito K: Mechanisms of glomerular macrophage infiltration in lipid-induced renal injury. Kidney Int 1999;71:S47–S50.
23 Lu J, Bankovic-Calic N, Ogborn M, Saboorian MH, Aukema HM: Detrimental effects of a high fat diet in early renal injury are ameliorated by fish oil in Han:SPRD-cy rats. J Nutr 2003;133:180–186.
24 Dominquez JH, Tang N, Xu W, Evan AP, Siakotos AN, Agarwal R, Walsh J, Deeg M, Pratt JH, March KL, Monnier VM, Weiss MF, Baynes JW, Peterson R: Studies of renal injury III: Lipid-induced nephropathy in type II diabetes. Kidney Int 2000;57:92–104.
25 Maddox DA, Alavi FK, Santella RN, Zawata ET Jr: Prevention of obesity-linked renal disease: age-dependent effects of dietary food restriction. Kidney Int 2002;62:208–219.
26 Kasiske BL, O'Donnell MD, Cleary MP, Keane WF: Treatment of hyperlipidemia reduces glomerular injury in obese Zucker rats. Kidney Int 1988;33:667–672.
27 Joles JA, Kunter U, Janssen U, Kriz W, Rabelink TJ, Koomans HA, Floege J: Early mechanisms of renal injury in hypercholesterolemic or hypertriglyceridemic rats. J Am Soc Nephrol 2000;11:669–683.
28 Han J, Hajjar DP, Tauras JM, Nicholson AC: Cellular cholesterol regulates expression of the macrophage type B scavenger receptor, CD36. J Lipid Res 1999;40:830–838.

29 Sun L, Halaihel N, Zhang W, Rogers T, Levi M: Role of sterol regulatory element-binding protein 1 in regulation of renal lipid metabolism and glomerulosclerosis in diabetes mellitus. J Biol Chem 2002;277:18919–18927.

30 O'Donnell MP: Mechanisms and clinical importance of hypertriglyceridemia in the nephrotic syndrome. Kidney Int 2001;59:380–382.

31 Lynn EG, Siow YL, O K: Very low-density lipoprotein stimulates the expression of monocyte chemoattractant protein-1 in mesangial cells. Kidney Int 2000;57:1472–1483.

32 Chait A, Brazg RL, Tribble DL, Krauss RM: Susceptibility of small, dense, low-density lipoproteins to oxidative modification in subjects with the atherogenic lipoprotein phenotype, pattern B. Am J Med 1993;94:350–356.

33 Kwiterovich PO Jr: Clinical relevance of the biochemical, metabolic, and genetic factors that influence low-density lipoprotein heterogeneity. Am J Cardiol 2002;90:30i–47i.

34 Oram JF, Heinecke JM: ATP-binding cassette transporter A1: A cell cholesterol exporter that protects against cardiovascular disease. Physiol Rev 2005;85:1343–1372.

35 Ribalta J, Vallve JC, Girona J, Masana L: Apolipoprotein and apolipoprotein receptor genes, blood lipids and disease. Curr Opin Clin Nutr Metab Care 2003;6:177–187.

36 Sesti G, Sciacqua A, Cardellini M, Marini MA, Maio R, Vatrano M, Succurro E, Lauro R, Federici M, Perticone F: Plasma concentration of IGF-I is independently associated with insulin sensitivity in subjects with different degrees of glucose tolerance. Diabetes Care 2005;28:120–125.

37 Juul A, Scheike T, Davidsen M, Gyllenborg J, Jorgensen T: Low serum insulin-like growth factor I is associated with increased risk of ischemic heart disease: A population-based case-control study. Circulation 2002;106:939–944.

38 Berfield AK, Chait A, Oram JF, Zager RA, Johnson AC, Abrass CK: IGF-1 induces rat glomerular mesangial cells to accumulate triglyceride. Am J Physiol Renal Physiol 2005;290:F138–F147.

39 Conti E, Carrozza C, Capoluongo E, Volpe M, Crea F, Zuppi C, Andreotti F: Insulin-like growth factor-1 as a vascular protective factor. Circulation 2004;110:2260–2265.

40 Karl M, Potier M, Schulman IH, Rivera A, Werner H, Fornoni A, Elliot SJ: Autocrine activation of the local insulin-like growth factor I system is up-regulated by estrogen receptor (ER)-independent estrogen actions and accounts for decreased ER expression in type 2 diabetic mesangial cells. Endocrinol 2005;146:889–900.

41 Koonen DPY, Glatz JFC, Bonen A, Luiken JJFP: Long-chain fatty acid uptake and FAT/CD36 translocation in heart and skeletal muscle. Biochim Biophys Acta 2005;1736:163–180.

42 Elliot SJ, Striker L, Hattori M, Yang CW, He CJ, Peten EP, Striker GE: Mesangial cells from diabetic NOD mice constitutively secrete increased amounts of insulin-like growth factor-I. Endocrinol 1993;133:1783–1788.

43 Green TL, Hunter DD, Chan W, Merlie JP, Sanes JR: Synthesis and assembly of the synaptic cleft protein s-laminin by cultured cells. J Biol Chem 1992;267:2014–2022.

44 Frystyk J, Ledet T, Møller N, Flyvberg A, Ørskov H: Cardiovascular disease and insulin-like growth factor I. Circulation 2002;106:893–895.

45 El Nahas AM, Sayed-Ahmed N: Insulin-like growth factor I and the kidney: friend or foe? Exp Nephrol 1993;4:205–217.

46 Knott RM: Insulin-like growth factor type 1 – friend or foe? Br J Ophthalmol 1998;82:719–720.

47 Tabas I: Consequences of cellular cholesterol accumulation: basic concepts and physiological implications. J Clin Invest 2002;110:905–911.

48 Hussain MM, Strickland DK, Bakillah A: The mammalian low-density lipoprotein receptor family. Annu Rev Nutr 1999;19:141–172.

49 Ruan XZ, Varghese Z, Powis SH, Moorhead JF: Dysregulation of LDL receptor under the influence of inflammatory cytokines: A new pathway for foam cell formation. Kidney Int 2001;60:1716–1725.

50 Podrez EA, Febbraio M, Sheibani N, Schmitt D, Silverstein RL, Hajjar DP, Cohen PA, Frazier WA, Hoff HF, Hazen SL: Macrophage scavenger receptor CD36 is the major receptor for LDL modified by monocyte-generated reactive nitrogen species. J Clin Invest 2000;105:1095–1108.

51 Li HY, Kotaka M, Kostin S, Lee SMY, Kok LDS, Chan KK, Tsui SKW, Schaper J, Zimmerman R, Lee CY, Fung KP, Waye MMY: Translocation of a human focal adhesion LIM-only protein, FHL2, during myofibrillogenesis and identification of LIM2 as the principal determinants of FHL2 focal adhesion localization. Cell Motil Cytoskeleton 2001;48:11–23.

52   Draude G, Lorenz RL: TGF-beta1 downregulates CD36 and scavenger receptor A but upregulates LOX-1 in human macrophages. Am J Physiol 2000;278:H1042–H1048.
53   Sniderman AD, Scantlebury T, Cianflone K: Hypertriglyceridemia hyperapoB: The unappreciated atherogenic dyslipoproteinemia in type 2 diabetes mellitus. Ann Intern Med 2001;135:447–459.
54   Wang Y, Oram JF: Unsaturated fatty acids phosphorylate and destabilize ABCA1 through a phospholipase D2 pathway. J Biol Chem 2005;280:35896–35903.
55   Takahashi S, Sakai J, Fujino T, Hattori H, Zenimaru Y, Suzuki J, Miyamori I, Yamamoto TT: The very low-density lipoprotein (VLDL) receptor: characterization and functions as a peripheral lipoprotein receptor. J Atheroscler Thromb 2004;11:200–208.
56   Chawla A, Repa JJ, Evans RM, Mangelsdorf DJ: Nuclear receptors and lipid physiology: Opening the X-files. Science 2001;294:1866–1870.
57   Ruan XZ, Moorhead JF, Fernando R, Wheeler DC, Powis SH, Varghese Z: PPAR antagonists protect mesangial cells from interleukin 1β-induced intracellular lipid accumulation by activating the ABCA1 cholesterol efflux pathway. J Am Soc Nephrol 2003;14:593–600.
58   Han J, Hajjar DP, Zhou X, Gotto AM, Jr., Nicholson AC: Regulation of peroxisome proliferator-activated receptor-γ-mediated gene expression. A new mechanism of action for high density lipoprotein. J Biol Chem 2002;277:23582–23586.
59   Barish GD, Narkar VA, Evans RM: PPARδ: a dagger in the heart of the metabolic syndrome. J Clin Invest 2006;116:590–597.
60   Blaschke F, Takata Y, Caglayan E, Law RE, Hsueh WA: Obesity, peroxisome proliferator-activated receptor, and atherosclerosis in type 2 diabetes. Arterioscler Thromb Vasc Biol 2006;26:28–40.
61   Lee C-H, Olson P, Hevener A, Mehl I, Chong LW, Olefsky JM, Gonzalez FJ, Ham J, Kang H, Peters JM, Evans RM: PPARδ regulates glucose metabolism and insulin sensitivity. Proc Natl Acad Sci USA 2006;103:3444–3449.
62   Lee C-H, Kang K, Mehl IR, Nofsinger F, Alaynick WA, Chong L-W, Rosenfeld JM, Evans RM: Peroxisome proliferator-activated receptor δ promotes very low-density lipoprotein-derived fatty acid catabolism in the macrophage. Proc Natl Acad Sci USA 2006;103:2434–2439.
63   Chisolm GM: Cytotoxicity of oxidized lipoproteins. Curr Opin Lipidol 1991;2:311–316.
64   de Villiers WJ, Smart EJ: Macrophage scavenger receptors and foam cell formation. J Leukoc Biol 1999;66:740–746.
65   Tabata T, Mine S, Kawahara C, Okada Y, Tanaka Y: Monocyte chemoattractant protein-1 induces scavenger receptor expression and monocyte differentiation into foam cells. Biochem Biophys Res Commun 2003;305:380–385.
66   Febbraio M, Hajjar DP, Silverstein RL: CD36:a calls B scavenger receptor involved in angiogenesis, atherosclerosis, inflammation, and lipid metabolism. J Clin Invest 2001;108:785–791.
67   Febbraio M, Podrez EA, Smith JD, Hajjar DP, Hazen SL, Hoff HF, Sharma K, Silverstein RL: Targeted disruption of the class B scavenger receptor CD36 protects against atherosclerotic lesion development in mice. J Clin Invest 2000;105:1049–1056.
68   Rubic T, Lorenz RL: Downregulated CD36 and oxLDL uptake and stimulated ABCA1/G1 and cholesterol efflux as anti-atherosclerotic mechanisms of interleukin-10. Cardiovasc Res 2006;69:527–535.
69   Li AC, Glass CK: The macrophage foam cell as a target for therapeutic intervention. Nat Med 2002;8:1236–1242.
70   Milosavljevic D, Kontush A, Griglio S, Le Naour G, Thillet J, Chapman MJ: VLDL-induced triglyceride accumulation in human macrophages is mediated by modulation of LPL lipolytic activity in the absence of change in LPL mass. Biochim Biophys Acta 2003;1631:51–60.
71   Glazier AM, Scott J, Aitman TJ: Molecular basis of the Cd36 chromosomal deletion underlying SHR defects in insulin action and fatty acid metabolism. Mamm Genome 2002;13:108–113.
72   Pravenec M, Kurtz TW: Genetics of CD36 and the hypertension metabolic syndrome. Semin Nephrol 2002;22:148–153.
73   Studer RK, Negrete H, Craven PA, DeRubertis FR: Protein kinase C signals thromboxane-induced increases in fibronectin synthesis and TGF-β bioactivity in mesangial cells. Kidney Int 1995;48:422–430.
74   Zager RA, Johnson A: Renal cortical cholesterol accumulation is an integral component of the systemic stress response. Kidney Int 2001;60:2299–2310.

75 Zager RA, Johnson A, Anderson K, Wright S: Cholesterol ester accumulation: An immediate consequence of acute in vivo ischemic renal injury. Kidney Int 2001;59:1750–1761.
76 Zager RA, Johnson A, Hanson S, dela Rosa V: Altered cholesterol localization and caveolin expression during the evolution of acute renal failure. Kidney Int 2002;61:1674–1683.
77 Johnson AC, Yabu JM, Hanson S, Shah VO, Zager RA: Experimental glomerulopathy alters renal cortical cholesterol, SR-B1, ABCA1, and HMG CoA reductase expression. Am J Pathol 2003;162:283–291.
78 Lee HS, Kim BC, Hong HK, Kim YS: LDL stimulates collagen mRNA synthesis in mesangial cells through induction of PKC and TGF-beta expression. Am J Physiol 1999;277:F369–F376.
79 Nishikawa T, Suematsu S, Saito J, Soyama A, Ito H, Kino T, Chrousos G: Human renal mesangial cells produce aldosterone in response to low-density lipoprotein (LDL). J Steroid Biochem Mol Biol 2005;96:309–316.
80 Ruan XZ, Varghese Z, Powis SH, Moorhead JF: Human mesangial cells express inducible macrophage scavenger receptor. Kidney Int 1999;56:440–451.
81 Kamanna VS, Bassa BV, Vaziri ND, Roh DD: Atherogenic lipoproteins and tyrosine kinase mitogenic signaling in mesangial cells. Kidney Int Suppl 1999;71:S70–S75.
82 Massy ZA, Kim Y, Guijarro C, Kasiske B, Keane WF, O'Donnell MP: Low-density lipoprotein-induced expression of interleukin-6, a marker of human mesangial cell inflammation: effects of oxidation and modulation by lovastatin. Biochem Biophys Res Commun 2000;267:536–540.
83 Berfield AK, Abrass CK: IGF-1 induces foam cell formation in rat glomerular mesangial cells. J Histochem Cytochem 2002;50:395–403.
84 Berfield AK, Andress DL, Abrass CK: IGF-1-induced lipid accumulation impairs mesangial cell migration and contractile function. Kidney Int 2002;62:1229–1237.
85 Berfield AK, Spicer D, Abrass CK: Insulin-like growth factor I (IGF-I) induces unique effects in the cytoskeleton of cultured rat glomerular mesangial cells. J Histochem Cytochem 1997;45:583–593.
86 Fowler SD, Greenspan P: Applications of nile red, a fluorescent hydrophobic probe, for the detection of neutral lipid deposits in tissue sections. J Histochem Cytochem 1985;33:833–836.
87 Kruth HS, Huang W, Ishii I, Zhang W-Y: Macrophage foam cell formation with native low density lipoprotein. J Biol Chem 2002;277:34573–34580.
88 Abrass CK, Berfield AK, Andress DL: Heparin binding domain of insulin-like growth factor binding protein-5 stimulates mesangial cell migration. Am J Physiol 1997;273:F899–F906.
89 Berfield AK, Andress DL, Abrass CK: IGFBP-5$^{201-218}$ stimulates Cdc42 aggregation and filopodia formation in migrating mesangial cells. Kidney Int 2000;57:1991–2003.
90 Bezaire V, Bruce CR, Heigenhauser GJF, Tandon NN, Glatz JFC, Luiken JJJF, Bonen A, Spriet LL: Identification of fatty acid translocase on human skeletal muscle mitochondrial membranes: essential role in fatty acid oxidation. Am J Physiol Endocrinol Metab 2006;290:E509–E515.
91 Hostetter TH, Rennke HG, Brenner BM: The case for intrarenal hypertension in the initiation and progression of diabetic and other glomerulopathies. Am J Med 1982;72:375–380.
92 Doi T, Striker LJ, Gibson CC, Agodoa LYC, Brinster RL, Striker GE: Glomerular lesions in mice transgenic for growth hormone and insulinlike growth factor-I. I. Relationship between increased glomerular size and mesangial sclerosis. Am J Pathol 1990;137:541–552.
93 Elliot SJ, Striker LJ, Hattori M, Yang CW, He CJ, Peten EP, Striker GE: Mesangial cells from diabetic NOD mice constitutively secrete increased amounts of insulin-like growth factor-I. Endocrinology 1993;133:1783–1788.
94 Oda H, Keane WF: Recent advances in statins and the kidney. Kidney Int 1999;71:S2–S5.

Christine K. Abrass, MD
University of Washington School of Medicine, UW Medicine South Lake Union
815 Mercer Street
Seattle, WA 98109 (USA)
Tel. +1 206 897 1966, Fax +1 206 897 1300, E-Mail cabrass@u.washington.edu

Role of the Renin–Angiotensin–Aldosterone System in the Metabolic Syndrome

Stefan Engeli

Franz Volhard Clinical Research Center, Charité Campus Buch, Berlin, Germany

Abstract

Clinical trials of angiotensin-converting enzyme inhibitors and angiotensin type 1-receptor blockers have demonstrated a significant reduction in cardiovascular, renal and new-onset diabetes risk. This finding raises the question on the pathophysiological role of the renin–angiotensin–aldosterone system (RAAS) in the Metabolic Syndrome. Inappropriate activation of the RAAS in the circulation but also on the tissue level, suggests profound changes in the regulation of this system in the Metabolic Syndrome. Possible physiological and molecular mechanisms that link the RAAS with the Metabolic Syndrome include interactions between insulin and angiotensin type 1-receptors, hemodynamic effects that change nutrient partition to tissues such as skeletal muscle and adipose tissue, inhibition of adipogenesis which may limit the storage capacity of adipose tissue and allows ectopic lipid storage which leads to lipotoxicity, influence on adipokine secretion, and influence on insulin secretion from pancreatic β-cells. Most of these mechanisms are linked with decreased insulin sensitivity. Convincing experimental evidence on these possible mechanisms, however, is rare, and several obscurities remain, which are partly due to species differences between rodent models and humans. Thus, data from clinical trials clearly suggest that the RAAS plays an important role for the pathophysiology of the Metabolic Syndrome, but the molecular mechanisms are not understood very well at present.

Copyright © 2006 S. Karger AG, Basel

Introduction

The renin–angiotensin–aldosterone system (RAAS) is a complex system with many components. In brief, the precursor angiotensinogen (AGT) is cleaved subsequently by renin and the angiotensin-converting enzyme (ACE) yielding angiotensin II (Ang II). The principle receptors for Ang II are the membrane-bound angiotensin type I and type II receptors (AT1 and AT2) that are responsible for most of the biological actions of Ang II, including stimulation of aldosterone secretion from the adrenal gland. New components are still discovered, but their physiological or pathophysiological importance is unclear in most cases (e.g. ACE2 and other enzymes that process angiotensin peptides such as chymase and cathepsins, several other angiotensin peptides, receptors such as AT4 and the apelin and renin receptors). Beside the well-known effects on sodium homeostasis and vascular tone, the RAAS is involved in physiological and pathophysiological processes such as tissue growth and hypertrophy, inflammation, and interferes with glucose, lipid and energy metabolism [1–3].

The complexity of the RAAS is also due to the fact that the circulating systemic and local tissue RAAS interact, and that RAAS regulation is tissue- and sometimes also species-specific [4]. Adipose tissue is of great importance for metabolic regulation and metabolic deteriorations in the Metabolic Syndrome [5], partly because a complete local RAAS is present in adipose tissue. Evidence is accumulating on upregulated RAAS activity in obesity and the Metabolic Syndrome, both on the systemic and local adipose-tissue level [3, 6].

Clinical trials of ACE inhibitors and AT1-receptor blockers have demonstrated significant reductions in cardiovascular and renal risk, as well as a reduction of new-onset diabetes risk [7]. This chapter will describe RAAS activation, which has predominantly studied with respect to obesity rather than to the Metabolic Syndrome, both on the systemic and tissue level. Possible physiological and molecular mechanisms that link the RAAS with the Metabolic Syndrome will be described. These discussions should help to understand the role of the RAAS for the Metabolic Syndrome and the beneficial metabolic effects described in clinical trials on RAAS blockade.

Dysregulation of the RAAS in Obesity and the Metabolic Syndrome

Systemic Renin–Angiotensin–Aldosterone System
Several studies found increased plasma levels of AGT, renin, ACE, and aldosterone in obese, sometimes hypertensive, humans and animals (see [3, 6] for extensive review). Obesity is associated with sodium retention and volume expansion [8], which normally would suppress renin and aldosterone levels. In fact, experimental salt loading in obese healthy males clearly demonstrated that renin and aldosterone are not appropriately suppressed, reflecting profound changes of RAAS regulation in obesity [9]. Increased renal sympathetic nerve activity may increase renin release [10]. Renal sympathetic nerve activity is stimulated by leptin in experimental models, and a selective leptin resistance was

demonstrated in animal models of obesity, with preserved activation of the renal sympathetic nervous system, but a resistance against metabolic effects [11]. In humans, a close relationship between whole body leptin release and renal norepinephrine spill-over was found [12].

Aldosterone release in the obese may be stimulated by EKODE (12,13-epoxy-9-keto-10(trans)-octadecenoic acid). EKODE is a potent stimulator of aldosterone secretion in vitro. EKODE was positively associated with plasma aldosterone levels in a small human study, and also with BMI in a subset of African-Americans [13]. Whether EKODE is the candidate to explain increased aldosterone levels in obese subjects, or merely reflects oxidative stress in obesity, needs to be determined in larger clinical trials. Conditioned media of human adipocytes increased aldosterone secretion in vitro as well, independently of potassium or Ang II [14]. The close proximity of adipocytes to the adrenal cells suggests the influence of adipose-derived secretagogues on aldosterone levels in obesity.

The reasons for increased circulating ACE activity in obesity are unclear, and the available data on Ang II are controversial, with increased levels in diet-induced obese rats [15] but similar levels between lean and obese women [16]. Some organs are particularly important sources for circulating RAAS components (liver, kidney, adrenals, adipose tissue). An interesting question is, whether these tissue-RAAS are differentially regulated in obesity.

Local Tissue Renin-Angiotensin Systems

Data (mostly) on mRNA levels in diet-induced obese animals clearly show that neither AGT in the liver, nor renin in the kidney or ACE in the lung are stimulated by high fat feeding and weight gain [15, 17]. Unchanged renin expression, however, was a surprising finding in the light of high circulating Ang II levels, which should suppress renin in the kidney [15]. This finding, again, points to profound changes in the regulation of the RAAS in obesity, and may be attributed to the counteracting effects of leptin on renal sympathetic activity [10, 11].

A recent human study demonstrated that obese human subjects present with markedly enhanced renal plasma flow responsiveness when treated with ACE-inhibitors or angiotensin-receptor blockers [18]. Also, renal vascular reactivity to Ang II infusions is blunted in obese human subjects [19]. These data suggest that intrarenal Ang II content is increased in obesity. This hypothesis would also help to understand why renal AT_1-receptor density was decreased in diet-induced obese rats. Nevertheless, increased levels of circulating Ang II may also have contributed to AT_1-receptor downregulation in this model [15]. In clear contrast to these findings, AT1-receptors were upregulated in the kidney of obese Zucker rats for unknown reasons [20].

All components of the RAAS including angiotensin peptides and mineralocorticoid receptors, have been identified in adipose tissue, and the expression of all the RAAS genes was demonstrated in isolated human adipocytes [3]. Local formation of Ang II in adipose tissue or by adipocytes has been demonstrated [21], and angiotensin receptors are also expressed on adipose cells. Although all RAAS components are invariably present in adipose tissue, there are notable differences in the regulation and physiology between humans and rodents. For example, the AT_1-receptor is the only receptor present on human preadipocytes and adipocytes, but both, AT_1- and AT_2-receptors, have been identified in rodent adipose cells [3].

Several studies examined the influence of obesity on the expression of RAAS genes in adipose tissue with differing findings (see [3, 6] for extensive review). In general, RAAS gene expression is higher in visceral adipose tissue compared to subcutaneous adipose tissue. With the exception of AGT, RAAS gene expression in human adipose tissue is dependent on the presence of hypertension: renin, ACE and AT1-receptor mRNA levels are only increased in adipose tissue of obese hypertensive subjects [16, 22–24]. AGT, however, is typically downregulated in adipose tissue of obese normotensive and hypertensive subjects [16, 22, 25]. AGT secretion from isolated human subcutaneous adipocytes was not different between lean and obese donors [26]. The influence of obesity on AGT expression in rodent adipose tissue is dependent on the underlying genotype [27, 28], but diet-induced obesity in 'wild type' animals typically leads to an upregulation of AGT gene expression [15, 29].

These data demonstrate that obesity has a model-dependent influence on RAAS regulation in several tissues. Nevertheless, available data do not support the idea that the liver, lung or kidney do significantly contribute to increased circulating levels of AGT, ACE or renin, respectively. A special case, however, must be made for the adipose-tissue RAAS. Especially in diet-induced obese rats, increased AGT gene expression in adipose tissue was associated with increased circulating AGT and Ang II levels [15]. This finding clearly raises the question, whether local secretion of AGT by adipocytes can contribute to circulating AGT levels.

Relationship between the Local Adipose-Tissue and Circulating RAAS

Targeted AGT expression in adipocytes of AGT-knockout mice lead to detectable circulating AGT levels, and near normal blood pressure. Targeted AGT expression in adipocytes of 'wild type' animals raised circulating AGT and blood pressure above normal levels [30]. No high-fat diet was given to these animals, thus the data represent a proof of principle, but do not reflect the situation in obesity. Targeted expression of 11β-hydroxysteroid dehydrogenase-1 in adipocytes increased local levels of corticosterone, induced a metabolic syndrome and was associated with increased blood pressure, plasma AGT, and adipose-tissue AGT expression [31]. In human isolated adipocytes, however, cortisol had no effect on AGT gene expression [22]. The effects of the dominant negative P467L mutation of PPARγ was studied in heterozygous mice presenting with increased subcutaneous adipose tissue, mild insulin resistance, and hypertension. Adipose expression of the AGT gene was again increased [32]. Thus, several mouse models present with the correlation of increased local expression of AGT, AGT plasma levels and hypertension.

We recently reported that the decrease of plasma AGT during weight loss is closely linked to the decrease in waist circumference, a surrogate for reduced body fat mass, and the decrease in adipose-tissue AGT expression [16]. This finding implies that AGT secretion from adipocytes may contribute to AGT plasma levels in humans as well. A recent study described for the first time that subcutaneous adipose tissue contributes significantly to circulating Ang II blood levels in humans, with 23% higher adipose-venous Ang II concentrations compared to arterial blood [33]. As in previous studies [16], no correlation was found in this study between BMI and systemic or adipose-venous Ang II levels.

Influence of Weight Loss on the RAAS

Reductions of plasma renin, aldosterone and ACE upon weight loss have occasionally been reported. Furthermore, high renin levels predicted the decline in blood pressure induced by weight loss [34–37]. Possible mechanisms that may increase renin in the obese, namely hyperleptinemia and sympathetic activation, are reduced by weight loss [38, 39]. The link between decreased circulating aldosterone and weight loss is unknown, but decreased renin activity may contribute per se, as well as the possible reduction of adipocyte products and oxidized fatty acid derivatives.

We have recently reported the influence of 5% body weight reduction on the circulating and adipose-tissue RAAS in 17 postmenopausal obese women [16]. Five percent body weight reduction in our study was achieved by a reduction in total calorie consumption, and was associated by 4 cm reduction in waist circumference, a decrease in daily mean ambulatory blood pressure by 7/2 mm Hg, and a slight decrease of fasting insulin. Weight reduction reduced levels of circulating AGT, renin, aldosterone and ACE, and decreased expression of the AGT gene in adipose tissue. We found a highly significant correlation between the decline in AGT plasma levels with the reduction in waist circumference, with the decrease of AGT gene expression in adipose tissue and with the reduction in systolic blood pressure. Thus, our data point to the link between adipose AGT and the circulating RAAS in obesity as described above. The influence of RAAS genotypes to the body weight–blood pressure relationship has been discussed elsewhere [6].

Evidence from Clinical Trials

The presence of one or two components of the Metabolic Syndrome doubles cardiovascular risk, whereas the co-occurrence of all components more than triples the risk [40, 41]. The Metabolic Syndrome is associated with a significant, graded relationship between the number of metabolic risk factors and the presence of kidney disease [42, 43]. Although many of the major trials on RAAS blockade did not specifically address obesity or the Metabolic Syndrome, mean BMI was $\geq 28\,kg/m^2$ in a number of clinical trials conducted in high-risk populations [44–47]. As patients with the Metabolic Syndrome have a high incidence of left ventricular hypertrophy, hyperfiltration, and microalbuminuria, the effect of RAAS blockade on cardiovascular and renal outcomes is important to consider [48].

Of similar importance is the finding that RAAS blockade significantly reduces the risk of new-onset diabetes compared with other anti-hypertensive classes [44–47]. In ALLHAT (mean BMI approximately $30\,kg/m^2$), treatment with the ACE inhibitor lisinopril was associated with a significantly lower rate of new-onset diabetes compared with the diuretic chlorthalidone [46]. A substudy of LIFE, which demonstrated a reduction in new-onset diabetes with losartan versus atenolol, suggests that preservation of insulin sensitivity may explain the reduced diabetes risk [45]. Results of VALUE demonstrated a significantly lower risk of new-onset diabetes with valsartan compared with amlodipine, an agent with neutral effects on glucose metabolism [47]. The ALPINE study randomized overweight patients with newly diagnosed hypertension to candesartan alone or with felodipine versus hydrochlorothiazide (HCTZ) alone or with atenolol [49]. HCTZ-based therapy was associated with increased serum insulin and glucose levels, whereas candesartan-based therapy had no effect on these parameters. The Metabolic Syndrome was present in similar numbers of patients in the two groups at baseline, but was significantly increased in the HCTZ group at 1 year. ALPINE also found a significant reduction in the rate of newly diagnosed diabetes with candesartan compared with HCTZ. A more detailed review of RAAS blockade, metabolic risk, and ongoing trials is given elsewhere [7, 50].

Mechanisms that Link the RAAS and the Metabolic Syndrome

Insulin resistance is one of the hallmarks of the Metabolic Syndrome [51]. The mechanisms by which Ang II/RAAS blockade may influence metabolic regulation are mostly related to the influence on insulin sensitivity. Signs of improved insulin sensitivity by RAAS blockade have been reported by several investigators, and decreased basal insulin levels or a diminished insulin response following a glucose load were measured as surrogates [52, 53]. RAAS inhibition prevents the insulin-desensitizing effect of Ang II on the molecular level [54]. On the other hand, insulin secretion by pancreatic β-cells is also influenced by Ang II/RAAS blockade, which may be of special interest in later stages of the Metabolic Syndrome [55, 56].

Insulin sensitivity is dependent on blood flow which regulates the delivery of nutrients to the tissue, the mass of insulin-sensitive cells in a given tissue, and the molecular events of insulin signaling. Cellular insulin sensitivity depends in part on extracellular signals such as adipokines or Ang II, which are also closely linked to inflammatory processes and oxidative stress. On the other hand, intracellular molecules that mediate the insulin signal must also function correctly (e.g. PPARγ). All these mechanisms may be influenced by Ang II, and thus also by RAAS blockade [57]. To make things even more complicated, beneficial effects of RAAS blockade are partly independent of Ang II or the AT1-receptor. ACE inhibitors improve glucose metabolism also through increased bradykinin availability [58], whereas angiotensin-receptor blockers interfere with other metabolic pathways and are more heterogenous in their influence on insulin sensitivity [59].

Endothelial function determines blood flow regulation. In general, ACE inhibitors improve endothelial function by a number of effects, including increased bradykinin availability and increased nitric oxide production. Angiotensin receptor blockers also appear to improve endothelial dysfunction by increasing endothelial-dependent relaxation and inducing anti-inflammatory, anti-coagulant, and anti-oxidant effects [60]. Interactions between vascular insulin and AT1-receptors decrease insulin-mediated vasodilation, which is an important mechanism to ensure nutrient partition to tissues such as skeletal muscle and adipose tissue under physiological conditions [61].

The data on the dysregulation of the adipose-tissue RAAS suggest enhanced metabolic and hemodynamic effects of locally produced Ang II. We tested this hypothesis in obese men by local application of Ang II with the microdialysis technique, and by treating these subjects with irbesartan. Both, RAAS blockade and Ang II had rather limited effects on adipose tissue blood flow, glucose handling and lipolysis in obese humans ([6, 62] and M. Boschmann, manuscript in revision). Thus, direct effects of Ang II/RAAS blockade on adipose-tissue metabolism appear to be of less importance in the pathophysiology of the Metabolic Syndrome.

Adipose tissue is known to produce hormones and cytokines that influence inflammation, lipid accumulation, and insulin resistance [63]. A very important adipokine in this respect is adiponectin. RAAS blockade increases adiponectin levels [64–66]. However, none of these studies investigated the mechanisms that increase circulating adiponectin. Furthermore, the issue is complicated by the fact that accompanying insulin resistance was always ameliorated by RAAS blockade, which may by itself increase circulating adiponectin. Experimentally, Ang II decreases adiponectin plasma levels, and the stimulation of radical oxygen species production in adipose tissue appears to be an important mechanism [67, 68]. Thus, inhibition of Ang II-mediated effects may raise adiponectin plasma levels, but direct effects of ACE inhibitors or angiotensin-receptor blockers may have contributed as well.

Adiponectin plasma levels in humans are clearly regulated by PPARγ agonists [69]. PPARγ-activation by some angiotensin-receptor blockers has recently been demonstrated in mouse cloncal cell lines [70, 71], resulting in increased adiponectin secretion [72]. Similar findings on the stimulation of adipogenesis and of PPARγ-target gene expression by irbesartan and telmisartan in human adipocytes was also found by our group ([73] and Jürgen Janke, manuscript in revision). Stimulation of adipogenesis increases the number of new, small, and insulin-sensitive adipocytes, which may represent a mechanism to increase adipose-tissue insulin sensitivity by itself [73]. As Ang II also inhibited human adipogenesis by AT1-receptor mediated effects, the use of AT1-receptor blockers may result in increased adipogenesis by blocking Ang II. But the use of certain AT1-receptor blockers may additionally increase adipogenesis by a different mechanism [73]. Thus, findings on increased cellular insulin sensitivity upon AT1-blocker treatment may be attributed to both, decreased Ang II signaling and PPARγ-activation by direct molecular interactions [74–77].

Knowledge on metabolic effects of aldosterone are even more limited. Aldosterone receptors are expressed at least in rodent adipocytes [78, 79], and aldosterone exposure inhibited insulin-stimulated glucose uptake and induced the expression of inflammatory cytokines [80]. In the liver, aldosterone increased gluconeogenesis, partly by suppression of insulin effects, but the nuclear receptor involved appears to be the glucocorticoid receptor [81]. Clinical data on aldosterone effects on glucose metabolism are missing with the exception for primary aldosteronism [82]. Thus, the pathophysiological role for aldosterone in the Metabolic Syndrome is currently restricted to its effects on sodium retention.

Conclusions

Inappropriate upregulation of the RAAS suggests profound changes in the regulation of this system in obese individuals. Increased renin secretion is linked to increased renal sympathetic activity, which may in turn be linked to hyperleptinemia. The contribution of adipose tissue to increased AGT and aldosterone plasma levels in obesity is likely due to the secretion of AGT and as yet unidentified aldosterone secretagogues. Modest weight reduction significantly reduced RAAS activity in humans, again also by influencing the adipose-tissue RAAS. Several recent clinical trials demonstrated beneficial effects of RAAS-blockade on metabolic complications in cardiovascular patients. There remain, however, several obscurities. Why the RAAS is upregulated in obesity/the Metabolic Syndrome, is not well understood, and whether RAAS-blockade acts by direct metabolic effects or by indirectly influencing local inflammation and blood flow is also not clear. The adipose-tissue RAAS may play a substantial role in the development of metabolic and cardiovascular abnormalities associated with obesity. Possible mechanisms include inhibition of adipogenesis which may limit the storage capacity of adipose tissue and allows ectopic lipid storage which leads to lipotoxicity. Nevertheless, the molecular mechanisms that link the RAAS and the Metabolic Syndrome are not understood very well at present.

References

1. Carey RM, Siragy HM: Newly recognized components of the renin-angiotensin system: potential roles in cardiovascular and renal regulation. Endocr Rev 2003;24:261–271.
2. Re RN: Mechanisms of disease: local renin-angiotensin-aldosterone systems and the pathogenesis and treatment of cardiovascular disease. Nat Clin Pract Cardiovasc Med 2004;1:42–47.
3. Engeli S, Schling P, Gorzelniak K, Boschmann M, Janke J, Ailhaud G, Teboul M, Massiera F, Sharma AM: The adipose-tissue renin-angiotensin-aldosterone system: role in the metabolic syndrome? Int J Biochem Cell Biol 2003;35:807–825.
4. Cassis LA, Huang J, Gong MC, Daugherty A: Role of metabolism and receptor responsiveness in the attenuated responses to Angiotensin II in mice compared to rats. Regul Pept 2004;117:107–116.
5. Blüher M: Transgenic animal models for the study of adipose tissue biology. Best Pract Res Clin Endocrinol Metab 2005;19:605–623.
6. Sharma AM, Engeli S: Obesity and the renin-angiotensin-aldosterone system. Expert Rev Endocrinol Metab 2006;1:255–264.
7. Scheen AJ: Renin-angiotensin system inhibition prevents type 2 diabetes mellitus. Part 1. A meta-analysis of randomised clinical trials. Diabetes Metab 2004;30:487–496.
8. Hall JE: The kidney, hypertension, and obesity. Hypertension 2003;41:625–633.
9. Licata G, Volpe M, Scaglione R, Rubattu S: Salt-regulating hormones in young normotensive obese subjects: Effects of saline load. Hypertension 1994;23(suppl):I20–I24.
10. Vaz M, Jennings G, Turner A, Cox H, Lambert G, Esler M: Regional sympathetic nervous activity and oxygen consumption in obese normotensive human subjects. Circulation 1997;96:3423–3429.
11. Rahmouni K, Morgan DA, Morgan GM, Mark AL, Haynes WG: Role of selective leptin resistance in diet-induced obesity hypertension. Diabetes 2005;54:2012–2018.
12. Eikelis N, Lambert G, Wiesner G, Kaye D, Schlaich M, Morris M, Hastings J, Socratous F, Esler M: Extra-adipocyte leptin release in human obesity and its relation to sympathoadrenal function. Am J Physiol Endocrinol Metab 2004;286:E744–E752.
13. Goodfriend TL, Ball DL, Egan BM, Campbell WB, Nithipatikom K: Epoxy-keto derivative of linoleic acid stimulates aldosterone secretion. Hypertension 2004;43:358–363.
14. Ehrhart-Bornstein M, Lamounier-Zepter V, Schraven A, Langenbach J, Willenberg HS, Barthel A, Hauner H, McCann SM, Scherbaum WA, Bornstein SR: Human adipocytes secrete mineralocorticoid-releasing factors. Proc Natl Acad Sci USA 2003;100:14211–14216.

15 Boustany CM, Bharadwaj K, Daugherty A, Brown DR, Randall DC, Cassis LA: Activation of the systemic and adipose renin-angiotensin system in rats with diet-induced obesity and hypertension. Am J Physiol Regul Integr Comp Physiol 2004;287:R943–R949.
16 Engeli S, Böhnke J, Gorzelniak K, Janke J, Schling P, Bader M, Luft FC, Sharma AM: Weight loss and the renin-angiotensin-aldosterone system. Hypertension 2005;45:356–362.
17 Barton M, Carmona R, Morawietz H, D'Uscio LV, Goettsch W, Hillen H, Haudenschild CC, Krieger JE, Munter K, Lattmann T, Lüscher TF, Shaw S: Obesity is associated with tissue-specific activation of renal angiotensin-converting enzyme in vivo: evidence for a regulatory role of endothelin. Hypertension 2000;35:329–336.
18 Ahmed SB, Fisher ND, Stevanovic R, Hollenberg NK: Body mass index and angiotensin-dependent control of the renal circulation in healthy humans. Hypertension 2005;46:1316–1320.
19 Hopkins PN, Lifton RP, Hollenberg NK, Jeunemaitre X, Hallouin MC, Skuppin J, Williams CS, Dluhy RG, Lalouel JM, Williams RR, Williams GH: Blunted renal vascular response to angiotensin II is associated with a common variant of the angiotensinogen gene and obesity. J Hypertens 1996;14:199–207.
20 Xu ZG, Lanting L, Vaziri ND, Li Z, Sepassi L, Rodriguez-Iturbe B, Natarajan R: Upregulation of angiotensin II type 1 receptor, inflammatory mediators, and enzymes of arachidonate metabolism in obese Zucker rat kidney: reversal by angiotensin II type 1 receptor blockade. Circulation 2005;111:1962–1969.
21 Schling P, Schäfer T: Human adipose tissue cells keep tight control on the angiotensin II levels in their vicinity. J Biol Chem 2002;277:48066–48075.
22 Gorzelniak K, Engeli S, Janke J, Luft FC, Sharma AM: Hormonal regulation of the human adipose-tissue renin-angiotensin system: relationship to obesity and hypertension. J Hypertens 2002;20:965–973.
23 Faloia E, Gatti C, Camilloni MA, Mariniello B, Sardu C, Garrapa GG, Mantero F, Giacchetti G: Comparison of circulating and local adipose tissue renin-angiotensin system in normotensive and hypertensive obese subjects. J Endocrinol Invest 2002;25:309–314.
24 Giacchetti G, Faloia E, Mariniello B, Sardu C, Gatti C, Camilloni MA, Guerrieri M, Mantero F: Overexpression of the renin-angiotensin system in human visceral adipose tissue in normal and overweight subjects. Am J Hypertens 2002;15:381–388.
25 Davis D, Liyou N, Lockwood D, Johnson A: Angiotensinogen genotype, plasma protein and mRNA concentration in isolated systolic hypertension. Clin Genet 2002;61:363–368.
26 Prat-Larquemin L, Oppert JM, Clement K, Hainault I, Basdevant A, Guy-Grand B, Quignard-Boulange A: Adipose angiotensinogen secretion, blood pressure, and AGT M235T polymorphism in obese patients. Obes Res 2004;12:556–561.
27 Tamura K, Umemura S, Yamakawa T, Nyui N, Hibi K, Watanabe Y, Ishigami T, Yabana M, Tanaka SI, Sekihara H, Murakami K, Ishii M: Modulation of tissue angiotensinogen gene expression in genetically obese hypertensive rats. Am J Physiol Regul Integr Comp Physiol 1997;272:R1704–R1711.
28 Hainault I, Nebout G, Turban S, Ardouin B, Ferre P, Quignard-Boulange A: Adipose tissue-specific increase in angiotensinogen expression and secretion in the obese (fa/fa) Zucker rat. Am J Physiol Endocrinol Metab 2002;282:E59–E66.
29 Rahmouni K, Mark AL, Haynes WG, Sigmund CD: Adipose depot-specific modulation of angiotensinogen gene expression in diet-induced obesity. Am J Physiol Endocrinol Metab 2004;286:E891–E895.
30 Massiera F, Bloch-Faure M, Ceiler D, Murakami K, Fukamizu A, Gasc JM, Quignard-Boulange A, Negrel R, Ailhaud G, Seydoux J, Meneton P, Teboul M: Adipose angiotensinogen is involved in adipose tissue growth and blood pressure regulation. FASEB J 2001;15:2727–2729.
31 Masuzaki H, Paterson J, Shinyama H, Morton NM, Mullins JJ, Seckl JR, Flier JS: A transgenic model of visceral obesity and the metabolic syndrome. Science 2001;294:2166–2170.
32 Tsai YS, Kim HJ, Takahashi N, Kim HS, Hagaman JR, Kim JK, Maeda N: Hypertension and abnormal fat distribution but not insulin resistance in mice with P465L PPARgamma. J Clin Invest 2004;114:240–249.
33 Harte A, McTernan P, Chetty R, Coppack S, Katz J, Smith S, Kumar S: Insulin-mediated upregulation of the renin angiotensin system in human subcutaneous adipocytes is reduced by rosiglitazone. Circulation 2005;111:1954–1961.

34 Tuck ML, Sowers J, Dornfeld L, Kledzik G, Maxwell M: The effect of weight reduction on blood pressure, plasma renin activity, and plasma aldosterone levels in obese patients. N Engl J Med 1981;304:930–933.
35 Blaufox MD, Lee HB, Davis B, Oberman A, Wassertheil Smoller S, Langford H: Renin predicts diastolic blood pressure response to nonpharmacologic and pharmacologic therapy. JAMA 1992;267:1221–1225.
36 Harp JB, Henry SA, DiGirolamo M: Dietary weight loss decreases serum angiotensin-converting enzyme activity in obese adults. Obes Res 2002;10:985–990.
37 Sowers JR, Nyby M, Stern N, Beck F, Baron S, Catania R, Vlachis N: Blood pressure and hormone changes associated with weight reduction in the obese. Hypertension 1982;4:686–691.
38 Grassi G, Seravalle G, Colombo M, Bolla G, Cattaneo BM, Cavagnini F, Mancia G: Body weight reduction, sympathetic nerve traffic, and arterial baroreflex in obese normotensive humans. Circulation 1998;97:2037–2042.
39 Straznicky NE, Lambert EA, Lambert GW, Masuo K, Esler MD, Nestel PJ: Effects of dietary weight loss on sympathetic activity and cardiac risk factors associated with the metabolic syndrome. J Clin Endocrinol Metab 2005;90:5998–6005.
40 Malik S, Wong ND, Franklin SS, Kamath TV, L'Italien GJ, Pio JR, Williams GR: Impact of the metabolic syndrome on mortality from coronary heart disease, cardiovascular disease, and all causes in United States adults. Circulation 2004;110:1245–1250.
41 Klein BE, Klein R, Lee KE: Components of the metabolic syndrome and risk of cardiovascular disease and diabetes in Beaver Dam. Diabetes Care 2002;25:1790–1794.
42 Chen J, Muntner P, Hamm LL, Jones DW, Batuman V, Fonseca V, Whelton PK, He J: The metabolic syndrome and chronic kidney disease in U.S. adults. Ann Intern Med 2004;140:167–174.
43 Palaniappan L, Carnethon M, Fortmann SP: Association between microalbuminuria and the metabolic syndrome: NHANES III. Am J Hypertens 2003;16:952–958.
44 Yusuf S, Sleight P, Pogue J, Bosch J, Davies R, Dagenais G: Effects of an angiotensin-converting-enzyme inhibitor, ramipril, on cardiovascular events in high-risk patients. The Heart Outcomes Prevention Evaluation Study Investigators. N Engl J Med 2000;342:145–153.
45 Lindholm LH, Ibsen H, Borch-Johnsen K, Olsen MH, Wachtell K, Dahlöf B, Devereux RB, Beevers G, De Faire U, Fyhrquist F, Julius S, Kjeldsen SE, Kristianson K, Lederballe-Pedersen O, Nieminen MS, Omvik P, Oparil S, Wedel H, Aurup P, Edelman JM, Snapinn S: Risk of new-onset diabetes in the Losartan Intervention For Endpoint reduction in hypertension study. J Hypertens 2002;20:1879–1886.
46 The ALLHAT Officers and Coordinators for the ALLHAT Collaborative Research Group: Major outcomes in high-risk hypertensive patients randomized to angiotensin-converting enzyme inhibitor or calcium channel blocker vs. diuretic: The Antihypertensive and Lipid-Lowering Treatment to Prevent Heart Attack Trial (ALLHAT). JAMA 2002;288:2981–2997.
47 Julius S, Kjeldsen SE, Weber M, Brunner HR, Ekman S, Hansson L, Hua T, Laragh J, McInnes GT, Mitchell L, Plat F, Schork A, Smith B, Zanchetti A: Outcomes in hypertensive patients at high cardiovascular risk treated with regimens based on valsartan or amlodipine: the VALUE randomised trial. Lancet 2004;363:2022–2031.
48 Sharma AM, Pischon T, Engeli S, Scholze J: Choice of drug treatment for obesity-related hypertension: where is the evidence? J Hypertens 2001;19:667–674.
49 Lindholm LH, Persson M, Alaupovic P, Carlberg B, Svensson A, Samuelsson O: Metabolic outcome during 1 year in newly detected hypertensives: results of the Antihypertensive Treatment and Lipid Profile in a North of Sweden Efficacy Evaluation (ALPINE study). J Hypertens 2003;21:1563–1574.
50 Sharma AM, Engeli S: The role of renin-angiotensin system blockade in the management of hypertension associated with the cardiometabolic syndrome. J Cardiometab Syndr 2006;1:29–35.
51 Eckel RH, Grundy SM, Zimmet PZ: The metabolic syndrome. Lancet 2005;365:1415–1428.
52 Jordan J, Engeli S, Boschmann M, Weidinger G, Luft FC, Sharma AM, Kreuzberg U: Hemodynamic and metabolic responses to valsartan and atenolol in obese hypertensive patients. J Hypertens 2005;23:2313–2318.
53 Masuo K, Mikami H, Ogihara T, Tuck ML: Weight reduction and pharmacologic treatment in obese hypertensives. Am J Hypertens 2001;14:530–538.

54 Folli F, Saad MJ, Velloso L, Hansen H, Carandente O, Feener EP, Kahn CR: Crosstalk between insulin and angiotensin II signaling systems. Exp Clin Endocrinol Diabetes 1999;107:133–139.

55 Ramracheya RD, Muller DS, Wu Y, Whitehouse BJ, Huang GC, Amiel SA, Karalliedde J, Viberti G, Jones PM, Persaud SJ: Direct regulation of insulin secretion by angiotensin II in human islets of Langerhans. Diabetologia 2006;49:321–331.

56 Chu KY, Lau T, Carlsson PO, Leung PS: Angiotensin II type 1 receptor blockade improves beta-cell function and glucose tolerance in a mouse model of type 2 diabetes. Diabetes 2006;55:367–374.

57 Scheen AJ: Renin-angiotensin system inhibition prevents type 2 diabetes mellitus. Part 2. Overview of physiological and biochemical mechanisms. Diabetes Metab 2004;30:498–505.

58 Carvalho CR, Thirone AC, Gontijo JA, Velloso LA, Saad MJ: Effect of captopril, losartan, and bradykinin on early steps of insulin action. Diabetes 1997;46:1950–1957.

59 Katovich MJ, Pachori A: Effects of inhibition of the renin-angiotensin system on the cardiovascular actions of insulin. Diabetes Obes Metab 2000;2:3–14.

60 Watanabe T, Barker TA, Berk BC: Angiotensin II and the endothelium: diverse signals and effects. Hypertension 2005;45:163–169.

61 Nakashima H, Suzuki H, Ohtsu H, Chao JY, Utsunomiya H, Frank GD, Eguchi S: Angiotensin II regulates vascular and endothelial dysfunction: recent topics of Angiotensin II type-1 receptor signaling in the vasculature. Curr Vasc Pharmacol 2006;4:67–78.

62 Boschmann M, Engeli S, Adams F, Franke G, Luft FC, Sharma AM, Jordan J: Influences of AT1 receptor blockade on tissue metabolism in obese men. Am J Physiol Regul Integr Comp Physiol 2006;290:R219–R223.

63 Berg AH, Scherer PE: Adipose tissue, inflammation, and cardiovascular disease. Circ Res 2005;96:939–949.

64 Nomura S, Shouzu A, Omoto S, Nishikawa M, Fukuhara S, Iwasaka T: Effect of valsartan on monocyte/endothelial cell activation markers and adiponectin in hypertensive patients with type 2 diabetes mellitus. Thromb Res 2006;117:385–392.

65 Furuhashi M, Ura N, Higashiura K, Murakami H, Tanaka M, Moniwa N, Yoshida D, Shimamoto K: Blockade of the renin-angiotensin system increases adiponectin concentrations in patients with essential hypertension. Hypertension 2003;42:76–81.

66 Yenicesu M, Yilmaz MI, Caglar K, Sonmez A, Eyileten T, Acikel C, Kilic S, Bingol N, Bingol S, Vural A: Blockade of the renin-angiotensin system increases plasma adiponectin levels in type-2 diabetic patients with proteinuria. Nephron Clin Pract 2005;99:c115–c121.

67 Ran J, Hirano T, Fukui T, Saito K, Kageyama H, Okada K, Adachi M: Angiotensin II infusion decreases plasma adiponectin level via its type 1 receptor in rats: an implication for hypertension-related insulin resistance. Metabolism 2006;55:478–488.

68 Hattori Y, Akimoto K, Gross SS, Hattori S, Kasai K: Angiotensin-II-induced oxidative stress elicits hypoadiponectinaemia in rats. Diabetologia 2005;48:1066–1074.

69 Yu JG, Javorschi S, Hevener AL, Kruszynska YT, Norman RA, Sinha M, Olefsky JM: The effect of thiazolidinediones on plasma adiponectin levels in normal, obese, and type 2 diabetic subjects. Diabetes 2002;51:2968–2974.

70 Schupp M, Janke J, Clasen R, Unger T, Kintscher U: Angiotensin type 1 receptor blockers induce peroxisome proliferator-activated receptor-gamma activity. Circulation 2004;109:2054–2057.

71 Benson SC, Pershadsingh HA, Ho CI, Chittiboyina A, Desai P, Pravenec M, Qi N, Wang J, Avery MA, Kurtz TW: Identification of telmisartan as a unique angiotensin II receptor antagonist with selective PPAR{gamma}-modulating activity. Hypertension 2004;43:993–1002.

72 Clasen R, Schupp M, Foryst-Ludwig A, Sprang C, Clemenz M, Krikov M, Thöne-Reineke C, Unger T, Kintscher U: PPARgamma-activating angiotensin type-1 receptor blockers induce adiponectin. Hypertension 2005;46:137–143.

73 Janke J, Engeli S, Gorzelniak K, Luft FC, Sharma AM: Mature adipocytes inhibit in vitro differentiation of human preadipocytes via angiotensin type 1 receptors. Diabetes 2002;51:1699–1707.

74 Furuhashi M, Ura N, Takizawa H, Yoshida D, Moniwa N, Murakami H, Higashiura K, Shimamoto K: Blockade of the renin-angiotensin system decreases adipocyte size with improvement in insulin sensitivity. J Hypertens 2004;22:1977–1982.

75 Fujimoto M, Masuzaki H, Tanaka T, Yasue S, Tomita T, Okazawa K, Fujikura J, Chusho H, Ebihara K, Hayashi T, Hosoda K, Nakao K: An angiotensin II AT1 receptor antagonist, telmisartan augments

glucose uptake and GLUT4 protein expression in 3T3-L1 adipocytes. FEBS Lett 2004;576: 492–497.
76 Shiuchi T, Iwai M, Li HS, Wu L, Min LJ, Li JM, Okumura M, Cui TX, Horiuchi M: Angiotensin II type-1 receptor blocker valsartan enhances insulin sensitivity in skeletal muscles of diabetic mice. Hypertension 2004;43:1003–1010.
77 Schupp M, Clemenz M, Gineste R, Witt H, Janke J, Helleboid S, Hennuyer N, Ruiz P, Unger T, Staels B, Kintscher U. Molecular characterization of new selective peroxisome proliferator-activated receptor gamma modulators with angiotensin receptor blocking activity. Diabetes 2005;54:3442–3452.
78 Rondinone CM, Rodbard D, Baker ME: Aldosterone stimulated differentiation of mouse 3T3-L1 cells into adipocytes. Endocrinology 1993;132:2421–2426.
79 Penfornis P, Viengchareun S, Le Menuet D, Cluzeaud F, Zennaro MC, Lombes M: The mineralocorticoid receptor mediates aldosterone-induced differentiation of T37i cells into brown adipocytes. Am J Physiol Endocrinol Metab 2000;279:E386–E394.
80 Kraus D, Jager J, Meier B, Fasshauer M, Klein J: Aldosterone inhibits uncoupling protein-1, induces insulin resistance, and stimulates proinflammatory adipokines in adipocytes. Horm Metab Res 2005;37:455–459.
81 Yamashita R, Kikuchi T, Mori Y, Aoki K, Kaburagi Y, Yasuda K, Sekihara H: Aldosterone stimulates gene expression of hepatic gluconeogenic enzymes through the glucocorticoid receptor in a manner independent of the protein kinase B cascade. Endocr J 2004;51:243–251.
82 Corry DB, Tuck ML: The effect of aldosterone on glucose metabolism. Curr Hypertens Rep 2003;5:106–109.

Dr. med. Stefan Engeli
Franz-Volhard-Centrum für Klinische Forschung (Haus 129)
Charité Campus Buch
Wiltberg Strasse 50
DE–13125 Berlin (Germany)
Tel. +49 30 9417 2271, Fax +49 30 9417 2265, E-Mail stefan.engeli@charite.de

Functional and Structural Renal Changes in the Early Stages of Obesity

Vashu Thakur, Stephen Morse, Efrain Reisin

Section of Nephrology, Department of Medicine, Louisiana State University School of Medicine, New Orleans, La., USA

Abstract

Obesity is an epidemic of colossal proportion, and the healthcare system has been singularly unable to abate this growing threat to society. Two-thirds of US adults are now overweight or obese. Also, the incidence of obesity is increasing in children and adolescents at an alarming rate. The mechanisms of kidney damage, can be both direct and consequential from hypertension and diabetes mellitus. Both hemodyanmic and hormonal effects of obesity are discussed. Benefits of weight loss and pharmacological treatment are evaluated.

Copyright © 2006 S. Karger AG, Basel

Introduction

Obesity and its ill effects have been described for centuries [1]. However, an epidemic of the current colossal proportion is unprecedented, and the healthcare system has been singularly unable to abate this growing threat to society. The prevalence of obesity in US adults has doubled from 15 to 30.5% in last 20 years, and two-thirds of US adults are now overweight or obese [2]. What is also disturbing is the rate at which the incidence of obesity is increasing in children and adolescents. Obesity in children and adolescents is associated with many of the same short- and long-term complications suffered by obese adults [3, 4]. These medical complications include hypertension (HTN), hypercholesterolemia and dyslipidemia, type 2 diabetes mellitus (T2DM), obstructive sleep apnea, nonalcoholic steatohepatitis, orthopedic abnormalities, and psychological problems. Many epidemiological studies have provided substantial evidence for the increased prevalence of HTN in obese children and adolescents using casual blood-pressure measurements [5–7]. Autopsy studies in adolescents and young adults have revealed associations between increased body mass index (BMI) and the presence of atherosclerotic lesions in the aorta and coronary arteries [8].

T2DM has turned into a significant public health problem, with escalating numbers of new cases. It was referred to as an 'epidemic' by the American Diabetes Association (ADA) [9, 10]. Moreover, the problem is not limited to North America; it also has been reported in children from Europe [11, 12], Asia [13], and Australia [14].

Mechanisms by which increased BMI causes kidney damage are subjects of intense study. Although the kidney damage caused by obesity may be secondary to the increased incidence of HTN and DM, there seems to be an independent deleterious effect of obesity. Obesity promotes kidney damage through both hemodynamic and hormonal effects. In studies of kidney pathology, glomerular injury was described in animal models of obesity in the absence of diabetes [15]. There are, however, similar pathobiological mechanisms that appear to underlie both nephrosclerosis and atherosclerosis.

Even being overweight confers an increased risk of damage to the kidneys and other body organs [16]. Early recognition provides increased opportunities for early intervention and prevention of the more serious complications of obesity. In many specific cases of organ damage related to obesity, there is perhaps a point of no return, e.g., dilated cardiomyopathy or cirrhosis of the liver. At the very least, obesity (with its attendant complications) is easier to reverse with early intervention. Although the focus of this paper is early obesity, overweight and obesity are considered to be a continuum of the same disease process.

Metabolic Syndrome as an Important Cause of CKD

Approximately 11% of adults (19.2 million Americans) have chronic kidney disease (CKD) based on decreased kidney function or persistent albuminuria [17]. Several clinical studies have linked CKD/microalbuminuria with metabolic syndrome [18–21]. Recently, Chen et al. [19] examined the correlation between CKD and the components of metabolic syndrome; they found not only that each component of metabolic syndrome was associated with increased prevalence of CKD and microalbuminuria but also that the more components are present, the higher the prevalence of CKD and microalbuminuria. In addition to hyperglycemia and HTN as risks for CKD, increased waist circumference was also significantly correlated with glomerular filtration rate (GFR) decline and microalbuminuria [19]. Insulin resistance and hyperinsulinemia were shown to predict CKD strongly and positively in nondiabetic adults [21].

Gelber et al. [16] demonstrated an increased in the risk of developing CKD 3 in subjects with a BMI as low as 22.7. The risk increased progressively with higher BMIs. It is noteworthy that in Gelber's study, the highest BMI of any group was >26.6, still shy of the technical definition of obesity. According to Framingham data, baseline BMI predicts subsequent kidney disease after a mean follow-up of 18.5 years [22]. Each unit increase in BMI SD was associated with a 1.2-fold increased risk for new-onset kidney disease, controlling for age, sex, baseline GFR, smoking, and diabetes status. Obesity-associated proteinuria, nephritic syndrome, and CKD are frequently seen in clinical practice and are well described [23–26]. An early autopsy study of kidneys from obese patients reported glomerulomegaly as a consistent feature and focal segmental glomerulosclerosis (FSGS) without predilection in the juxtamedullary sites [24]. A recent renal biopsy study found a 10-fold increase in the incidence of obesity-related glomerulopathy over 10 years [25]. All subjects had glomerulomegaly and proteinuria, and 48% of the proteinuria was in the nephrotic range. The subjects with obesity-related FSGS demonstrated the concomitant presence of glomerulomegaly; fortunately, they also exhibited less severe proteinuria, less cholesterol elevation, less clinical nephrotic syndrome, less podocyte injury, and more indolent progression than those with idiopathic FSGS [25]. Therefore, metabolic syndrome may not only be an independent cause of CKD but may also contribute to the increasing number of Americans suffering from CKD [26].

The Impact of Obesity on the Kidney: Hemodynamic and Mechanical Effects

Even though the total sodium excretion is normal or modestly elevated (Tuck), the pressure natriuresis normally seen with increased blood pressure is blunted [27, 28]. These investigators studied dogs with diet-induced obesity. They were able to demonstrate sodium retention at the level of loop of Henle that may be caused by insulin resistance and hyperinsulinemia, increased sympathetic activity, activation of renin–angiotensin–aldosterone system (RAAS), and/or higher renal interstitial fluid hydrostatic pressure. Microscopic examination of the animal kidneys revealed an increase in the interstitial cells and expansion of the extracellular matrix between tubules in the renal medulla. They concluded that obesity-HTN is associated with a shift of pressure natriuresis toward a higher blood pressure (fig. 1).

Reisin et al. [29] have proved that obesity in humans is associated with increased renal blood flow. Other investigators have documented a decreased renal vascular resistance and increased GFR [30]. The number of nephrons does not increase with obesity [31]; rather, each nephron filters more. Consequently, obese subjects have hyperperfusion and hyperfiltration. Whether obesity-associated hyperfiltration is due to increased hydrostatic pressure or increased

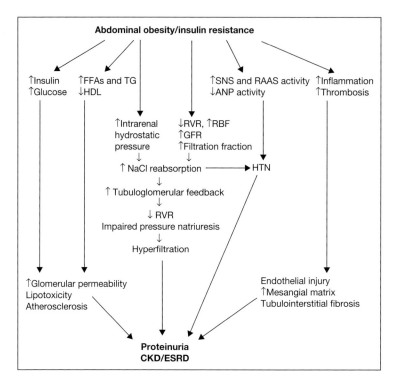

Fig. 1. Mechanisms of kidney injury. ANP = Atrial natriuretic peptide; CKD = chronic kidney disease; ESRD = end-stage renal disease; FFA = free fatty acid; GFR = glomerular filtration rate; HDL = high-density lipoprotein; HTN = hypertension; RAAS = renin–angiotensin–aldosterone system; RBF = renal blood flow; RVR = renal vascular resistance; SNS = sympathetic nervous system; TG = triglycerides. Available from reference [98].

permeability (Kf) is not settled. Chagnac et al. [32] attempted to resolve the issue using different sized pores and pressure. They employed sophisticated modeling to analyze their data and concluded that increase in hydrostatic pressure was responsible for hyperfiltration in the obese. They further showed that the GFR increased more than renal plasma flow (51% and 31%, respectively), resulting in an increased filtration fraction in nondiabetic, nonhypertensive obese subjects. Abnormal transmission of elevated arterial pressure to the intraglomerular capillaries through dilated afferent arterioles could account for the increased transcapillary hydraulic pressure difference, which leads to a high filtration fraction. A high filtration fraction raises the postglomerular oncotic pressure in peritubular arterioles, which increases NaCl reabsorption and decreases NaCl delivery to the macula densa. This, in turn, leads to activation of the tubuloglomerular feedback and maintenance of hyperfiltration [32, 33].

This view is supported indirectly by the microscopic findings of vasodilation of afferent arterioles and increased cellularity in the renal medulla [34].

Excess Excretory Load and Thrifty Phenotype

Obesity is associated with excess metabolic and excretory load from increased body mass. Glomerular hyperperfusion and hyperfiltration with increased filtration fraction have been well documented by several studies of obese animals and patients [23, 32, 33, 35]. These hemodynamic changes can cause glomerulomegaly and eventual glomerulosclerosis. The first clinical evidence of renal injury is microalbuminuria in obese subjects, and weight loss can effectively reduce the microalbuminuria [35, 36]. Nephron overwork and intraglomerular HTN in obesity can be exaggerated if the nephron number is already reduced, such as occurs after uninephrectomy or in offspring with low-birth weight [37, 38].

Clinical studies have observed a consistent relationship between low birth weight and increased risk of developing adult metabolic syndrome [39, 40]. Early intrauterine growth restriction may cause a programmed increase in both appetite and efficiency in the utilization of calories, which in turn lead to rapid compensatory childhood growth when food is available. This 'thrifty phenotype' generates body size that exceeds relative organ capacity and promotes abdominal obesity [39, 40]. Fetal nutrient deprivation and growth restriction also impair renal development and reduce nephron number. No additional nephron can be formed postnatally. In subjects with a lifestyle of physical inactivity and dietary excess, the thrift phenotype promotes obesity and creates an imbalance between the excess metabolic/excretory load and the nephron deficit/reduced renal capacity. In one clinical study, Lackland et al. [38] found that low-birth weight indeed increased the relative risk of early-onset end-stage renal disease.

Hall [41] observed that the interstitial fluid hydrostatic pressure was elevated to 19 mm Hg in obese dogs, compared to only 9–10 mm Hg in lean dogs. The elevated interstitial hydrostatic pressure reduces medullary blood flow (vasa recta) and causes tubular compression, which slows the tubular flow rate and increases fractional tubular reabsorption. Increased NaCl reabsorption reduces macula densa NaCl delivery, which leads to a feedback mediated renal vascular dilatation, elevation of GFR, and stimulation of RAAS, despite the volume expansion. These compensatory responses aim to overcome the increased tubular reabsorption and to maintain sodium balance; however, persistent glomerular hyperfiltration, working along with other components of metabolic syndrome, will cause proteinuria and progressive renal injury [23, 33, 41].

It is interesting to note that babies who are small for gestation do not have fully developed kidneys; the total number of nephrons is much less than normal. These babies are prone to developing obesity, which is also linked to other conditions associated with rapid growth. Since obesity is a hyperfiltrable condition, it exposes the smaller number of nephrons to much higher pressures and thus may be especially deleterious to the kidneys. In addition to hemodynamics, several hormones play an important part in the development of kidney damage in the obese. These are discussed below.

The Impact of Obesity on the Kidney: Hormonal Effects

Ghrelin, a gut peptide, is the endogenous ligand for the growth hormone secretagogue receptor [42]. It increases appetite and induces obesity in rodents. Wortley et al. [43] suggested that ghrelin plays a pivotal role in the control of obesity by demonstrating a lack of weight gain in ghrelin $-/-$ mice even after exposing them to a high-fat diet. Other genetic studies in mice question the role of endogenous ghrelin in regulating energy balance [44, 45]. The variability in results may be due to the constitutive activity of GSHR, and some activity may occur even in the ghrelin-deficient mice [46]. The ghrelin pathway is a potential target for the development of anti-obesity pharmaceutical agents; further studies are needed.

Leptin, Adiponectin, and other Adipocyte-Related Hormones and Cytokines

Leptin is a 167-amino acid hormone secreted by adipocytes. It decreases caloric intake by interacting with leptin receptors in the hypothalamus [47]. Its circulating concentration reflects the amount of body fat. Cultured cell and animal studies found that the kidney is a target organ for leptin action. The kidney expresses both isoforms of leptin receptors (Ob-Ra and Ob-Rb). Leptin can stimulate TGF-β1 synthesis, type 4 collagen production, and cellular proliferation in glomerular endothelial cells. It also upregulates TGF-β type 2 receptor and stimulates type 1 collagen production [35, 48, 49]. The infusion of leptin into normal rats causes proteinuria and FSGS [49]. On the other hand, leptin administration was shown partially to reverse insulin resistance and lipoprotein abnormalities in patients with lipodystrophy [50].

Increased adipocyte mass has been associated with increased expression of angiotensinogen, tumor necrosis factor-α (TNF-α), resistin, leptin, and PAI-1 [36–38, 48, 49, 51]. Together, leptin and cytokines in the adipocytes seem to be responsible for kidney damage associated with obesity; CRP attaches to the plasma membrane of damaged cells and causes cell death through activation of the complement cascade. Increased fibrinogen contributes to a prothrombotic state. Visceral fat also has an activated RAAS that has both paracrine and systemic effects. IL-6, leptin, and angiotensin-2 can directly cause endothelial cell injury and inflammation, and both changes could contribute to the atherosclerotic process [48, 49].

High levels of resistin are associated with insulin-resistance in mice. Also, resistin administration led to increased insulin resistance [52]. Adiponectin is a protein expressed by adipocytes. Circulating adiponectin concentrations are decreased in people who are obese as well as in those who have T2DM [50]. Loss of adiponectin may lead to insulin resistance and atherosclerosis. Thiazolidinediones are effective in raising adiponectin levels in plasma via their effect on adiponectin gene expression. This, in turn, leads to improved insulin sensitivity.

Free Fatty Acids and Renal Injury

Lipotoxicity is a cytotoxic process that is associated with the progression of metabolic syndrome and can cause multi-organ injury [39]. High-circulating free fatty acid (FFA) from abdominal fat mass drives the cellular uptake of more fatty acid (FA), which inhibits the secretion of adiponectin and reduces the mitochondrial uptake and oxidation of FA. Despite the increased release of leptin, which normally enhances FA oxidation, tissue resistance to leptin in obesity further promotes cytosolic FA build-up. The excess intracellular FA is shunted toward the production of reactive intermediates, such as fatty acyl CoA, diacylglycerol, and ceramide. These reactive compounds are cytotoxic and capable of inducing cell apoptosis and organ damage [39, 53, 54].

High circulating FFA also enhances the number of FA moieties bound to albumin and increases FA load to proximal tubules in patients with proteinuria. Lipotoxicity from the accumulation of albumin-bound FA in proximal tubular cells contributes to the tubulointerstitial inflammation and fibrosis [53]. In the presence of intraglomerular HTN and hyperfiltration, the mesangial cells can be at particular risk for exposure to the circulating FFA. Oxidative stress may also be involved in this cytotoxic process as reactive oxygen scavengers block FFA-induced apoptosis [54]. Therefore, high FFA per se in metabolic syndrome may be an important linkage of abdominal obesity, microalbuminuria, and renal injury [39, 51, 53].

Prevention and Treatment

A recent editorial in the *Lancet* suggests that the exploding epidemic of obesity in developed countries may reverse many gains made by improved diagnosis and treatment [55]. It lamented the fact that physicians and societies do very little to curb this most preventable disease state. Obesity, however, is not just a problem of developed countries. It has global implications for health and well-being (The Center for Global Health and Economic Development. A race against time: the challenge of cardiovascular disease in developing economies. New York: The Center, Columbia University, 2004).

An Australian perspective calls for more government involvement and the creation of a Public Health Ministry to deal with the endemic [56]; however, the factors that will transform medical practice and opinion as well as unambiguous goals for practitioners to follow must first be clearly defined. Maximal attention and efforts must be focused on those who are at a higher risk of being overweight, e.g., Mexican Americans [57, 58] and the middle aged [59]. Also, early intervention is necessary when obesity is more likely to be reversible.

Resolution of Microalbuminuria as a Goal of Clinical Intervention

Microalbuminuria not only indicates the development of early renal injury, but more importantly, it is a surrogate marker of generalized systemic endothelial dysfunction. Microalbuminuria is now established as a modifiable independent risk of increased CVD morbidity and mortality. It is also increasingly recognized as a component of metabolic syndrome. Of the individual components of metabolic syndrome, microalbuminuria confers the strongest risk of CVD mortality. Reduction of microalbuminuria may reflect the restoration of endothelial function and adequate control of the underlying pathogenic process; therefore, the therapeutic goal should be to reduce overall CVD risk, including CKD.

Clinical studies have found associations between microalbuminuria and central obesity, insulin resistance, glucose intolerance, dyslipidemia, and HTN, as well as LVH [23, 60–62]. Certain interventions, including improving insulin sensitivity, controlling hyperglycemia or dyslipidemia, reducing BP, and reducing weight have been shown to decrease microalbuminuria [35, 36, 61]. Reduction of microalbuminuria correlated significantly with weight loss [61]. A study of insulin resistance in atherosclerosis found that increasing insulin sensitivity is associated with decreasing microalbuminuria [60]. Statins improve endothelial function and reduce microalbuminuria, thereby providing beneficial effects beyond lipid control [63]. ACE inhibitors and angiotensin type 1 receptor blockers (ARBs) also have an insulin-sensitizing effect and can prevent the development of diabetes in addition to supplying BP control and cardiovascular and renal protective benefits [64, 65].

Weight Loss

Reisin et al. and others have documented the benefits of weight loss [66, 67]. They showed reduction in blood pressure despite normal salt intake. Even modest weight reduction (5–10 kg) was effective in 75% of the subjects. In addition to blood pressure control, weight loss induces reduction in insulin levels, reduces sympathetic activity, and possibly reduces renin, aldosterone levels, and intracellular sodium levels. Cardiac benefits of weight loss are manifest in the decrease in interventricular septal thickness, posterior wall thickness, and total left ventricular mass.

Sinaiko et al. [68] reviewed the diets of children and observed that increasing fiber increased insulin sensitivity with increasing grain intake and lowered fasting insulin level. Eating the same number of calories as fat can result in more efficient energy storage and more weight gain. One should note, however, that although the historical 38% fat content in the US diet has decreased to 32%, rates of obesity have increased. There are a number of studies of weight loss on Atkin's (high fat, low carbs) diet, one of which has implicated a high protein intake in excessive weight gain and insulin sensitivity [69]. A diet rich in fiber and low in fat is generally recommended, along with increased physical activity.

Despite increasing societal awareness of the ill effects of obesity, incidence of obesity is rising because weight loss is not an easy goal to attain. Most weight-loss programs report dropout rates of 50–70%, and even fewer subjects are able to sustain the weight loss. A study of Finnish children demonstrated modest results in girls despite early intervention. Furthermore, there was no benefit to the boys [70].

Research suggests that prenatal characteristics, particularly race, ethnicity, maternal smoking during pregnancy, and maternal prepregnancy obesity, exert influence on the child's weight states through an early tendency toward overweight, which then is perpetuated as the child ages. These findings provide additional clues to the genesis of childhood overweight and suggest that overweight prevention may need to begin in early childhood and perhaps before birth of the baby [71].

The Effect of ACE-I and ARBs on the Kidneys

In a previous study with obese Zucker rats treated with the ACE-I quinapril, kidney protection with inhibition of the immunohistochemical expression of alpha smooth muscle actin in the glomerulus and intertitium has been shown [72]. In other animal models, ACE-I reduced glomerular capillary HTN and slowed the progression of renal disease [73].

In human studies, ACE-I has been shown to reduce markers of vascular oxidative stress, a benefit that may attenuate the progression of the cardiovascular and renal alterations described in subjects with the metabolic syndrome [74]. Also, the treatment of humans with some ARBs may induce peroxisome proliferator-activated receptors (PPAR)-γ activation that may be protective in the obese subjects through the insulin-sensitizing anti-diabetic effect [75].

In the first large, prospective, multicenter, double-blind trial performed in obese, hypertensive patients, the therapeutic effect of hydrochlorothiazide was compared with lisinopril. This study has shown that more than half of the patients controlled with lisinopril only need a low dose (10 mg/day), whereas 46% of the patients controlled with hydrochlorothiazide required a high dose (50 mg/day). Consequently, the authors suggest that the use of ACE-I in obese hypertensives may generate a more rapid rate of response in the blood-pressure control with fewer side effects [76]. Other authors have shown that in patients with T2DM, ARBs improve kidney and patient survival [77, 78]. Because ACE-I and ARBs control blood pressure and induce hemodynamic, metabolic, and anti-inflammatory benefits, some investigators have proposed that these anti-hypertensives should be the first line of therapy for patients with obesity and/or the metabolic syndrome [35, 79].

Peroxisome Proliferator-Activated Receptor Agonists and Renal Protection

PPAR are nuclear hormone-activated receptors and transcription factors. Cumulating data demonstrate the importance of PPAR in the pathogenesis of metabolic syndrome, including adipogenesis, dyslipidemia, insulin sensitivity, obesity, inflammation, atherosclerosis, HTN, and microalbuminuria [80]. Three isoforms, PPAR-α, PPAR-β/$\varnothing$, and PPAR-γ, have been cloned. Their ligands or agonists have been considered as potential therapeutic agents for preventing and treating metabolic syndrome and its related renal damage [80–83]. In the presence of their ligands, these isoforms form heterodimers with retinoid X receptor-α, which bind to PPAR-response element in the promoter regions of PPAR-driven genes and regulate these gene transcriptions [80–82]. Therapeutic implications of different classes of PPAR are discussed below.

The fibric acid derivative class of lipid-lowering drugs, such as fenofibrate, gemfibrozil, and clofibrate, are PPAR-α agonists. Since both PPAR-α and PPAR-γ are expressed in renal proximal tubule and mesangial cells [80], it is not surprising that a study of diabetic animals found that fenofibrate inhibits TGF-β1 and TGF-β receptor expression and reduces collagen deposition in glomeruli [83]. Yet other studies have suggested a beneficial effect of fibrate therapy on diabetic nephropathy [84].

The thiazolidinedione class of PPAR-γ agonists, such as rosiglitazone, troglitazone, and pioglitazone, has been used for treating insulin resistance and type 2 diabetes. There are more animal and clinical data supporting the renal protective effects of this class. In both type 1 and type 2 diabetic animal models, PPAR-γ agonist troglitazone significantly inhibited TGF-β expression and renal matrix protein, ameliorated mesangial expansion, and reduced glomerular hyperfiltration and urinary protein excretion [85, 86]. Recall that both PPAR-α and PPAR-γ are expressed in renal proximal tubule and mesangial cells. Rosiglitazone was shown to improve GFR and reduce glomerulosclerosis, tubulointerstitial fibrosis and albuminuria [87]. Clinical studies also found that PPAR-γ agonists can significantly decrease microalbuminuria in patients with type 2 diabetes [88]. Compared with other oral hypoglycemic agents (including metformin, glyburide, glibenclamide), PPAR-γ agonists (rosiglitazone, troglitazone, and pioglitazone) achieve similar glucose control but provide superior renal protection [89, 90, 91]. Cultured mesangial and proximal tubular cell studies have further demonstrated that PPAR-γ agonists have anti-proliferative, anti-fibrotic, and anti-inflammatory effects [92, 93]. Therefore, PPAR-γ agonists may be beneficial in slowing the progression of glomerulosclerosis and tubulointerstitial fibrosis.

Interestingly, recent animal and cell model studies noted that the ARB telmisartan can function as a partial agonist of PPAR-γ, regulate the expression of PPAR-γ target genes, and reduce glucose, insulin, and TG levels when tested at the low plasma concentrations typically achieved with conventional oral doses [75, 94]. This may explain the anti-diabetic effect of ARB that has been reported in clinical trials.

PPAR-β/ø is also expressed widely along the nephron, but its role there remains to be understood. There are data suggesting that PPAR-β/ø may serve as a 'survival factor' for medullary interstitial cells in the hypertonic condition of renal medulla [95, 59].

There has been great interest in finding newer agents that have dual- or pan-PPAR activating properties since each of the PPAR isoforms seems to play a distinct and important role in the pathogenesis of metabolic syndrome and kidney damage. For example, Tesaglitazar was recently found to be a novel dual PPAR-α and PPAR-γ agonist [96].

In summary, the pharmacologic intervention should include medications that are beneficial for both retarding/preventing kidney damage and preventing metabolic syndrome. ACEIs and ARBs fit this role nicely; in addition to providing cardiovascular and renal benefits, ACE inhibitors and ARBs can increase insulin sensitivity, prevent the onset of diabetes, and ameliorate microalbuminuria [35, 64, 65]. The use of fibrates and statins in the treatment of hyperlipidemia (PPAR antagonism and anti-inflammatory effect) and thiazolidinediones in the management of T2DM is encouraged. Additionally, Metformin has been shown to reduce the development of diabetes by 31% in prediabetic individuals [97]. New, safe, effective, and inexpensive drugs will be discovered to combat obesity and protect the kidneys. In the meantime, the medical community should attempt to slow the epidemic down by educating the public about the ill effects of increasing weight. A concerted effort must be made to change the overfeeding and the sedentary lifestyle that are rampant in this society.

Conclusion

The evidence reviewed suggests that the hyper-hemodynamic status of obesity, impaired pressure natriuresis from macula–densa compensatory reflex, excess excretory load, insulin resistance, increased FFA, chronic inflammatory and prothrombotic status, and endothelial dysfunction all initiate renal injury in early obesity, both individually and interdependently. Late, overt diabetes, HTN, and/or dyslipidemia additionally damage injured kidneys. CKD then becomes irreversible and eventually progresses to end-stage renal disease. Future studies are needed to identify the novel therapeutic interventions to prevent renal injury in the early stage of metabolic syndrome.

Current treatment of early obesity and its complications requires a multifactorial and comprehensive approach. Lifestyle changes include diet control, exercise, weight loss, and smoking cessation. Pharmacological intervention includes insulin sensitizers, tight glycemic control, lipid-lowering agents, aggressive BP control (with RAAS blockers, BP goal < 130/80 mm Hg), and anti-inflammatory and anti-thrombotic therapy with aspirin. Developing novel agonists with dual- or pan-PPAR activities may have great promise in treating metabolic syndrome and renal complications.

There is an ongoing epidemic of obesity in the United States and the world. While we're getting better at treating the complications of obesity, e.g., cardiovascular and renal disease, the real challenge is to prevent early kidney injury before the onset of irreversible damage. Multi-faceted approach is needed: It should include focused government policies, education of general practitioners and other health care professionals, education of the general public by these health care professionals.

Acknowledgement

We thank Ms. Michelle Burke for her editorial assistance.

References

1 Abdel-Halim R: Obesity: 1000 years ago. Lancet 2005;366:204.
2 Flegal K, et al: Prevalence and trends in obesity among US adults, 1999–2000. JAMA 2002;288: 1723–1727.
3 American Academy of Pediatrics, Committee on Nutrition: Prevention of pediatric overweight and obesity. Pediatrics 2003;112:424–430.
4 Ogden C, et al: Prevalence and trends in overweight among US children and adolescents, 1999–2000. JAMA 2002;288:1728–1732.
5 Macedo M, Trigueiros D, de Freitas F: Prevalence of high blood pressure in children and adolescents: Influence of obesity. Rev Port Cardiol 1997;16:27–28.

6   Freedman D, et al: The relation of overweight to cardiovascular risk factors among children and adolescents: the Bogalusa Heart Study. Pediatrics 1999;103:1175–1182.
7   Sorof J, et al: Overweight, ethnicity, and the prevalence of hypertension in school-aged children. Pediatrics 2004;113:475–482.
8   Kortelainen M: Adiposity, cardiac size and precursors of coronary atherosclerosis in children: a prospective study of 210 violent deaths. Int J Obes Relat Metab Disord 1997;21:691–697.
9   American Diabetes Association: Type 2 diabetes in children and adolescents. Diabetes Care 2000;23:381–389.
10 Fagot-Campagna A, et al: Type 2 diabetes among North American children and adolescents: an epidemiologic review and a public health perspective. J Pediatr 2000;136:664–672.
11 Holl R, et al: Prevalence and clinical characteristics of patients with non-type 1 diabetes in the pediatric age range: analysis of a multicenter database including 20,410 patients from 148 center in Germany and Austria. Diabetologia 2003;46(suppl 2):A26.
12 Ehtisham S, et al: Prevalence of type 2 diabetes in children in Birmingham. BMJ 2001;322:1428.
13 Wei J, et al: National surveillance for type 2 diabetes mellitus in Taiwanese children. JAMA 2003;290:1345–1350.
14 Braun B, et al: Diabetes Care 1996;19:472–479.
15 Henegar J, et al: Functional and structural changes in the kidney in the early stages of obesity. J Am Soc Nephrol 2001;12:1211–1217.
16 Gelber R, et al: Association between body mass index and CKD in apparently healthy men. Am J Kidney Dis 2005;46:871–880.
17 Coresh J, et al: Prevalence of chronic kidney disease and decreased kidney function in the adult US population: Third national health and nutrition examination survey. Am J Kidney Dis 2003; 41:1–12.
18 Hoehner C, et al: Association of the insulin resistance syndrome and microalbuminuria among nondiabetic native Americans. The Inter-Tribal Heart Project. J Am Soc Nephrol 2000;13:1626–1634.
19 Chen J, et al: The metabolic syndrome and chronic kidney disease in US adults. Ann Intern Med 2004;140:167–174.
20 Palaniappan L, Carnethon M, Fortmann S: Association between microalbuminuria and the metabolic syndrome: NHANES III. Am J Hypertens 2003;16:952–958.
21 Chen J, et al: Insulin resistance and risk of chronic kidney disease in nondiabetic US adults. J Am Soc Nephrol 2003;14:469–477.
22 Fox C, et al: Predictors of new-onset kidney disease in a community-based population. JAMA 2004;291:844–850.
23 Zhang R, Reisin E: Obesity-hypertension: The effects on cardiovascular and renal systems. J Hypertens 2000;13:1308–1314.
24 Verani R: Obesity-associated focal segmental glomerulosclerosis: pathological features of the lesion and relationship with cardiomegaly and hyperlipidemia. Am J Kidney Dis 1992;20:629–634.
25 Kambham N, et al: Obesity-related glomerulopathy: an emerging epidemic. Kidney Int 2001; 59:1498–1509.
26 USRDSU: 15th Annual Data Report: Atlas of End-Stage Renal Disease in the United States. Bethesda, MD, National Institutes of Health, National Institute of Diabetes and Digestive and Kidney Diseases, 2003.
27 Hall J: Renal and cardiovascular mechanisms of hypertension in obesity. Hypertension 1994;23:381–394.
28 Hall J, et al: Obesity-induced hypertension: Renal function and systemic hemodynamics. Hypertension 1993;22:292–299.
29 Reisin E, et al: Renal hemodynamic studies in obesity hypertension. J Hypertens 1987;5:397–400.
30 Ribstein J, Cailar G, Mirman A: Combined renal effects of overweight and hypertension. Hypertension 1995;26:610–615.
31 Nyengaard J, Bendtsen T: Glomerular number and size in relation to age, kidney weight, and body surface area in normal man. Anat Rec 1992;232:194–201.
32 Chagnac A, et al: Glomerular hemodynamics in severe obesity. Am J Physiol Renal Phsiol 2000;278:f817–f822.

33 Reisin E: Obesity and the kidney connection. Am J Kidney Dis 2001;38:1129–1134.
34 Arnold M, et al: Obesity associated renal medullary changes. Lab Invest 1994;70:156A.
35 Zhang R, Thakur V, Reisin E: Renal and cardiovascular considerations for the nonpharmacological and pharmacological therapies of obesity-hypertension. J Human Hypertens 2002;16:819–827.
36 Morales E, et al: Beneficial effects of weight loss in overweight patients with chronic proteinuric nephropathies. Am J Kidney Dis 2003;41:319–327.
37 Praga M, et al: Influence of obesity on the appearance of proteinuria and renal insufficiency after unilateral nephrectomy. Kidney Int 2000;58:2111–2118.
38 Lackland D, et al: Low birth weights contribute to high rates of early-onset chronic renal failure in the southeastern United States. Arch Intern Med 2000;160:1472–1476.
39 Bagby S: Obesity-initiated metabolic syndrome and the kidney: a recipe for chronic kidney disease? J Am Soc Nephrol 2004;15:2775–2791.
40 Osmond C, Barker D: Fetal, infant, and childhood growth are predictors of coronary heart disease, diabetes, and hypertension in adult men and women. Environ Health Perspect 2000;108(suppl 3): 545–553.
41 Hall J: Mechanism of abnormal renal sodium handing in obesity hypertension. Am J Hypertens 1997;10:49S–55S.
42 Kojima M, et al: Ghrelin is a growth-hormone-releasing acylated peptide from stomach. Nature 1999;402:656–660.
43 Wortley KE, et al: Absence of ghrelin protects against early-onset obesity. J Clin Invest 2005;115:3573–3578.
44 Sun Y, Ahmed S, Smith RG: Deletion of ghrelin impairs neither growth nor appetite. Mol Cell Biol 2003;23:7973–7981.
45 Sun Y, et al: Ghrelin stimulation of growth hormone release and appetite is mediated through the growth hormone secretagogue receptor. Proc Natl Acad Sci USA 2004;101:4679–4684.
46 Zigman J, et al: Mice lacking ghrelin receptors resist the development of diet-induced obesity. J Clin Invest 2005;115:3564–3572.
47 Schwartz M, et al: Identification of targets of leptin action in rat hypothalamus. J Clin Invest 1996;98:1101–1106.
48 Wisse B: The inflammatory syndrome: the role of adipose tissue cytokines in metabolic disorders linked to obesity. J Am Soc Nephrol 2004;15:2792–2800.
49 Wolf G, et al: Leptin and renal disease. Am J Kidney Dis, 2002;39:1–11.
50 Havel P: Control of energy homeostasis and insulin action by adipocyte hormones: leptin, acylation stimulating protein, and adiponectin. Curr Opin Lipidol 2002;13:51–59.
51 Mulyadi L, et al: Body fat distribution and total body fat as risk factors for microalbuminuria in the obese. Ann Nutr Metab 2001;45:67–71.
52 Steppan C, et al: The hormone resistin links obesity to diabetes. Nature 2001;409:307–312.
53 Kamijo A, et al: Urinary free fatty acids bound to albumin aggravate tubulointerstitial damage. Kidney Int 2002;62:1628–1637.
54 Unger R: Minireview: weapons of lean body mass destruction: the role of ectopic lipids in the metabolic syndrome. Endocrinology 2003;144:5159–5165.
55 The catastrophic failure of public health [editorial]. Lancet 2004;363:745.
56 Corbett S: A Ministry for the Public's Health: an imperative for disease prevention in the 21st century? Med J Aust 2005;183:254–257.
57 Batey L, et al: Summary measures of the insulin resistance syndrome are adverse among Mexican-American versus non-Hispanic white children: the Corpus Christi Child Heart Study. Circulation 1997;96:4319–4325.
58 Reaven P, et al: Cardiovascular disease insulin risk in Mexican-American and Anglo-American children and mothers. Pediatrics 1998;101:E12.
59 Kwon H, et al: Prevalence and clinical characteristics of the metabolic syndrome in middle-aged Korean adults. Korean J Intern Med 2005;20:310–316.
60 Mykkanen L, et al: Microalbuminuria is associated with insulin resistance in nondiabetic subjects: the insulin resistance atherosclerosis study. Diabetes 1998;47:793–800.
61 Castro J, et al: Cardiometabolic syndrome: pathophysiology and treatment. Curr Hypertens Rep 2003;5:393–401.

62   Mangrum A, Bakris G: Predictors of renal and cardiovascular mortality in patients with non-insulin-dependent diabetes: a brief overview of microalbuminuria and insulin resistance. J Diabetes Complications 1997;11:352–357.
63   McFarlane S, Banerji M, Sowers J: Insulin resistance and cardiovascular disease. J Clin Endocrinol Metab 2001;86:713–718.
64   Yusuf S, et al: Effects of an angiotensin-converting enzyme inhibitor, ramipril, on cardiovascular events in high-risk patients. The Heart Outcomes Prevention Evaluation Study Investigators. N Engl J Med 2000;342:145–153.
65   Lindholm L, et al: Risk of new-onset diabetes in the losartan intervention for endpoint reduction in hypertension study. J Hypertens 2002;20:1879–1886.
66   Langsford H, et al: Dietary therapy slows the return of hypertension after stopping prolonged medication. JAMA 1985;253:657–669.
67   Reisin E: Obesity hypertension: nonpharmacologic and pharmacologic therapeutic modalities; in Laragh JH, Brenner B (eds): Hypertension: Pathophysiology, Diagnosis, and Management New York, Raven Press, 1995, 2683–2691.
68   Sinaiko A, et al: Insulin resistance syndrome in childhood: associations of the euglycemic insulin clamp and fasting insulin with fatness and other risk factors. J Pediatr 2001;139:700–707.
69   Gunther A, Buyken A, Kroke A: The influence of habitual protein intake in early childhood on BMI and age at adiposity rebound: results from the DONALD Study. Int J Obes (Lond) 2006 [Epub ahead of print].
70   Hakanen M, et al: Development of overweight in an atherosclerosis prevention trial starting in early childhood. The STRIP study. Int J Obes (Lond) 2006;30:618–626.
71   Oken E, et al: Associations of maternal prenatal smoking with child adiposity and blood pressure. Obes Res 2005;13:2021–2028.
72   Richards R, et al: Effects of dehydroepiandrosterone and quinapril on nephropathy in obese Zucker rats. Kidney Int 2001;59:37–43.
73   Zatz R, et al: Prevention of diabetic glomerulopathy by pharmacological amelioration of glomerular capillary hypertension. J Clin Invest 1986;77:1925–1930.
74   Khan B, et al: Quinapril, an ACE inhibitor, reduces markers of oxidative stress in the metabolic syndrome. Diabetes Care 2004;27:1712–1715.
75   Schupp M, et al: Angiotensin type 1 receptor blockers induce peroxisome proliferator-activated receptor-alpha activity. Circulation 2004;109:2054–2057.
76   Reisin E, et al: Lisinopril versus hydrochlorothiazide in obese hypertensive patients: a multicenter placebo-controlled trial. Treatment in Obese Patients With Hypertension (TROPHY) Study Group. Hypertension 1997;30:140–145.
77   Lewis E, et al: Collaborative Study Group: renoprotective effect of the angiotensin-receptor antagonist irbesartan in patients with nephropathy due to type 2 diabetes. N Engl J Med 2001;345:851–860.
78   Pohl M, et al: Independent and additive impact of blood pressure control and angiotensin II receptor blockade on renal outcomes in the irbesartan diabetic nephropathy trial: clinical implications and limitations. J Am Soc Nephrol 2005;16:3027–3037.
79   Julius S, Majahalme S, Palatini P: Antihypertensive treatment of patients with diabetes and hypertension. Am J Hypertens 2001;14(pt 2):310S–316S.
80   Guan Y: PPAR family and its relationship to renal complications of the metabolic syndrome. J Am Soc Nephrol 2004;15:2801–2815.
81   Guan Y, Breyer, MD: Peroxisome proliferator-activated receptors (PPARs): novel therapeutic targets in renal disease. Kidney Int 2001;60:14–30.
82   Guan Y, Breyer M: Targeting peroxisome proliferator-activated receptors (PPARs) in kidney and urologic disease. Minerva Urol Nephrol 2002;54:65–79.
83   Park C, et al: A PPAR alpha agonist improves diabetic nephropathy in db/db mice. Am Soc Nephrol 2003;14:393A.
84   Fried L, Orchard T, Kasiske B: Effect of lipid reduction on the progression of renal disease: a meta-analysis. Kidney Int 2001;59:260–269.
85   Isshiki K, et al: Thiazolidinedione compounds ameliorate glomerular dysfunction independent of their insulin-sensitizing action in diabetic rats. Diabetes 2000;49:1022–1032.

86   McCarthy K, et al: Troglitazone halts diabetic glomerulosclerosis by blockade of mesangial expansion. Kidney Int 2000;58:2341–2350.
87   Baylis C, et al: Peroxisome proliferator-activated receptor γ agonist provides superior renal protection versus angiotension-converting enzyme inhibition in a rat model of type 2 diabetes with obesity. J Pharmacol Exp Ther 2003;307:854–860.
88   Bakris G, et al: Rosiglitazone reduces urinary albumin excretion in type II diabetes. J Hum Hypertens 2003;17:7–12.
89   Nakamura T, et al: Comparative effects of pioglitazone, glibenclamide, and voglibose on urinary endothelin-1 and albumin excretion in diabetes patients. J Diabetes Complications 2000;14:250–254.
90   Wolffenbuttel B, et al: Addition of low-dose rosiglitazone to sulphonylurea therapy improves glycaemic control in type 2 diabetic patients. Diabet Med 2000;17:40–47.
91   Imano E, et al: Effect of troglitazone on microalbuminuria in patients with incipient diabetic nephropathy. Diabetes Care 1998;21:2135–2139.
92   Guo B, et al: Peroxisome proliferator-activated receptor-γ ligands inhibit TGF-β1-induced fibronectin expression in glomerular mesangial cells. Diabetes 2004;53:200–208.
93   Chana R, Lewington A, Brunskill N: Differential effects of peroxisome proliferator activated receptor-gamma (PPAR gamma) ligands in proximal tubular cell: thiazolidinediones are partial PPAR gamma agonist. Kidney Int 2004;65:2081–2090.
94   Benson S, et al: Identification of telmisartan as a unique angiotensin 2 receptor antagonist with selective PPAR-modulating activity. Hypertension 2004;43:993–1002.
95   Hao C, et al: Peroxisome proliferator-activated receptor delta activation promotes cell survival following hypertonic stress. J Biol Chem 2002;277:21341–21345.
96   Hegarty B, et al: PPAR activation induces tissue-specific effects on fatty acid uptake and metabolism in vivo – a study using the novel PPAR (alpha)/(gamma) agonist Tesaglitazar. Endocrinology 2004;145:3158–3164.
97   Knowler W, et al: Reduction in the incidence of type 2 diabetes with lifestyle intervention or metformin. N Engl J Med 2002;346:393–403.
98   Zhang R, et al: Kidney disease and the metabolic syndrome. Am J Med Sci 2005;330:319–325.

Efrain Reisin, MD
Professor of Medicine and Chief of Section of Nephrology
1542 Tulane Avenue, Room 354
New Orleans, LA 70112 (USA)
E-Mail ereisi@lsuhsc.edu

Leptin as a Proinflammatory Cytokine

Graham M. Lord

Department of Nephrology and Transplantation, 5th Floor Thomas Guy House, King's College London, Guy's Hospital, London, UK

Abstract

Leptin is a 16-kDa protein produced mainly by adipocytes. Animal models demonstrate that leptin is required for control of bodyweight and reproduction, since mice defective in leptin or the leptin receptor are obese, hyperphagic insulin resistant and infertile. Our initial series of observations lead us to propose that leptin also had significant effects on human type I proinflammatory immune responses. In support of this hypothesis, leptin deficient mice are resistant to a wide range of autoimmune diseases and display features of immune deficiency. Subsequent work has confirmed that leptin has a pleiotrophic role on the immune response and can rightly be considered, both structurally and functionally, as a proinflammatory cytokine.

Copyright © 2006 S. Karger AG, Basel

Introduction

The field of leptin and the immune system has expanded rapidly since the first paper described the interrelationship of this cytokine, initially thought of as simply a weight reducing hormone, with the optimal functioning of the immune system in vitro and in vivo [1]. This review starts with an introduction to the biology of leptin in order to illustrate how it compares with other cytokine systems. This section also highlights what may be the main physiological role of leptin; namely that it acts as a signal of failing energy reserves. A brief discussion of the in vivo models of leptin deficiency follows, which leads to a thorough review of the data which suggest that leptin has marked and specific effects on the proinflammatory immune response. The consequences of leptin acting as a proinflammatory cytokine in the context of insulin resistance and obesity are clearly of importance in understanding the growing mortality associated with obesity related pathophysiology.

The Biology of the Action of Leptin

The Obese Gene and Its Protein Product, Leptin

In 1994, the mouse *obese* (*ob*) gene and its human homologue were isolated by positional cloning [2]. In humans, the gene is positioned at 7q31. The mouse *ob* gene resides on chromosome 6 and consists of 3 exons and 2 introns and encodes a 4.5 kilobase (kb) mRNA. The coding sequence is contained within exons 2 and 3. The wildtype *ob* gene encodes a 167 amino acid protein named leptin, expressed almost exclusively in white adipose tissue, although recently its expression has been found at lower levels in the placenta, stomach brain and T cells [3–6]. The homozygous mutant mouse *ob/ob* has a mutation in the *ob* gene, either due to a 5 kb ETn transposon inserted into the first intron of *ob* (*ob²ʲ* mice) or due to a non-sense mutation causing protein truncation [7]. Leptin (*leptos* (Greek): thin) is a 16 kDa class I cytokine, with structural similarity to Interleukin (IL)-2, the paradigm T cell growth factor, although there is no sequence similarity [8]. It consists of 167 amino acids and is a helical cytokine, belonging to the family of haemopoietic cytokines, which includes IL-2, IL-3, IL-4 and GM-CSF. Leptin circulates in the bloodstream bound to plasma proteins and soluble leptin receptor (ObRe) [9]. The ratio of free:bound leptin in serum increases with increasing obesity [10]. Its serum levels are proportional to body fat mass [11] and are dynamically regulated, being increased by inflammatory mediators such as TNF-α, IL-1 and LPS [12–14] and being rapidly reduced by starvation [15, 16]. The release of leptin is pulsatile [17] and is inversely related to ACTH and cortisol secretion [18]. The fact that obesity is associated with high leptin levels rather than vice versa, as would be predicted by the obese phenotype of the *ob/ob* mouse, has lead to the hypothesis that appetite suppression is not the main physiological role of leptin [19]. It is excreted via the kidneys and is therefore elevated in patients with end-stage renal failure [20]. The regulation of leptin gene expression is highly complex as it involves multiple mediators whose relative importance is, as yet, undetermined [21]. A high serum leptin level reduces leptin gene expression and changes in nutrient availability result in rapid alterations of gene expression [22]. Other important regulatory factors are glucocorticoids, insulin and thyroid hormones. Corticosteroids enhance gene expression and protein secretion, whereas insulin causes release of leptin from an intracellular pool [23]. Thyroid hormones inhibit leptin gene expression [24] and sex steroids such as oestrogen increase leptin mRNA levels [25]. This latter fact may underlie the sexual dimorphism observed in serum leptin concentration, such that for a given body mass index, females have significantly higher leptin levels than males [20, 26].

Role of Leptin in the Regulation of Body Weight

Chronic administration of recombinant *ob* protein has been shown to produce a significant reduction in body weight in *ob/ob* and normal mice due to a reduction in food intake but also an increase in energy expenditure [27–29]. Centrally administered leptin (into the lateral or third cerebral ventricles) has been shown to be particularly effective in promoting anorexia and weight loss at doses which when administered peripherally were without effect on feeding behaviour [27]. This suggests that leptin acts on receptors within the central nervous system, probably at the level of the hypothalamus and clearly implicate leptin as an important factor in the regulation of body weight in rodents [30]. Further evidence that implicates leptin as a regulator of body weight is the fact that total deficiency of leptin in mouse and man causes obesity [30, 31], which is reversed by leptin treatment [28, 31, 32]. However, as mentioned above, leptin levels are high in rodent and human models of obesity [11, 33] leading to the hypothesis that either there is resistance to the actions of leptin akin to insulin resistance in type 2 diabetes mellitus [34], or that the main function of leptin is a signal of starvation and not body weight regulation [15]. There is some experimental evidence for the occurrence of leptin resistance since leptin induces the expression of SOCS3 (suppressor of cytokine signalling), which subsequently prevents that cell from responding to further leptin [35, 36]. In vivo proof of this concept has recently been provided by the finding that mice with haploinsufficiency of the *SOCS3* gene are more leptin sensitive and resistant to diet induced obesity [37].

Role of Leptin as a Signal of Starvation

In 1996, Flier proposed a role for leptin as a signal of energy deficiency [15]. Circulating leptin levels fall rapidly in response to starvation when energy intake is limited and energy stores (fat) are declining. It was suggested that leptin may have evolved to signal, the shift between sufficient and insufficient energy stores [19]. The hypothesis that reduced circulating leptin levels signal nutrient deprivation is supported by the demonstration that prevention of the starvation-induced fall in plasma leptin levels by exogenous replacement is able to prevent the starvation-induced delay in ovulation in female mice. In addition, such a regime of leptin replacement was also shown partially to prevent the fasting-induced rise in plasma corticosterone levels and the fall in plasma thyroid hormone (total T4) levels [15]. It is likely that leptin exerts these effects at the level of the central nervous system [19].

Peripheral Effects of Leptin

In addition to the centrally mediated effects of leptin, many peripheral effects have been reported. Leptin receptor expression has been demonstrated in the pancreas [38, 39], haemopoietic stem cells [40], endothelial cells [41] and reproductive organs [40], where leptin has been shown to have direct effects in vitro. Thus, leptin can influence ovarian hormone synthesis, can induce proliferation and differentiation of bone marrow cells [42] and can act as an angiogenic factor by its action on endothelial cells [41]. Leptin can also affect insulin secretion from pancreatic β-cells by inhibiting glucose stimulated insulin release both in vitro and in vivo [38, 39]. Metabolic effects of leptin are partially peripheral and partly central, via the sympathetic nervous system [43]. Leptin directly promotes glucose uptake and glycogen synthesis in skeletal muscle [44, 45] and promotes lipolysis with no increase in free fatty acids [46]. The relative importance of these central and peripheral effects of leptin still remains to be determined. It will be difficult to dissect observed responses completely, however, because many effects are likely to be both peripherally and centrally mediated [21].

Leptin Receptor Genetics and Structure

The receptor for leptin was identified by expression cloning in 1995. Sequencing of the original mouse cDNA demonstrated that it coded for a single membrane spanning protein of the class I cytokine receptor family residing on chromosome 4 (human – 1p31). It consists of 20 exons, of which the first 2 are non-coding [47]. This family includes gp130, G-CSF, IL-6 and leukaemia inhibitory factor receptors among its members and is consistent with predictions that leptin itself was a helical cytokine structurally homologous to growth hormone and IL-2. There are several isoforms of the leptin receptor (ObRa–e) in both mice and humans arising from alternative RNA splicing at the most C-terminal coding exon of the leptin receptor gene. All isoforms have a common extracellular domain but differ in the length of their intracytoplasmic domain [47–49]. One isoform (ObRe) is predicted to be a soluble receptor as it lacks the transmembrane domain [49]. The 'long isoform' of the leptin receptor (ObRb) has a long intracellular domain of about 303 amino acids and contains sequence motifs enabling it to employ the JAK-STAT pathway for signal transduction [50, 51]. Mice and humans that have mutant receptors with impaired or no signal transducing capacity have a phenotype of obesity and insulin resistance identical to that of *ob* gene mutations [48, 52].

Distribution of the Leptin Receptor

The short leptin receptor isoforms appear to be ubiquitously expressed, although their function remains to be established [51]. It has been suggested that the high levels of ObRa observed in the choroid plexus may play a role in transporting leptin from the bloodstream into the cerebrospinal fluid from where it can access centres in the hypothalamus involved in energy homeostasis [53]. The long isoform of the leptin receptor, ObRb, is expressed at high levels in the hypothalamic nuclei known to be of importance in the regulation of body weight [54, 55]. However, ObRb mRNA expression has also been detected in rodent peripheral tissues such as pancreatic beta cells [39] and lymph nodes [29, 51] suggesting that leptin may have a broader physiological role. We and others have shown that ObRb is expressed in T cells [1, 56] and there is good evidence to suggest that it is also expressed by monocytes/macrophages [54, 57].

In vivo Models of Leptin Deficiency

The ob/ob and db/db Mouse

Recessive mutations in the mouse *diabetes* (*db*) and *obese* (*ob*) genes have long been recognised to cause a syndrome of obesity and diabetes resembling morbid obesity in humans [33, 58]. These mutant strains of mice are phenotypically identical, each weighing three times more than normal mice with a 5-fold increase in body fat content when compared to its wild-type litter mate control. Many years ago, Coleman and Hummel [59, 60] showed that the obesity in the *ob/ob* mouse was due to the lack of a circulating satiety factor and that the phenotype of the *db/db* mouse was probably due to a receptor defect for that factor. The *ob/ob* phenotype has since been found to be due to two different mutations in two different strains of mice. The mutation in ob^{2j} mice prevents synthesis of leptin due to a mutation that results in the insertion of a retroviral-like transposon in the first intron of the *ob* gene. This insertion contains several splice acceptor and polyadenylation sites, which leads to the production of chimeric RNAs in which the first exon is spliced to sequences in the transposon, thus preventing the production of mature mRNA [7]. In the C57BL/6 *ob/ob* mutant, a non-sense mutation results in the production of a truncated inactive leptin [2].

The *db/db* mouse has been shown to have a missense mutation (G–T transversion) within an exon encoding the extreme C-terminus and 3′ untranslated region of ObRa, the predominant short isoform. This generates a new splice donor site that results in the truncation of the intracellular domain of what would have been the long isoform splice variant of the leptin receptor and instead generates a transcript encoding for a protein that is identical to the major short isoform, ObRa [49]. The demonstration that the defect in the *db/db* mouse is in the leptin receptor gene confirmed the importance of the ObRb receptor in body weight regulation [48]. In vivo, the *db/db* mouse is resistant to the actions of exogenous leptin administration [28, 29].

All of these mice strains are obese and have insulin resistance, which leads to frank diabetes later in life. The mice also exhibit the development of raised circulating corticosterone, which is partly responsible for the insulin resistance [58]. Male and female mice are sterile due a variety of reproductive disturbances.

This sterility can be reversed in *ob/ob* mice with recombinant exogenous leptin but is, of course, without effect on the *db/db* mouse [61]. An additional phenotypic feature of these mice is that they have impaired immunity, as will be discussed below.

Immunological Abnormalities of the ob/ob *and* db/db *Mouse*

Prior to the discovery of leptin, impaired cellular immune function had been noted in both *ob/ob* and *db/db* mice [62–66]. Leptin deficient *ob/ob* mice and receptor defective *db/db* mice had been found to exhibit defective cell-mediated immunity and lymphoid atrophy analogous to that observed in chronic human undernutrition where leptin levels are low. In one study, skin graft rejection from a fully allogeneic mouse strain was delayed when grafted onto *db/db* mice compared with wild-type controls [63]. These mutant mouse strains also show increased susceptibility to pathogens, most notably to coxsackie virus. In one study, infection of *db/db* mice with coxsackie virus caused 100% mortality as opposed to less than 10% mortality in the wild-type control group [66]. Of note, there seemed to be a discrepancy between the in vivo and in vitro results in that cell-mediated immune responses were significantly impaired in vivo but in vitro T cell responses were less affected [63]. Further defects were found in thymic and splenic cellularity [62, 63]. All of these defects were considered to be the result of some unidentified aspect of obesity [62].

The Effects of Leptin on the Immune System

The Innate Immune Response

As mentioned above, the pattern of leptin release during an acute phase response mirrors other cytokine gene expression, particularly IL-6 [12–14]. It has been shown that LPS, IL-1, TNF-α and other inflammatory stimuli increase gene expression and serum concentration of leptin as early as 6 h after the initial stimulus. The induction of gene expression makes leptin an ideal candidate to be a key player in an immune/inflammatory response. It is of particular interest that LPS binds to a Toll-like receptor (TLR-4) on adipocytes and induces adipocyte expression of TLR-2 and secretion of leptin and other proinflammatory cytokines [67]. TLR engagement provides an elegant mechanism, whereby the production of leptin is induced at the start of an immune response before cognate recognition of a foreign antigen has occurred, which will in turn upregulate Th1 cognate immune responses if appropriate. This further illustrates the well-accepted concept of a critical interplay between non-cognate and cognate immune responses, which serves to coordinate the host response to any foreign or 'dangerous' antigen [68]. Leptin itself does not cause induction of other acute phase proteins, which distinguishes it from other members of the IL-6 family [69]. However, it has recently been shown that leptin binds to C-reactive protein, a major acute phase protein providing another mechanism whereby leptin resistance may occur [70]. It was initially thought that the induction of leptin expression during an acute phase response was responsible for the anorexia of infection [12]. However, it has since been shown that LPS-induced anorexia occurs in the absence of leptin [71], indicating that other factors are probably responsible for this phenomenon.

Leptin induces proinflammatory or Th1 type cytokines from a variety of cell types, including T cells. It has been clearly shown that macrophages express the leptin receptor and respond to added exogenous leptin by increasing the synthesis and release of TNF-α, IL-6 and IL-12 [57, 72]. It is important to note that in most studies, the basal synthesis of these cytokines is not affected by leptin. It is only the stimulated production that is enhanced (e.g. after addition of LPS), indicating that leptin does not initiate immune responses per se, but rather influences their outcome. There is some in vitro and in vivo data to suggest that leptin enhances the phagocytic and bactericidal activity of neutrophils and macrophages [73]. One study showed that leptin deficient mice were unable to clear *Escherichia coli* as efficiently as normal mice and that phagocytosis by peritoneal macrophages from these mice was significantly impaired [57]. Other experiments show that in vitro, leptin enhances the phagocytic activity of macrophages against *Leishmania major* and *Candida parapsilopsis* [74]. In contrast with this data, there is evidence that leptin can induce IL-1Ra production from monocytes in vitro, which acts as an anti-inflammatory protein [75].

There is a body of data relating to the responses of leptin deficient *ob/ob* mice to septic shock, which indicates that these mice are more sensitive to LPS or TNF-α induced injury [76, 77]. It has been suggested that *ob/ob* mice have increased numbers of circulating monocytes and that this may underlie their hypersensitivity to septic shock [78]. The cytokine responses in *ob/ob* mice in these models do show marked dysregulation, but as yet there is not a clear mechanism to explain these findings. It may be well, that in the total absence of leptin these mice have a chronic undiagnosed infection that renders them susceptible to LPS induced shock, such as occurs in normal mice primed with *Propionibacterium acnes* [79]. Alternatively, alterations in lymphocyte apoptosis may underlie this phenomenon, since enhanced lymphocyte apoptosis is associated with impaired survival in LPS induced septic shock and leptin has been shown to have anti-apoptotic properties [80]. Whatever the mechanism, it is clear that in the absence of leptin, either congenitally in the *ob/ob* mouse, or acutely during starvation [81], sensitivity to septic shock is markedly increased and that this hypersensitivity is reversed by exogenous leptin treatment [82].

The Cognate Proinflammatory Immune Response

As mentioned above, the innate and cognate immune systems are interlinked, so that many of the effects of leptin discussed in the previous section will have implications for the behaviour of responding B and T lymphocytes. It is also clear that an organism mounts a coordinated immune response to an infectious pathogen that initially comprises innate immunity and then evolves, if appropriate to involve the cognate immune system. An important point of communication between the innate and cognate arms of the immune systems lies with the macrophage and dendritic cell. These cells produce cytokines that polarise and activate T cell responses and act as professional antigen presenting cells by presenting peptide bound to MHC molecules to T cells along with high levels of costimulation. As leptin can effect the production of proinflammatory cytokines from macrophages [57] and dendritic cells [83] and also increase expression of MHC molecules and costimulatory molecules [72], it can be seen how this will affect T cell immune responses. Interestingly, leptin treatment of dendritic cells induces other type 1 cytokines (such as IL-12 and TNF-α) which serves to licence naïve CD4+ T cells for Th1 priming [83].

There are a number of papers regarding responses to specific microbial pathogens in leptin deficient hosts. Studies by Loffreda et al. [57] show that *ob/ob* mice display impaired clearance of *E. coli* and a paper by Webb et al. [66] demonstrates that *db/db* mice are startlingly susceptible to coxsackie virus infection. Recently, leptin deficient mice have been shown to exhibit impaired responses to tuberculosis, which fits in with the clinical observations of a marked impairment of resistance to tuberculosis in malnourished and thus hypoleptinaemic patients [84]. Also, resistance to listeria is profoundly suppressed in the absence of leptin [85].

Ob/ob mice have been shown to be completely resistant to a Con A induced T cell mediated hepatitis. This resistance was abrogated by exogenous leptin replacement. It was also demonstrated that equally obese mice that had their endogenous leptin levels raised by a hypothalamic lesion induced by an injection of gold thioglucose had worse hepatitis that lean wild type mice [78]. This would suggest that the key cytokine in this T cell dependent model was leptin. Similar findings have been reported in a Th1 dependent model of colitis, which shows elevated leptin levels during the induction of colitis that correlate with the severity of the colonic inflammation [86]. T cell expression of the long form of the leptin receptor was shown to be important for the development of colitis [87], although leptin expression itself by T cells is not critical for disease progression [88] as had been suggested in another autoimmune disease model [6].

Probably the best-studied model of T cell mediated autoimmune disease is experimental autoimmune encephalomyelitis (EAE), a murine model of multiple sclerosis in humans. This disease is critically dependent on antigen-specific

CD4+ Th1 cell that produce IFN-γ since it can be adoptively transferred by these cells into a naïve host. Antigen-specific CD4+ T cells that have a Th2 phenotype and produce IL-4 do not cause disease and under some circumstances can be protective. *Ob/ob* mice are totally resistant to induction of EAE, but develop disease to the same extent as wild-type mice when given leptin exogenously [89]. This disease protection is caused by skewing of the immune response towards a Th2 phenotype. When leptin is given to *ob/ob* mice, Th1 CD4+ T cells are generated normally and produce similar amounts of IFN-γ in vitro to wild-type mice. Associated with this is impaired antibody isotype switching. In untreated *ob/ob* mice, all the antigen specific antibody produced is IgG1 (Th2 dependent antibody). In wild type and leptin treated mice, the antigen-specific antibody is mainly IgG2a, which is a Th1 dependent isotype. These findings have been essentially replicated in a murine arthritis model [90]. To extend these findings further, we looked at EAE prone and resistant normal mice strains. Male SJL mice, which have lower endogenous leptin levels, are resistant to EAE, whereas females are susceptible. Exogenous leptin rendered the male mice susceptible to EAE and worsened disease in females [91]. These data may well be informing us of an additional factor in the sexual dimorphism of autoimmune disease [92, 93]. Additional data suggest that serum leptin levels peak before the onset of EAE and that leptin itself is produced by pathogenic T cells [6], although the relevance of this is unclear [88]. Importantly, blocking leptin action also leads to amelioration of disease severity in this model [93].

Findings in vitro support the above in vivo data. Essentially, T cells respond to exogenous leptin by increased proliferation and increased Th1 type cytokine production in the context of the mixed lymphocyte reaction in both humans and mice. This is associated with suppression of Th2 cytokine production [1, 56]. In the case of polyclonal stimulation, memory T cell (CD45RO+) proliferation is inhibited and IFN-γ production is increased and naïve T cell (CD45RA+) proliferation is increased [94]. Therefore in two different in vitro systems, leptin seems to have differential proliferative effects on naïve and memory T cell proliferation, but consistently upregulates Th1 cytokine production. These differential effects have been confirmed using naïve T cells from cord blood as responder cells [1, 94]. Furthermore, leptin has marked effects on CD4+ CD8+ thymocytes, a developmental stage immediately preceding naïve T cells [95].

Conclusions

Initially, leptin was implicated in the regulation of bodyweight and reproductive function [29]. However, recent reports of its effect on the immune system both in vitro and in vivo indicate that it has a potentially important role in the modulation of a wide variety of inflammatory and immune responses. Indeed, it is entirely feasible to classify leptin as a proinflammatory cytokine. We have previously proposed that one of the functions of falling leptin levels is to preserve limited energy supplies, as immune responses use up significant amounts of energy [96]. This evolutionary adaptive response can also be seen to have deleterious consequences if leptin levels are low for a significant length of time. However, manipulation of the leptin axis may provide an opportunity to design novel treatments for inflammatory and immune-mediated diseases, which are a cause of significant morbidity and mortality.

References

1 Lord GM, Matarese G, Howard JK, Baker RJ, Bloom SR, Lechler RI: Leptin modulates the T cell immune response and reverses starvation induced immunosuppression. Nature 1998;394: 897–901.
2 Zhang Y, Proenca R, Maffei M, Barone M, Leopold L, Friedman JM: Positional cloning of the mouse obese gene and its human homologue. Nature 1994;372:425–432.
3 Bado A, Levasseur S, Attoub S, Kermorgant S, Laigneau JP, Bortoluzzi MN, Moizo L, Lehy T, GuerreMillo M, LeMarchandBrustel Y, Lewin MJM: The stomach is a source of leptin. Nature 1998;394:790–793.
4 Masuzaki H, Ogawa Y, Sagawa N, Hosoda K, Matsumoto T, Mise H, Nishimura H, Yoshimasa Y, Tanaka I, Mori T, Nakao K: Nonadipose tissue production of leptin: leptin as a novel placenta-derived hormone in humans. Nat Med 1997;3:1029–1033.
5 Morash B, Li A, Murphy PR, Wilkinson M, Ur E: Leptin gene expression in the brain and pituitary gland. Endocrinology 1999;140:5995–5998.
6 Sanna V, Di Giacomo A, La Cava A, Lechler RI, Fontana S, Zappacosta S, Matarese G: Leptin surge precedes onset of autoimmune encephalomyelitis and correlates with development of pathogenic T cell responses. J Clin Invest 2003;111:241–250.
7 Moon BC, Friedman JM: The molecular basis of the obese mutation in ob2J mice. Genomics 1997;42:152–156.
8 Zhang F, Basinski MB, Beals JM, Briggs SL, Churgay LM, Clawson DK, DiMarchi RD, Furman TC, Hale JE, Hsiung HM, Schoner BE, Smith DP, Zhang XY, Wery JP, Schevitz RW: Crystal structure of the obese protein leptin-E100. Nature 1997;387:206–209.
9 Sinha MK, Opentanova I, Ohannesian JP, Kolaczynski JW, Heiman ML, Hale J, Becker GW, Bowsher RR, Stephens TW, Caro JF: Evidence of free and bound leptin in human circulation. Studies in lean and obese subjects and during short-term fasting. J Clin Invest 1996;98: 1277–1282.
10 Houseknecht KL, Mantzoros CS, Kuliawat R, Hadro E, Flier JS, Kahn BB: Evidence for leptin binding to proteins in serum of rodents and humans: modulation with obesity. Diabetes 1996;45:1638–1643.
11 Considine RV, Sinha MK, Heiman ML, Kriauciunas A, Stephens TW, Nyce MR, Ohannesian JP, Marco CC, McKee LJ, Bauer TL, et al: Serum immunoreactive-leptin concentrations in normal-weight and obese humans. N Engl J Med 1996;334:292–295.
12 Grunfeld C, Zhao C, Fuller J, Pollack A, Moser A, Friedman J, Feingold KR: Endotoxin and cytokines induce expression of leptin, the ob gene product, in hamsters. J Clin Invest 1996;97: 2152–2157.
13 Janik JE, Curti BD, Considine RV, Rager HC, Powers GC, Alvord WG, Smith JW, Gause BL, Kopp WC: Interleukin 1 alpha increases serum leptin concentrations in humans. J Clin Endocrinol Metab 1997;82:3084–3086.

14 Sarraf P, Frederich RC, Turner EM, Ma G, Jaskowiak NT, Rivet DJ, Flier JS, Lowell BB, Fraker DL, Alexander HR: Multiple cytokines and acute inflammation raise mouse leptin levels: potential role in inflammatory anorexia. J Exp Med 1997;185:171–175.

15 Ahima RS, Prabakaran D, Mantzoros C, Qu D, Lowell B, Maratos Flier E, Flier JS: Role of leptin in the neuroendocrine response to fasting. Nature 1996;382:250–252.

16 Boden G, Chen X, Mozzoli M, Ryan I: Effect of fasting on serum leptin in normal human subjects. J Clin Endocrinol Metab 1996;81:3419–3423.

17 Sinha MK, Sturis J, Ohannesian J, Magosin S, Stephens T, Heiman ML, Polonsky KS, Caro JF: Ultradian oscillations of leptin secretion in humans. Biochem Biophys Res Commun 1996;228: 733–738.

18 Licinio J, Mantzoros C, Negrao AB, Cizza G, Wong ML, Bongiorno PB, Chrousos GP, Karp B, Allen C, Flier JS, Gold PW: Human leptin levels are pulsatile and inversely related to pituitary-adrenal function. Nat Med 1997;3:575–579.

19 Flier JS: What's in a name? In search of leptin's physiologic role. J Clin Endocrinol Metab 1998;83:1407–1413.

20 Howard JK, Lord GM, Clutterbuck EJ, Ghatei MA, Pusey CD, Bloom SR: Plasma immunoreactive leptin concentration in end stage renal disease. Clin Sci 1997;93:119–126.

21 Friedman JM, Halaas JL: Leptin and the regulation of body weight in mammals. Nature 1998;395: 763–770.

22 Wang J, Liu R, Liu L, Chowdhury R, Barzilai N, Tan J, Rossetti L: The effect of leptin on Lep expression is tissue-specific and nutritionally regulated. Nat Med 1999;5:895–899.

23 Bradley RL, Cheatham B: Regulation of ob gene expression and leptin secretion by insulin and dexamethasone in rat adipocytes. Diabetes 1999;48:272–278.

24 Kristensen K, Pedersen SB, Langdahl BL, Richelsen B: Regulation of leptin by thyroid hormone in humans: studies in vivo and in vitro. Metabolism 1999;48:1603–1607.

25 Brann DW, De Sevilla L, Zamorano PL, Mahesh VB: Regulation of leptin gene expression and secretion by steroid hormones. Steroids 1999;64:659–663.

26 Saad MF, Damani S, Gingerich RL, Riad Gabriel MG, Khan A, Boyadjian R, Jinagouda SD, el Tawil K, Rude RK, Kamdar V: Sexual dimorphism in plasma leptin concentration. J Clin Endocrinol Metab 1997;82:579–584.

27 Campfield LA, Smith FJ, Guisez Y, Devos R, Burn P: Recombinant mouse OB protein: evidence for a peripheral signal linking adiposity and central neural networks. Science 1995;269:546–549.

28 Halaas JL, Gajiwala KS, Maffei M, Cohen SL, Chait BT, Rabinowitz D, Lallone RL, Burley SK, Friedman JM: Weight-reducing effects of the plasma protein encoded by the obese gene. Science 1995;269:543–546.

29 Pelleymounter MA, Cullen MJ, Baker MB, Hecht R, Winters D, Boone T, Collins F: Effects of the obese gene product on body weight regulation in ob/ob mice. Science 1995;269:540–543.

30 Friedman JM: Leptin, leptin receptors and the control of body weight. Eur J Med Res 1997;2:7–13.

31 Montague CT, Farooqi IS, Whitehead JP, Soos MA, Rau H, Wareham NJ, Sewter CP, Digby JE, Mohammed SN, Hurst JA, Cheetham CH, Earley AR, Barnett AH, Prins JB, O'Rahilly S: Congenital leptin deficiency is associated with severe early-onset obesity in humans. Nature 1997;387:903–908.

32 Farooqi IS, Jebb SA, Langmack G, Lawrence E, Cheetham CH, Prentice AM, Hughes IA, McCamish MA, O'Rahilly S: Effects of recombinant leptin therapy in a child with congenital leptin deficiency. N Engl J Med 1999;341:879–884.

33 Friedman JM, Leibel RL: Tackling a weighty problem. Cell 1992;69:217–220.

34 Friedman JM: The alphabet of weight control. Nature 1997;385:119–120.

35 Bjorbaek C, Elmquist JK, Frantz JD, Shoelson SE, Flier JS: Identification of SOCS-3 as a potential mediator of central leptin resistance. Mol Cell 1998;1:619–625.

36 Bjorbaek C, El-Haschimi K, Frantz JD, Flier JS: The role of SOCS-3 in leptin signaling and leptin resistance. J Biol Chem 1999;274:30059–30065.

37 Howard JK, Cave BJ, Oksanen LJ, Tzameli I, Bjorbaek C, Flier JS: Enhanced leptin sensitivity and attenuation of diet-induced obesity in mice with haploinsufficiency of Socs3. Nat Med 2004;10:734–738.

38   Emilsson V, Liu YL, Cawthorne MA, Morton NM, Davenport M: Expression of the functional leptin receptor mRNA in pancreatic islets and direct inhibitory action of leptin on insulin secretion. Diabetes 1997;46:313–316.
39   Kulkarni RN, Wang ZL, Wang RM, Hurley JD, Smith DM, Ghatei MA, Withers DJ, Gardiner JV, Bailey CJ, Bloom SR: Leptin rapidly suppresses insulin release from insulinoma cells, rat and human islets and, in vivo, in mice. J Clin Invest 1997;100:2729–2736.
40   Cioffi JA, Shafer AW, Zupancic TJ, Smith Gbur J, Mikhail A, Platika D, Snodgrass HR: Novel B219/OB receptor isoforms: possible role of leptin in hematopoiesis and reproduction. Nat Med 1996;2:585–589.
41   Sierra HM, Nath AK, Murakami C, Garcia CG, Papapetropoulos A, Sessa WC, Madge LA, Schechner JS, Schwabb MB, Polverini PJ, Flores RJ: Biological action of leptin as an angiogenic factor. Science 1998;281:1683–1686.
42   Bennett BD, Solar GP, Yuan JQ, Mathias J, Thomas GR, Matthews W: A role for leptin and its cognate receptor in hematopoiesis. Curr Biol 1996;6:1170–1180.
43   Haque MS, Minokoshi Y, Hamai M, Iwai M, Horiuchi M, Shimazu T: Role of the sympathetic nervous system and insulin in enhancing glucose uptake in peripheral tissues after intrahypothalamic injection of leptin in rats. Diabetes 1999;48:1706–1712.
44   Berti L, Gammeltoft S: Leptin stimulates glucose uptake in C2C12 muscle cells by activation of ERK2. Mol Cell Endocrinol 1999;157:121–130.
45   Kamohara S, Burcelin R, Halaas JL, Friedman JM, Charron MJ: Acute stimulation of glucose metabolism in mice by leptin treatment. Nature 1997;389:374–377.
46   Wang MY, Lee Y, Unger RH: Novel form of lipolysis induced by leptin. J Biol Chem 1999;274:17541–17544.
47   Tartaglia LA, Dembski M, Weng X, Deng N, Culpepper J, Devos R, Richards GJ, Campfield LA, Clark FT, Deeds J, et al: Identification and expression cloning of a leptin receptor, OB-R. Cell 1995;83:1263–1271.
48   Chen H, Charlat O, Tartaglia LA, Woolf EA, Weng X, Ellis SJ, Lakey ND, Culpepper J, Moore KJ, Breitbart RE, Duyk GM, Tepper RI, Morgenstern JP: Evidence that the diabetes gene encodes the leptin receptor: identification of a mutation in the leptin receptor gene in db/db mice. Cell 1996;84:491–495.
49   Lee GH, Proenca R, Montez JM, Carroll KM, Darvishzadeh JG, Lee JI, Friedman JM: Abnormal splicing of the leptin receptor in diabetic mice. Nature 1996;379:632–635.
50   Baumann H, Morella KK, White DW, Dembski M, Bailon PS, Kim H, Lai CF, Tartaglia LA: The full-length leptin receptor has signaling capabilities of interleukin 6-type cytokine receptors. Proc Natl Acad Sci USA 1996;93:8374–8378.
51   Ghilardi N, Ziegler S, Wiestner A, Stoffel R, Heim MH, Skoda RC: Defective STAT signaling by the leptin receptor in diabetic mice. Proc Natl Acad Sci USA 1996;93:6231–6235.
52   Clement K, Vaisse C, Lahlou N, Cabrol S, Pelloux V, Cassuto D, Gourmelen M, Dina C, Chambaz J, Lacorte JM, Basdevant A, Bougneres P, Lebouc Y, Froguel P, Guy Grand B: A mutation in the human leptin receptor gene causes obesity and pituitary dysfunction. Nature 1998;392:398–401.
53   Tartaglia LA: The leptin receptor. J Biol Chem 1997;272:6093–6096.
54   Fei H, Okano HJ, Li C, Lee GH, Zhao C, Darnell R, Friedman JM: Anatomic localization of alternatively spliced leptin receptors (Ob-R) in mouse brain and other tissues. Proc Natl Acad Sci USA 1997;94:7001–7005.
55   Mercer JG, Hoggard N, Williams LM, Lawrence CB, Hannah LT, Trayhurn P: Localization of leptin receptor mRNA and the long form splice variant (Ob-Rb) in mouse hypothalamus and adjacent brain regions by in situ hybridization. FEBS Lett 1996;387:113–116.
56   Martin-Romero C, Santos-Alvarez J, Goberna R, Sanchez-Margalet V: Human leptin enhances activation and proliferation of human circulating T lymphocytes. Cell Immunol 2000;199:15–24.
57   Loffreda S, Yang SQ, Lin HZ, Karp CL, Brengman ML, Wang DJ, Klein AS, Bulkley GB, Bao C, Noble PW, Lane MD, Diehl AM: Leptin regulates proinflammatory immune responses. FASEB J 1998;12:57–65.
58   Coleman DL: Obese and diabetes: two mutant genes causing diabetes-obesity syndromes in mice. Diabetologia 1978;14:141–148.

59 Coleman DL, Hummel KP: Effects of parabiosis of normal with genetically diabetic mice studies with the mutation, diabetes, in the mouse. Am J Physiol 1967;3:238–248.
60 Coleman DL, Hummel KP: Effects of parabiosis of normal with genetically diabetic mice. Am J Physiol 1969;217:1298–1304.
61 Mounzih K, Lu R, Chehab FF: Leptin treatment rescues the sterility of genetically obese ob/ob males. Endocrinology 1997;138:1190–1193.
62 Chandra RK: Cell-mediated immunity in genetically obese C57BL/6J (ob/ob) mice. Am J Clin Nutr 1980;33:13–16.
63 Fernandes G, Handwerger BS, Yunis EJ, Brown DM: Immune response in the mutant diabetic C57BL/Ks-dt+ mouse. Discrepancies between in vitro and in vivo immunological assays. J Clin Invest 1978;61:243–250.
64 Mandel MA, Mahmoud AA: Impairment of cell-mediated immunity in mutation diabetic mice (db/db). J Immunol 1978;120:1375–1377.
65 Meade CJ, Sheena J, Mertin J: Effects of the obese (ob/ob) genotype on spleen cell immune function. Int Arch Allergy Appl Immunol 1979;58:121–127.
66 Webb SR, Loria RM, Madge GE, Kibrick S: Susceptibility of mice to group B coxsackie virus is influenced by the diabetic gene. J Exp Med 1976;143:1239–1248.
67 Lin Y, Lee H, Berg AH, Lisanti MP, Shapiro L, Scherer PE: The lipopolysaccharide-activated Toll-like receptor (TLR)-4 induces synthesis of the closely related TLR-2 in adipocytes. J Biol Chem 2000;275:24255–24263.
68 Janeway CAJ, Medzhitov R: Introduction: the role of innate immunity in the adaptive immune response. Semin Immunol 1998;10:349–350.
69 Agnello D, Meazza C, Rowan CG, Villa P, Ghezzi P, Senaldi G: Leptin causes bodyweight loss in the absence of in vivo activities typical of cytokines of the IL-6 family. Am J Physiol 1998;275:R1518–R1525.
70 Chen K, Li F, Li J, Cai H, Strom S, Bisello A, Kelley DE, Friedman-Einat M, Skibinski GA, McCrory MA, Szalai AJ, Zhao AZ: Induction of leptin resistance through direct interaction of C-reactive protein with leptin. Nat Med 2006;12:425–432.
71 Faggioni R, Fuller J, Moser A, Feingold KR, Grunfeld C: LPS-induced anorexia in leptin-deficient (ob/ob) and leptin-receptor deficient (db/db) mice. Am J Physiol 1997;273:R181–R186.
72 Santos-Alvarez J, Goberna R, Sanchez-Margalet V: Human leptin stimulates proliferation and activation of human circulating monocytes. Cell Immunol 1999;194:6–11.
73 Caldefie-Chezet F, Poulin A, Tridon A, Sion B, Vasson MP: Leptin: a potential regulator of polymorphonuclear neutrophil bactericidal action? J Leukocyte Biol 2001;69:414–418.
74 Gainsford T, Willson TA, Metcalf D, Handman E, McFarlane C, Ng A, Nicola NA, Alexander WS, Hilton DJ: Leptin can induce proliferation, differentiation and functional activation of hemopoietic cells. Proc Natl Acad Sci USA 1996;93:14564–14568.
75 Gabay C, Dreyer M, Pellegrinelli N, Chicheportiche R, Meier CA: Leptin directly induces the secretion of interleukin 1 receptor antagonist in human monocytes. J Clin Endocrinol Metab 2001;86:783–791.
76 Faggioni R, Fantuzzi G, Gabay C, Moser A, Dinarello CA, Feingold KR, Grunfeld C: Leptin deficiency enhances sensitivity to endotoxin-induced lethality. Am J Physiol 1999;276:R136–R142.
77 Takahashi N, Waelput W, Guisez Y: Leptin is an endogenous protective protein against the toxicity exerted by tumor necrosis factor. J Exp Med 1999;189:207–212.
78 Faggioni R, Jones-Carson J, Reed DA, Dinarello CA, Feingold KR, Grunfeld C, Fantuzzi G: Leptin-deficient (ob/ob) mice are protected from T cell-mediated hepatotoxicity: role of tumor necrosis factor alpha and IL-18. Proc Natl Acad Sci USA 2000;97:2367–2372.
79 Guebre-Xabier M, Yang EK, Lin HZ, Schwenk R, Krzych U, Diehl AM: Altered hepatocyte lymphocyte subpopulations in obesity-related murine fatty livers: potential mechanisms for sensitization to liver damage. Hepatology 2000;31:633–640.
80 Hotchkiss RS, Tinsley KW, Swanson PE, Chang KC, Cobb JP, Buchman TG, Korsmeyer SJ, Karl IE: Prevention of lymphocyte cell death in sepsis improves survival in mice. Proc Natl Acad Sci USA 1999;96:14541–14546.
81 Faggioni R, Moser A, Feingold KR, Grunfeld C: Reduced leptin levels in starvation increase susceptibility to septic shock. Am J Pathol 2000;156:1781–1787.

82   Waelput W, Brouckaert P, Broekaert D, Tavernier J: A role for leptin in the systemic inflammatory response syndrome (SIRS) and in immune response, an update. Curr Med Chem 2006;13:465–475.
83   Mattioli B, Straface E, Quaranta MG, Giordani L, Viora M: Leptin promotes differentiation and survival of human dendritic cells and licenses them for Th1 priming. J Immunol 2005;174: 6820–6828.
84   Wieland CW, Florquin S, Chan ED, Leemans JC, Weijer S, Verbon A, Fantuzzi G, van der Poll T: Pulmonary Mycobacterium tuberculosis infection in leptin-deficient ob/ob mice. Int Immunol 2005;17:1399–1408.
85   Ikejima S, Sasaki S, Sashinami H, Mori F, Ogawa Y, Nakamura T, Abe Y, Wakabayashi K, Suda T, Nakane A: Impairment of host resistance to Listeria monocytogenes infection in liver of db/db and ob/ob mice. Diabetes 2005;54:182–189.
86   Siegmund B, Lehr HA, Fantuzzi G: Leptin: a pivotal mediator of intestinal inflammation in mice. Gastroenterology 2002;122:2011–2025.
87   Siegmund B, Sennello JA, Jones-Carson J, Gamboni-Robertson F, Lehr HA, Batra A, Fedke I, Zeitz M, Fantuzzi G: Leptin receptor expression on T lymphocytes modulates chronic intestinal inflammation in mice. Gut 2004;53:965–972.
88   Fantuzzi G, Sennello JA, Batra A, Fedke I, Lehr HA, Zeitz M, Siegmund B: Defining the role of T cell-derived leptin in the modulation of hepatic or intestinal inflammation in mice. Clin Exp Immunol 2005;142:31–38.
89   Matarese G, Di Giacomo A, Sanna V, Lord GM, Howard JK, Di Tuoro A, Bloom SR, Lechler RI, Zappacosta S, Fontana S: Requirement for leptin in the induction and progression of autoimmune encephalomyelitis. J Immunol 2001;166:5909–5916.
90   Busso N, So A, Chobaz-Peclat V, Morard C, Martinez-Soria E, Talabot-Ayer D, Gabay C: Leptin signaling deficiency impairs humoral and cellular immune responses and attenuates experimental arthritis. J Immunol 2002;168:875–882.
91   Matarese G, Sanna V, Di Giacomo A, Lord GM, Howard JK, Bloom SR, Lechler RI, Fontana S, Zappacosta S: Leptin potentiates experimental autoimmune encephalomyelitis in SJL female mice and confers susceptibility to males. Eur J Immunol 2001;31:1324–1332.
92   Matarese G, La Cava A, Sanna V, Lord GM, Lechler RI, Fontana S, Zappacosta S: Balancing susceptibility to infection and autoimmunity: a role for leptin? Trends Immunol 2002;23:182–187.
93   De Rosa V, Procaccini C, La Cava A, Chieffi P, Nicoletti GF, Fontana S, Zappacosta S, Matarese G: Leptin neutralization interferes with pathogenic T cell autoreactivity in autoimmune encephalomyelitis. J Clin Invest 2005;116:447–455.
94   Lord GM, Matarese G, Howard JK, Bloom SR, Lechler RI: Leptin inhibits the anti-CD3 driven proliferation of peripheral blood T cells but enhances the production of pro-inflammatory cytokines. J Leukoc Biol 2002;72:330–338.
95   Howard JK, Lord GM, Matarese G, Vendetti S, Ghatei MA, Ritter MA, Lechler RI, Bloom SR: Leptin protects mice from starvation-induced lymphoid atrophy and increases thymic cellularity in ob/ob mice. J Clin Invest 1999;104:1051–1059.
96   Lord GM, Matarese G, Howard JK, Lechler RI: The bioenergetics of the immune system. Science 2001;292:855–856.

Graham M. Lord, MA, MRCP, PhD
Professor of Medicine and Honorary Consultant Physician
Department of Nephrology and Transplantation, 5th Floor Thomas Guy House
King's College London, Guy's Hospital
London SE1 9RT (UK)
Tel. +44 020 7188 3053, Fax +44 020 7188 6413, E-Mail Graham.lord@kcl.ac.uk

Adipose Tissue and Inflammation in Chronic Kidney Disease

Jonas Axelsson, Olof Heimbürger, Peter Stenvinkel

Department of Clinical Science, Intervention and Technology, Karolinska Institutet, Division of Renal Medicine, Stockholm, Sweden

Abstract

Cardiovascular disease remains a major cause of morbidity and mortality in end-stage renal disease patients. As traditional risk factors cannot alone explain the unacceptable high prevalence and incidence of cardiovascular disease in this high-risk population, inflammation (interrelated to insulin resistance, oxidative stress, wasting and endothelial dysfunction) has been suggested to be a significant contributor. Recent studies show that the adipose tissue is a complex organ with functions far beyond the mere storage of energy. Indeed, it has been shown that fat tissue secretes a number of adipokines including leptin, adiponectin and visfatin, as well as a cytokines (here defined as signaling proteins mainly secreted by other cells present in adipose tissue, but sometimes also to a lesser degree by adipocytes *per se*), such as resistin, tumor-necrosis factor-α and interleukin-6. Adipokine serum levels are markedly elevated in chronic kidney disease, probably due to decreased renal excretion. Evidence suggests that they may have pro-inflammatory effects as well as contribute to metabolic derangements. Much research is thus still needed to elucidate the likely complex interactions between different fat tissue depots, muscle tissue and its' effects on inflammation, vascular health and outcome in this high-risk population.

Copyright © 2006 S. Karger AG, Basel

Introduction

The lifespan of chronic kidney disease (CKD) patients is markedly reduced, and cardiovascular disease (CVD) accounts for a premature death in the majority of dialysis patients in Europe and North America [1]. Despite more than 30 years of improvements in dialysis technology, most of maintenance dialysis patients die within 5 years of starting therapy – a survival worse than that of the majority of patients with cancer disease. Using data from large epidemiological studies extrapolated from the general population, much work has been done in optimizing

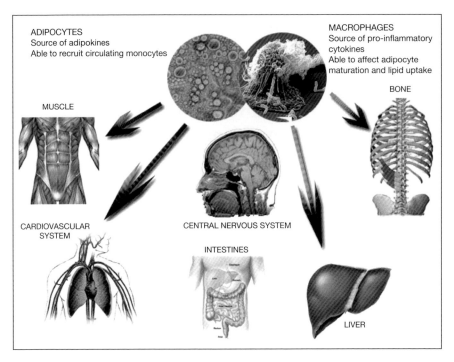

Fig. 1. Adipose tissue is the body's largest endocrine organ, secreting a number of adipokines that have pleiotropic effects on many tissues. Furthermore, macrophages resident in adipose tissue is an important source of cytokines.

traditional risk factors such as hypertension, diabetes mellitus and dyslipidemia. This is by no means unwarranted, as traditional risk factors seem to be the major contributor to cardiovascular mortality in elderly persons with mild-moderate CKD [2]. On the other hand, in moderate CKD the Atherosclerosis Risk In Communities (ARIC) data suggests that traditional and also a number of non-traditional risk factors are of equal importance [3], and studies in both hemodialysis (HD) [4] and peritoneal dialysis (PD) [5] patients suggest that novel (i.e. non-traditional) risk factors are both far more prevalent in this population than in the general population and play an important role in predicting outcome. As novel risk factors and traditional risk factors do not operate in separate rigid compartments, a rigid separation into traditional and novel factors are both difficult and unphysiological [6]. However, it is clear that among a number of factors, such as oxidative stress and hyperhomocysteinemia, that are both highly prevalent and strong predictors of outcome in end-stage renal disease (ESRD) persistent low-grade inflammation, a well-recognized independent risk factor also in other populations [7], may be the most important (fig. 1).

Systemic Inflammation Is Associated with CKD and a Predictor of Outcome

Evidence suggest that persistent inflammation (and oxidative stress) starts early in the process of a failing kidney function [2]. Indeed, recent studies have reported that CKD is associated with low-grade elevation of CRP even among patients with moderate renal impairment [8]. Since inflammatory and pro-thrombotic markers predict change in kidney function [9], interventions that reduce inflammation has been suggested to confer not only cardiovascular, but also renal benefits. Numerous studies have shown that elevated CRP predicts both all-cause and cardiovascular mortality in CKD [10] as well as ESRD patients treated by HD [11–13] or PD [14]. Moreover, persistent, rather than occasional, inflammation predicts death in dialysis patients [15] and elevated CRP concentrations after a HD session is associated with both cardiac hypertrophy [16] and a higher mortality risk [17]. Also other inflammatory markers, such as high interleukin (IL)-6 [18–20] and fibrinogen [21], have been shown to predict mortality in this population. Actually, Panichi et al. [20] have demonstrated that IL-6 has a stronger predictive value than CRP in a group of HD-patients. Their data are corroborated by own findings showing that out of four putative biochemical risk markers (CRP, IL-6, S-albumin and fetuin-A), IL-6 may be the most reliable predictor of CVD and mortality in ESRD [22].

Although evidence suggests that ESRD is a state of chronic low-grade inflammation, we do not yet fully understand the exact cause(s) of inflammation in this patient group. However, it appears likely that elevated plasma levels of inflammatory biomarkers, such as IL-6, in ESRD are caused by a combination of factors including the uremic syndrome *per se*, age, chronic heart failure and persistent infections, as well as bioincompatibility of the dialyzer membrane and endotoxin absorption from contaminated dialysate or from the gut [23]. As studies show that the adipose tissue is a complex organ with functions far beyond the mere storage of energy, obesity may be another factor that may contribute to inflammation in CKD. Indeed, it has recently been shown that fat tissue secretes a number of adipokines (here defined as signaling proteins mainly secreted by adipocytes), including leptin, adiponectin and visfatin, as well as cytokines (here defined as signaling proteins mainly secreted by other cells present in adipose tissue, but sometimes also to a lesser degree by adipocytes *per se*), such as resistin, tumor-necrosis factor (TNF)-α and IL-6. All of these may have important endocrine functions in CKD patients, and may be important contributors to systemic inflammation in this patient group [24]. In dialysis patients this may be of significance, as the initiation of PD is often associated with an increase in fat mass that is likely to be, at least partly, related to glucose absorption from the dialysis fluid [25], while HD may contribute to local ischemia in the adipose tissue, stimulating the recruitment of immunocompetent cells from the circulation.

Adipose Tissue as a Source of Inflammatory Cytokines

The new role of adipose tissue as not only an inert storage depot but also a source of adipokines has opened up a brand new area for research. We know that adipose tissue signaling may act as autogenic regulators of body fat depots, modulating gastrointestinal activities, metabolic changes and central nervous mechanisms, which have been speculated to play a central role in the development of complications, such as insulin resistance, CVD and sarcopenia (loss of muscle mass) [26]. Furthermore, there are intimate links between adipokines and pro-inflammatory cytokines, such as IL-6, as well as between fat and muscle tissue [27]. Considering the dramatic effect the loss of renal function has on the clearance of cytokines and adipokines [28–30], the systemic effects of these proteins may be greater in CKD patients than in the general population.

Interleukin-6 and Tumor Necrosis Factor-α
Although most of systemic circulating IL-6 is secreted from activated lymphocytes, other tissues, such as adipose tissue with resident macrophages, may contribute to about 20–30% of the production of IL-6 [31]. Furthermore, TNF-α is expressed by adipocytes, and this expression is markedly increased in obesity and has been linked to insulin sensitivity in non-renal populations [27]. Indeed, a recent study demonstrated a rapid beneficial effect of anti-TNF-α blockade (infliximab) on insulin resistance in patients with rheumatoid arthritis [32]. Clinical studies in other patient groups, such as obese individuals, demonstrated a correlation between fat mass and pro-inflammatory cytokines, while weight loss was associated with a reduction in circulating levels of inflammatory biomarkers, such as IL-6. It has also been demonstrated that the omental adipose tissue, most affected by PD, releases 2–3 times more IL-6 than subcutaneous fat tissue [31]. Indeed, in a cross-sectional study of ESRD patients, we have demonstrated that whereas truncal (i.e. visceral) fat mass correlated significantly to circulating levels of IL-6, no significant correlation was demonstrated between non-truncal fat mass and IL-6 [33].

Leptin
So far, relatively few studies have investigated the impact of adipokines on metabolic and inflammatory aspects of CKD. Leptin, the first adipokine described (1994), was shown to correlate strongly with total body fat mass and to modulate feeding behavior in rats [34]. While leptin signaling is more

complex in humans, loss of renal function leads to inappropriately elevated serum concentrations of leptin [35]. Leptin signaling in the CNS has been shown to be an important cause of anorexia in uremic rats [36]. As experimental uremic cachexia can be ameliorated by blockade of leptin signaling through the hypothalamic melanocortin-4 receptor, melanocortin receptor antagonism may provide a novel therapeutic strategy for inflammation-associated wasting [37]. In PD patients, we have shown that serum leptin levels increase with initiation of PD and are inversely related to inflammation and predict longitudinal changes in lean body mass [38]. In accordance, most [29, 39], but not all [40], studies have demonstrated an association between inflammatory biomarkers and leptin in CKD suggesting that free bioactive leptin may be associated to inflammatory-associated wasting. These data are corroborated by a study in mice showing that the total amount of bioactive leptin is increased during acute inflammation, suggesting that leptin participates in the host response to inflammation [41]. Indeed, in non-renal patients, leptin has been shown to be capable of initiating the recruitment and activation of immunocompetent cells [42], while leptin production may in turn be regulated by adipose tissue TNF-α levels [42]. Notably, serum leptin levels also appear to be an independent predictor of epoetin requirements in uremia even after adjustment for inflammation [43].

Adiponectin

Adiponectin is another adipokine exclusively secreted from adipocytes found in the circulation. In contrast to other adipokines, low circulating levels of adiponectin are generally found in populations at enhanced risk of atherosclerotic CVD [44]. Thus, reduced adiponectin levels predispose healthy individuals to insulin resistance [44]. Although plasma adiponectin levels are generally elevated in ESRD patients [45, 46] it has been reported that ESRD patients with low adiponectin levels have an increased mortality rate [46]. Also, a reciprocal relationship between CRP and adiponectin has been demonstrated in ESRD [45]. As the exact mechanisms whereby adiponectin mediates its' effects are still unclear, further studies will surely lead to novel physiological and pathological roles for this intriguing molecule. Indeed, Qi et al. [47] recently demonstrated that adiponectin, like leptin, acts on the central nervous system, where it reduces food intake. Also, in cultured myocytes, adiponectin inhibits apoptosis and TNF-α production after ischemia [48].

Resistin

Resistin may serve as an example of the close links band classification problems arising between adipokines and cytokines. Resistin was initially described as 'Found In Inflammatory Zone' (FIZZ)-1, and described as a cytokine [49]. Later, animal studies implicated it in obesity-induced insulin resistance and type-2 diabetes mellitus, and it was reclassified as an adipokine [50]. However, despite the initial excitement surrounding this adipokine, the true pathophysiological role of resistin in human disease remains unknown, but it now seems clear that many different cell types secrete this peptide, with a predominance of immuno-competent cells [51]. Indeed, Axelsson et al. [30] and Filippidis et al. [52] have found that increased circulating resistin levels are not associated with insulin resistance or fat mass, but rather correlate closely with inflammatory biomarkers and residual renal function.

Macrophages and the Link between Inflammation and Metabolism

As the biology and physiology of the adipose tissue is re-examined in the light of recent findings, much attention is being drawn to the close and interdependent signaling pathways of inflammation and metabolic control expressed there. Indeed, gene expression is highly similar between the two types of cells – adipocytes and macrophages – most common in adipose tissue. Thus, macrophages express many of the adipocytokines, such as PPAR-γ and resistin, while adipocytes can express many 'macrophage' proteins, such as TNF-α and IL-6 [27]. These two cells also share functional capabilities, i.e. macrophages can take up and store lipids to become atherosclerotic foam cells, while preadipocytes under some conditions can exhibit phagocytic and anti-microbial properties and appear to even be able to differentiate into macrophages in the right environment. This suggests a potential immunological role for preadipocytes [42].

Is It Good to Be Fat If You Are on Dialysis?

Although the relationship between obesity and CKD has been established in may studies few epidemiologic studies have examined whether obesity is an independent risk factor for ESRD. However, in a recent study by Hsu et al. [53], based on a cohort of 320,252 historical members of Kaiser Permanente, it was demonstrated that higher baseline BMI remained an independent predictor of ESRD even after adjustment for blood pressure and diabetes mellitus. However (and in contrast to findings in the general population), a number of studies have documented that when a CKD patient have reached ESRD a high BMI is actually associated with a better outcome [54, 55], but not necessarily better health status, quality of life and physical function [56, 57]. When these epidemiological data are interpreted it should be noted that BMI is not a very precise parameter of nutritional status and does not reflect body composition (e.g. to differentiate between muscle mass and fat mass). In a study of 70,028 patients that initiated dialysis in the Unites States from 1995 to 1999, it was demonstrated that a protective effect from a high BMI is only present in patients with a normal or high muscle mass [55]. In accordance, a study of 344 incident dialysis patients showed that a BMI $> 25\,\text{kg/m}^2$ was only associated with a better survival if the mid-arm muscle circumference was at least 90% of normal, while patients with a BMI $> 25\,\text{kg/m}^2$ and a lower muscle mass had the worst survival [58]. Furthermore, part of the positive relation between BMI and survival in dialysis patients may be due to time discrepancies among competitive risk factors. Even if obesity may induce a similar effects in dialysis patients, many patients with more important risk factors (such as CVD, diabetes and inflammation) may not live long enough for elevated body fat mass to be a significant outcome predictor. A long-term outcome study in 116 Japanese HD patients with low comorbidity, a BMI $> 19.0\,\text{kg/m}^2$ was associated with increased 12-year mortality, which was progressively higher with increasing BMI [59]. On the other hand, recent data based on a cohort of 553 North-American HD-patients demonstrated that fat loss over time is independently associated with higher mortality even after adjustment for demographics and surrogates of muscle mass and inflammation [57]. Thus, further mechanistic studies are needed to ascertain if CKD patients, like many other patients groups, may indeed have adverse metabolic consequences of an increased fat mass, or if the pathophysiology of CKD makes an increased adipokine level more beneficial than the presumed detrimental effect of an increase in systemic inflammation associated with increased fat mass. It could also be hypothesized that in the uremic milieu fat tissue secretes one or more substances with beneficial cardiovascular effects.

Conclusion

Human adipose tissue has recently been shown to be a hormonally active system that secretes various adipokines and cytokines (a clear distinction is currently not always possible), both of which likely play an important role in the development of metabolic complications associated with obesity, such as insulin resistance, type-2 diabetes and premature atherosclerosis. Adipokines are of special interest in CKD patients, where they may both increase systemic inflammation and contribute to metabolic derangements – something that paradoxically has yet to be proven to be detrimental to their outcome. Much research is thus still needed to elucidate the likely complex interactions between different fat tissue depots, muscle tissue and its' effects on inflammation, vascular health and outcome in this high-risk population.

Acknowledgements

This manuscript was in part supported by grants from the Swedish Kidney Foundation (JA), Swedish Medical Research Council (PS) and Söderbergs Foundation (PS).

References

1 Foley RN, Parfrey PS, Sarnak MJ: Clinical epidemiology of cardiovascular disease in chronic renal failure. Am J Kidney Dis 1998;32(suppl 5):S112–S119.
2 Shlipak MG, Fried LF, Cushman M, Manolio TA, Peterson D, Stehman-Breen C, et al: Cardiovascular mortality risk in chronic kidney disease: comparison of traditional and novel risk factors. JAMA 2005;293:1737–1745.
3 Muntner PHJ, Astor BC, Folsom AR, Coresh J: Traditional and nontraditional risk factors predict coronary heart disease in chronic kidney disease: results from the atherosclerosis risk in communities study. J Am Soc Nephrol 2005;16:529–538.
4 Cheung AK, Sarnak MJ, Yan G, Dwyer JT, Heyka RJ, Rocco MV, et al: Atherosclerotic cardiovascular disease risks in chronic hemodialysis patients. Kidney Int 2000;58:353–362.
5 Zoccali C, Enia G, Tripepi G, Panuccio V, Mallamaci F: Clinical epidemiology of major nontraditional risk factors in peritoneal dialysis patients. Perit Dial Int 2005;25(suppl 3):S84–S87.
6 Beddhu S, Kimmel PL, Ramkumar N, Cheung AK: Associations of metabolic syndrome with inflammation in CKD: results from the third national health and nutrition examination survey (NHANES III). Am J Kidney Dis 2005;46:577–586.
7 Ridker PM, Koenig W, Fuster V: C-reactive protein and coronary heart disease. N Engl J Med 2004;351:296–297.
8 Stenvinkel P, Ketteler M, Johnson RJ, Lindholm B, Pecoits-Filho R, Riella M, et al: Interleukin-10, IL-6 and TNF-a: important factors in the altered cytokine network of end-stage renal disease – the good, the bad and the ugly. Kidney Int 2005;67:1216–1233.
9 Fried L, Solomon C, Shlipak M, Seligerm S, Stehman-Breen C, Bleyer AJ, et al: Inflammatory and prothrombotic markers and the progression of renal disease in elderly individuals. J Am Soc Nephrol 2004;15:3184–3191.
10 Menon V, Greene T, Wang X, Pereira AA, Marcovina SM, Beck GJ, et al: C-reactive protein and albumin as predictors of all-cause and cardiovascular mortality in chronic kidney disease. Kidney Int 2005;68:766–772.
11 Yeun JY, Levine RA, Mantadilok V, Kaysen GA: C-reactive protein predicts all-cause and cardiovascular mortality in hemodialysis patients. Am J Kidney Dis 2000;35:469–476.
12 Zimmermann J, Herrlinger S, Pruy A, Metzger T, Wanner C: Inflammation enhances cardiovascular risk and mortality in hemodialysis patients. Kidney Int 1999;55:648–658.
13 Iseki K, Tozawa M, Yoshi S, Fukiyama K: Serum C-reactive (CRP) and risk of death in chronic dialysis patients. Nephrol Dial Transplant 1999;14:1956–1960.
14 Ducloux D, Bresson-Vautrin C, Kribs M, Abdelfatah A, Chalopin J-M: C-reactive protein and cardiovascular disease in peritoneal dialysis patients. Kidney Int 2002;62:1417–1422.
15 Nascimento MM, Pecoits-Filho R, Qureshi AR, Hayashi SY, Manfro RC, Pachaly MA, et al: The prognostic influence of fluctuating levels of C-reactive protein in Brazilian hemodialysis patients: a prospective study. Nephrol Dial Transplant 2004;19:2803–2809.
16 Park CW, Shin YS, Kim CM, Lee SY, Yu SE, Kim SY, et al: Increased C-reactive protein following hemodialysis predicts cardiac hypertrophy in chronic hemodialysis patients. Am J Kidney Dis 2002;40:1230–1239.
17 Korevaar JC, van Manen JG, Dekker FD, de Waart DR, Boeschoten EW, Krediet RT: Effect of an increase in CRP level during a hemodialysis session on mortality. J Am Soc Nephrol 2004;15:2916–2922.
18 Bologa RM, Levine DM, Parker TS, Cheigh JS, Seur D, Stenzel KH, et al: Interleukin-6 predicts hypoalbuminemia, hypocholesterolemia, and mortality in hemodialysis patients. Am J Kidney Dis 1998;32:107–114.

19 Pecoits-Filho R, Barany B, Lindholm B, Heimbürger O, Stenvinkel P: Interleukin-6 and its receptor is an independent predictor of mortality in patients starting dialysis treatment. Nephrol Dial Transplant 2002;17:1684–1688.
20 Panichi V, Maggiore U, Taccola D, Migliori M, Rizza GM, Consani C, et al: Interleukin-6 is a stronger predictor of total and cardiovascular mortality than C-reactive protein in haemodialysis patients. Nephrol Dial Transplant 2004;19:1154–1160.
21 Zoccali C, Mallamaci F, Tripepi G, Cutrupi S, Parlongo S, Malatino LS, et al: Fibrinogen, mortality and incident cardiovascular complications in end-stage renal failure. J Intern Med 2003;254:132–139.
22 Honda H, Qureshi AR, Heimbürger O, Barany P, Wang K, Pecoits-Filho R, et al: Serum albumin, C-reactive protein, interleukin-6 and fetuin-A as predictors of malnutrition, cardiovascular disease and mortality in patients with end-stage renal disease. Am J Kidney Dis 2005;47:139–148.
23 Stenvinkel P. Inflammation in end-stage renal disease – a fire that burns within. Contrib Nephrol 2005;149:185–199.
24 Axelsson J, Heimbürger O, Lindholm B, Stenvinkel P: Adipose tissue and its relation to inflammation: the role of adipokines. J Ren Nutr 2005;15:131–136.
25 Nordfors L, Heimbürger O, Lonnqvist F, Lindholm B, Helmrich J, Schalling M, et al: Fat tissue accumulation during peritoneal dialysis is associated with a polymorphism in uncoupling protein 2. Kidney Int 2000;57:1713–1719.
26 Nawrocki A, Scherer PE: The delicate balance between fat and muscle: adipokines in metabolic disease and musculoskeletal inflammation. Curr Opin Pharm 2004;4:281–289.
27 Wellen KE, Hotamisligil GS: Inflammation, stress and diabetes. J Clin Invest 2005;115:1111–1119.
28 Pecoits-Filho R, Heimbürger O, Bárány P, Suliman M, Fehrman-Ekholm I, Lindholm B, et al: Associations between circulating inflammatory markers and residual renal function in CRF patients. Am J Kidney Dis 2003;41:1212–1218.
29 Nordfors L, Lönnqvist F, Heimbürger O, Danielsson A, Schalling M, Stenvinkel P: Low leptin gene expression and hyperleptinemia in chronic renal failure. Kidney Int 1998;54:1267–1275.
30 Axelsson J, Bergsten A, Qureshi AR, Heimbürger O, Barany P, Lönnqvist F, et al: Elevated resistin levels in chronic kidney disease are associated with decreased glomerular filtration rate and inflammation, but not with insulin resistance. Kidney Int 2006;69:596–604.
31 Mohamed-Ali V, Goodrick S, Rawesh A, Katz DR, Miles JM, Yudkin JS, et al: Subcutaneous adipose tissue releases interleukin-6, but not tumour necrosis factor-a, in vivo. J Clin Endocrin Metab 1997;82:4196–4200.
32 Gonzales-Gay MA, De Matias JM, Gonzalez-Juanatey C, Garcia-Porrua C, Sanchez-Andrade A, Martin J, et al: Anti-tumor necrosis factor-alpha blockade improves insulin resistance in patients with rheumatoid arthritis. Clin Exp Rheumatol 2006;24:83–86.
33 Axelsson J, Qureshi AR, Suliman ME, Honda H, Pecoits-Filho R, Heimbürger O, et al: Truncal fat mass as a contributor to inflammation in end-stage renal disease. Am J Clin Nutr 2004;80:1222–1229.
34 Zhang Y, Proenca R, Maffei M, Barone M, Lori L, Friedman JM: Positional cloning of the mouse gene and its human homolouge. Nature 1994;372:425–432.
35 Heimbürger O, Lönnqvist F, Danielsson A, Nordenstrom J, Stenvinkel P: Serum immunoreactive leptin concentration and its relation to the body fat content in chronic renal failure. J Am Soc Nephrol 1997;8:1423–1430.
36 Cheung W, Yu PX, Little BM, Cone RD, Marks DL, Mak RH: Role of leptin and melanocortin signaling in uremia-associated cachexia. J Clin Invest 2005;115:1659–1665.
37 Mak RH, Cheung W, Cone RD, Marks DL: Leptin and inflammation-associated cachexia in chronic kidney disease. Kidney Int 2006;69:794–797.
38 Stenvinkel P, Lindholm B, Lönnqvist F, Katzarski K, Heimbürger O: Increases in serum leptin during peritoneal dialysis are associated with inflammation and a decrease in lean body mass. J Am Soc Nephrol 2000;11:1303–1309.
39 Pecoits-Filho R, Nordfors L, Heimbürger O, Lindholm B, Anderstam B, Marchlewska A, et al: Soluble leptin receptors and serum leptin in end-stage renal disease: relationship with inflammation and body composition. Eur J Clin Invest 2002;32:811–817.

40 Don BR, Rosales LM, Levine NW, Mitch W, Kaysen GA: Leptin is a negative acute phase protein in chronic hemodialysis patients. Kidney Int 2001;59:1114–1120.
41 Voegeling S, Fantuzzi G: Regulation of free and bound leptin and soluble leptin receptors during inflammation in mice. Cytokine 2001;14:97–103.
42 Wellen KE, Hotamisligil GS: Inflammation, stress, and diabetes. J Clin Invest 2005;115:1111–1119.
43 Axelsson J, Qureshi AR, Heimbürger O, Lindholm B, Stenvinkel P, Barany P: Body fat mass and serum leptin levels influence epoetin sensitivity in patients with ESRD. Am J Kidney Dis 2005;46:628–634.
44 Stenvinkel P, Pecoits-Filho R, Lindholm B: Leptin, ghrelin and proinflammatory cytokines: compounds with nutritional impact in chronic kidney disease. Adv Renal Repl Ther 2003;10:332–345.
45 Stenvinkel P, Marchlewska A, Pecoits-Filho R, Heimbürger O, Zhang Z, Hoff C, et al: Adiponectin in renal disease: relationship to phenotype and genetic variation in the gene encoding adiponectin. Kidney Int 2004;65:274–281.
46 Zoccali C, Mallamaci F, Panuccio V, Tripepi G, Cutrupi S, Parlongo S, et al: Adiponectin is markedly increased in patients with nephrotic syndrome and is related to metabolic risk factors. Kidney Int 2003;63(suppl 84):S98–S102.
47 Qi Y, Takahashi N, Hileman SM, Patel HR, Berg AH, Pajvani UB, et al: Adiponectin acts in the brain to decrease body weight. Nat Med 2004;10:524–529.
48 Shibata R, Sato K, Pimentel DR, Takemura Y, Kihara S, Ohashi K, et al: Adiponectin protects against myocardial ischemia-reperfusion injury through AMPK- and COX-2-dependent mechanisms. Nat Med 2005;11:1096–1103.
49 Holcomb IN, Kabakoff RC, Chan B, Baker TW, Gurney A, Henzel W, et al: FIZZ1, a novel cysteine-rich secreted protein associated with pulmonary inflammation, defines a new gene family. EMBO J 2000;19:4046–4055.
50 Rea R, Donnelly R: Resistin: an adipocyte-derived hormone. Has it a role in diabetes and obesity? Diab Obes Metab 2004;6:163–170.
51 Osawa H, Onuma H, Ochi M, Murakami A, Yamauchi J, Takasuka T, et al: Resistin SNP-420 determines its monocyte mRNA and serum levels inducing type 2 diabetes. Biochem Biophys Res Commun 2005;335:596–602.
52 Filippidis G, Liakopoulos V, Mertens PR, Kiropoulos T, Stakias N, Verikouki C, et al: Resistin serum levels are increased but not correlated with insulin resistance in chronic hemodialysis patients. Blood Purif 2005;23:421–428.
53 Hsu CY, McCulloch CE, Iribarren C, Darbinian J, Go AS: Body mass index and risk for end-stage renal disease. Ann Intern Med 2006;144:21–28.
54 Kalantar-Zadeh K, Abbott KC, Salahudeen AK, Kilpatrick RD, Horwich TB: Survival advantages of obesity in dialysis patients. Am J Clin Nutr 2005;81:543–554.
55 Beddhu S, Pappas LM, Ramkumar N, Samore MH: Effects of body size and body composition on survival in hemodialysis patients. J Am Soc Nephrol 2003;14:2366–2372.
56 Johansen KL, Kutner NG, Young B, Chertow GM: Association of body size with health status in patients beginning dialysis. Am J Clin Nutr 2006;83:543–549.
57 Kalantar-Zadeh K, Kuwae N, Wu DY, Shantouf RS, Fouque D, Anker SD, et al: Associations of body fat and its changes over time with quality of life and prospective mortality in hemodialysis patients. Am J Clin Nutr 2006;83:202–210.
58 de Araújo I KM, Draibe S, Canziani M, Manfredi S, Avesani C, Sesso R, Cuppari L: Nutritional parameters and mortality in incident heomdialysis patients: effect of body mass index, muscle mass, and dietary intake. J Ren Nutr 2006;16:27–35.
59 Kaizu Y, Tsunega Y, Yoneyama T, Sakao T, Hibi I, Miyaji K, et al: Overweight as another nutritional risk factor for the long-term survival of non-diabetic hemodialysis patients. Clin Nephrol 1998;50:44–50.

Peter Stenvinkel, MD, PhD
Department of Renal Medicine, K56, Karolinska University Hospital at Huddinge
SE–141 86 Stockholm (Sweden)
Tel. +46 8 58582532, Fax +46 8 7114742, E-Mail peter.stenvinkel@ki.se

Leptin and Renal Fibrosis

Gunter Wolf[a], *Fuad N. Ziyadeh*[b]

[a]Department of Internal Medicine III, Friedrich-Schiller-University, Jena, Germany;
[b]Renal Electrolyte and Hypertension Division, Department of Medicine, University of Pennsylvania, Philadelphia, Pa., USA

Abstract

Leptin is a peptide hormone that is mainly, but not exclusively, produced in adipose tissue and plays a pivotal role in regulating food intake and energy expenditure. Besides its effects on regulation of body weight, appetite and energy expenditure, leptin exhibits influence on the immune system and may contribute to the deterioration of renal function. These direct and indirect renal effects of leptin could partly explain obesity-associated kidney disease and may be also relevant for diabetic nephropathy in type 2 diabetes. Leptin is primarily metabolized in the kidney, presumably by binding to megalin, a multiligand receptor in the proximal tubule, tubular uptake and endocytosis. The kidney expresses abundant concentrations of the small isoform of the leptin receptor (Ob-Ra). In cultured renal rat endothelial cells and mesangial cells obtained from *db/db* mice, leptin can signal through the Ob-Ra receptor isoform. The peptide stimulates proliferation of glomerular endothelial cells, increases TGF-β1 synthesis, and collagen type IV production. In contrast, leptin did not influence TGF-β1 production in mesangial cells, but the peptide stimulates glucose transport in these cells, increased collagen type I synthesis, and lead to an upregulation of surface TGF-β type II receptors through signal transduction pathways involving phosphatidylinositol-3-kinase. Leptin also stimulates hypertrophy, but not proliferation in cultured rat mesangial cells. Infusion of leptin for 3 weeks into normal rats fosters development of glomerulosclerosis and proteinuria. In addition, transgenic mice with leptin overexpression demonstrated a increase in collagen type IV and fibronectin mRNA in the kidney. Additional previously described direct and indirect effects of leptin on the kidney include natriuretic effects, an increase in sympathetic nervous activity, and stimulation of reactive oxygen species. These findings collectively suggest that the kidney is a target organ for leptin and that this hormone might play an important role in renal pathophysiology.

Copyright © 2006 S. Karger AG, Basel

Introduction

Two recessive mutations, obese (*ob*) and diabetes (*db*), have been described in mice that both lead to hyperphagia, reduction in energy expenditure, early onset obesity, and eventually development of diabetes [1]. Many years later, the product of the *ob* gene was named leptin (from the Greek *leptos* = thin) and functions as an afferent signal in control of body weight. The mouse *ob* gene encodes a 4.5 kilo base transcript resulting in a 16 kDa protein [2]. The gene consists of three exons and has been localized on chromosome 6 in mouse and 7 in humans. Leptin is mainly produced in adipocytes of white fat tissue [2], but more recently evidence accumulated that there are several other sites of synthesis. Leptin is secreted, circulates as a monomer freely in the plasma, and also binds to other proteins [3]. Homozygous mutations of the *ob* gene cause leptin deficiency [2]. These mice are hyperphagic, hypothermic, exhibit several endocrine abnormalities such as infertility, and develop morbid obesity. Exogenous leptin replacement corrects all these abnormalities and induce weight loss that is restricted to adipose tissue with sparing of the lean body mass. Mutation of the *ob* gene does occur in humans, but is very rare [4].

The leptin receptor (Ob-R) belongs to the family of class 1 cytokine receptors and shares common features with the interleukin 6 receptor [5]. Different receptors result from alternative splicing have been described [6]. The long-form of the receptor (Ob-Rb) with a transmembrane domain is the leptin receptor principally present in the hypothalamus [7]. Binding of leptin to this receptor induces diverisation of the receptor and activation of the JAK (Janus kinase)/STAT (signal transducers and activators of transcription) pathway resulting in phosphorylation of STAT. Phosphorylated STATs translocate into the nucleus and bind to their putative elements in the DNA. Additional signal transduction pathways such as MAP kinases (Erk 1,2) and phosphatidylinositol 3-kinase are also activated [8–10]. In *db/db* mice, the transcript of the long form of the receptor is abnormal and has an insertion at the junction where the long- and short-form transcripts diverge [11]. This insertion contains a premature stop codon that leads to the synthesis of a truncated receptor that replace the Ob-Rb isoform with the Ob-Ra isoform [11]. Since *db/db* mice are resistant to the central actions of leptin, it has been assumed that the long Ob-Rb receptor is the only isoform capable of transmitting signals [12]. The abundant expression of the shorter receptor isoforms, particular Ob-Ra, in many peripheral tissues including the kidney have been originally contributed to a 'clearance' function of these receptors [11]. However, there is increasing evidence that the Ob-Ra isoform is active in signaling and may activate STATs [13–15]. This finding could be very important in understanding changes in *db/db* mice that have high circulating leptin levels because leptin may activate pathophysiological effects in peripheral tissue through Ob-Ra activation [16–21]. The hypothalamic melanocortin system with the melanocortin-4-receptors is pivotal in mediating the effect of leptin on appetite and metabolism [16].

Multiple Effects of Leptin

Accumulating data form the last years provided evidence that leptin is a multifunctional hormone with diverse actions clearly extending those involved in regulation of appetite and body weight. Some of these effects such as the maturation of the reproductive axis by stimulating gonadotropin release may be viewed as signal to linked the adequacy of existing energy (fat) stores to reproduction [17]. Leptin transduces signals including activation of STATs in intestinal and hepatic cells [19]. In the liver, contradictory results regarding a modulatory effect on insulin signaling have been described, and leptin probably has insulin-like as well as anti-insulin actions [20]. However, leptin inhibits insulin secretion in pancreatic islets [21]. A role of leptin in the inhibition of bone formation through hypothalamic influences has been recently suggested [22]. In addition, leptin also influences bone metabolism and enhances bone density by stimulating the transformation of progenitor cells into osteoblasts [23].

Interestingly, current studies demonstrated that leptin promotes angiogenesis [24, 25]. Leptin stimulates proliferation of endothelial cells and also promoted the formation of capillary-like tubes. Thus, by providing a local angiogenic signal, leptin may improve energy expenditure by increasing blood flow facilitating heat dissipation and lipid oxidation [24, 25]. Leptin induces proliferation, differentiation and function of hemopoietic cells including T-cells [26, 27]. However, leptin inhibits neutrophil migration in response to various chemoattractans [28]. In sera from patients with end-stage renal disease, a strict correlation between serum leptin concentrations and serum ability to suppress neutrophil locomotion was found indicating a crucial role of leptin in suppressing neutrophils in host defense in uremia [28]. These effects may explain how leptin levels in uremia causes a proinflammatory environment with nevertheless an increases risk of infections.

Leptin and the Kidney

As a small peptide hormone, leptin is principally cleared by the kidney, presumably involving a mechanism of megalin-mediated tubular uptake followed by endocytosis [29], see also chapter by Kumar Sharma. Consequently, plasma leptin concentrations are increased in end-stage renal failure and in patients undergoing hemodialysis [30–32]. Hyperleptinemia in patients with end-stage renal disease contributes to malnutrion, but rather by proinflammatory effects than directly suppressing appetite [33]. Experimental evidence suggests that uremic cachexia is attenuated in *db/db* mice [34]. Moreover, blocking leptin signaling through the hypothalamic melanocortin-4-receptors resulted in a resisted uremia-induced loss of lean body mass and maintained normal basal metabolic rates [34, 35].

The kidney express small amount of Ob-Rb receptors, but the small isoform Ob-Ra is much more abundant [36]. Leptin mediates sympathetic nerve activation [37, 38]. Infusion of leptin into rats produced increases renal sympathetic nervous activity [37]. Systolic blood pressures and urinary catecholamine excretion were also elevated in transgenic skinny mice that have elevated plasma leptin concentrations [38]. Although acute leptin application had no effect on blood pressure in rats, chronic infusion significantly increased arterial pressure and heart rate [39, 40]. Intracerebrovascular injection of leptin increases mean arterial pressure, probable by proopiomelanocortin products [41]. Interestingly, leptin has natriuretic and diuretic activity in the rat [42]. These effects are blunted in spontaneously hypertensive rats (SHR) with intact nerves [42, 43]. This observation suggests that leptin-mediated stimulation of sympathetic nervous system counteracts the natriuretic effects under normal conditions [44]. An increased sympathetic nerve activity has been recently associated with progression of renal disease [45]. For example, in the model of subtotally nephrectomized rats, a centrally sympathicoplegic agent significantly reduced glomerulosclerosis and proteinuria without affecting blood pressure [45]. These findings indicate blood pressure-independent effects of sympathetic overactivity on glomerulosclerosis. Indeed, we have previously found that catecholamines stimulate proliferation of renal cells in vitro and in vivo [46]. Thus, leptin may stimulate profibrotic actions in the kidney by sympathetic overactivity.

However, we have identified a more direct role for leptin in promoting renal pathophysiological changes [15, 47]. We discovered that leptin stimulates proliferation of cultured rat glomerular endothelial cells, but had no proliferative effect on mesangial cells [15]. Glomerular endothelial cells expressed high-affinity receptors of the Ob-Ra isoform and leptin induced phosphorylation of STA1α in these cells suggesting that the short-form of the receptor is involved in signaling [15]. In addition, leptin stimulates protein and mRNA expression of TGF-β1 in glomerular endothelial cells [15]. To further investigate a role of leptin signaling through the Ob-Ra, we studied mesangial cells from *db/db* mouse [48]. In accordance with observations obtained from rat mesangial cells, leptin did not increase TGF-β1 synthesis in *db/db* mesangial cells [15, 48], but upregulated dose-dependently transcription of the TGF-β type II receptor [48]. Moreover, leptin increased glucose uptake and stimulated collagen type I synthesis [48]. These effects were suppressed by a phosphatidylinositol3-kinase inhibitor indicating a major role of this enzyme in signal transduction [48]. The leptin-mediated upregulation of surface TGF-β type II receptors on *db/db* mesangial cells had functional consequences because addition of both exogenous TGF-β1 and leptin increased type I collagen production more that addition of either TGF-β1 or

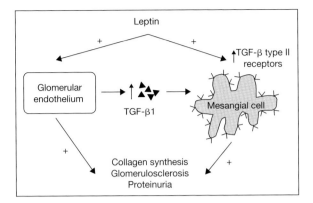

Fig. 1. Paracrine TGF-β pathways between glomerular endothelial and mesangial cells mediated by leptin. Leptin binds to short Ob-Ra receptors expressed on glomerular endothelial and mesangial cells. Whereas leptin increased synthesis of TGF-β1 in endothelial cells, it upregulates TGF-β type II receptors on mesangial cells without influencing TGF-β1 synthesis in these cells. TGF-β1 produced in endothelial cells may easily translocate to mesangial cells and induce an amplified response because of the already upregulated TGF-β receptors. Leptin also stimulates the synthesis of collagen type I in mesangial cells and of type IV in glomerular endothelial cells. The consequence of this paracrine TGF-β system is an increase in extracellular matrix deposition, proteinuria, and eventually glomerulosclerosis. Modified from reference [47].

leptin alone [48]. To test a potential role of leptin in glomerular pathology in vivo, we infused recombinant leptin for either 72 h or 3 weeks in normal rats [15]. Short-term infusion of leptin induces glomerular TGF-β expression and also increased the total number of proliferating cells as detected by PCNA-staining. At the end of 3-week infusion period, glomerular expression of collagen type IV expression was significantly enhanced in leptin-infused animals. These rats were also proteinuric [15]. However, leptin infusion had no effect on systolic blood pressure as measured by tail plethysmography. Thus, our data suggest a leptin cross-talk between glomerular endothelial and mesangial cells as depicted in figure 1. Leptin binds to Ob-Ra receptors expressed on glomerular endothelial and mesangial cells [47]. However, the binding has different biological effects. Whereas leptin increased synthesis of TGF-β1 in endothelial cells, it upregulates TGF-β type II receptors on mesangial cells without influencing TGF-β1 synthesis in these cells. TGF-β1 produced in endothelial cells may easily reach mesangial cells and induce an amplified response because of upregulated TGF-β receptors. Furthermore, leptin stimulates the synthesis of collagen type I in mesangial cells and of type IV in glomerular endothelial cells. The activation of this paracrine TGF-β system by leptin eventually induces extracellular matrix deposition, glomerulosclerosis, and proteinuria [49]. An important role of TGF-β1 in mediating profibrogenic effects is well-appreciated in various renal diseases [50]. Leptin may also have profibrotic effects on the mesangial cell independent of the TGF-β pathway since inhibition of phosphatidylinositol 3-kinase completely abolishes the leptin-induced increase in type I collagen and glucose uptake [48]. Furthermore, we found that 100 ng/ml of leptin was equipotent to 2 ng/ml of TGF-β1 in raising type I collagen protein production. Together, exogenous leptin and TGF-β1 at these particular concentrations exerted an additive effect on type I collagen production. These data may suggest that leptin and TGF-β1 promote mesangial extracellular matrix expression by different mechanisms.

In agreement with our findings, Suganami et al. [51] found an increase in renal collagen type IV and fibronectin mRNA expression in transgenic mice with leptin overexpression. A direct effect of leptin on cultured mesangial cells was also described by Lee et al. [52]. This group showed that leptin increased hypertrophy of mesangial cells with a 33% increase in cell size and a 40% increase in protein synthesis [52]. The hypertrophic effect of leptin was mediated via phosphatidylinositol 3-kinase and Erk 1,2 [52]. In contrast, leptin had no effect on proliferation and apoptosis of mesangial cells.

A role of the short-form leptin receptor has been also demonstrated in a model of non-alcoholic steatohepatitis [53]. Leptin stimulated in cultured hepatocytes from *db/db* mice that only expressed Ob-Ra osteopontin, a glycosylated phosphoprotein involved in inflammatory and profibrogenic diseases [53].

It has been recently described that leptin induces oxidative stress in human endothelial cells [54]. Since an increased formation of reactive oxygen species also occurs in the kidney under various conditions and mediates growth responses as well as proinflammatory effects [55, 56], it is possible that leptin further enhances renal oxidative stress.

In which pathophysiological situations could leptin contribute to renal pathology? An increased incidence of glomerulosclerosis has been described in individuals suffering from massive obesity [57–59]. Because such obese individuals exhibit a marked increase in serum leptin concentrations, it is possible that this hormone may contribute to the development of glomerulosclerosis in this population. Such effects may be even more exaggerated in situations with loss of functioning nephrons. Praga et al. [59] reported that obese patients are at high risk for developing proteinuria and chronic renal failure after unilateral nephrectomy. Among 14 obese patients with a body mass index $>30 \text{ kg/m}^2$, 13 developed proteinuria and renal insufficiency after unilateral nephrectomy whereas only 12% developed these complications when the body mass index was $<30 \text{ kg/m}^2$ [59] Since leptin is cleared by the kidney, nephrectomy may lead to higher serum levels further contributing to the development of glomerulosclerosis.

Pathophysiological effects of leptin may be more relevant in patients with type 2 diabetes that have high serum leptin concentrations [60, 61]. Although a fibrogenic role for leptin in human renal diseases remains to be established, the current clues are suggestive.

Acknowledgments

Original work in the authors' laboratories was supported by grants from the Deutsche Forschungsgemeinschaft and the National Institutes of Health.

References

1. Ingalls AM, Dickie MM, Snell GD: Obesity, a new mutation in the house mouse. J Hered 1950;14:317–318.
2. Auwerx J, Staels B: Leptin. Lancet 1998;351:737–742.
3. Diamond FB, Eichler DC, Duckett G, Jorgensen EV, Shulman D, Root AW: Demonstration of a leptin binding factor in human serum. Biochem Biophys Res Commun 1997;233:818–822.
4. Rau H, Reavens BJ, O'Rahilly S, Whitehead JP: Truncated human leptin (?133) associated with extreme obesity undergoes proteosomal degradation after defective intracellular transport. Endocrinology 1999;140:1718–1723.
5. Nakashima K, Narazaki M, Taga T: Leptin receptor (OB-R) oligomerizes with itself but not with its closely related cytokine signal transducer gp 130. FEBS Lett 1997;403:79–82.
6. Tartaglia LA: The leptin receptor. J Biol Chem 1997;272:6093–6096.
7. Fei H, Okano HJ, Li C, Lee GH, Zhao C, Dranell R, Friedman JM: Anatomic localization of alternatively spliced leptin receptors (Ob-R) in mouse brain and other tissues. Proc Natl Acad Sci USA 1997;94:7001–7005.
8. Wang MY, Zhou YT, Newgard CB, Unger RH: A novel leptin receptor isoform in rat. FEBS Lett 1996;392:87–90.
9. Cohen B, Novick D, Rubinstein M: Modulation of insulin activities by leptin. Science 1996;274: 1185–1188.
10. Berti L, Kellerer M, Capp E, Häring HU: Leptin stimulates glucose transport and glycogen synthesis in C2C12 myotubes: evidence for a PI3-kinase mediated effect. Diabetologia 1997;40: 606–609.
11. Chen H, Charlat O, Tartaglia LA, Woolf EA, Weng X, Ellis SJ, Lakey ND, Culpepper J, Moore KJ, Breitbart RE, Duyk GM, Tepper RI, Morgenstern JP: Evidence that the diabetes gene encodes the leptin receptor: identification of a mutation in the leptin receptor gene in *db/db* mice. Cell 1996;84:491–495.
12. Fei H, Okano HJ, Li C, Lee GW, Zhao C, Darnell R, Friedman JM: Anatomic localization of alternatively spliced leptin receptors (Ob-R) in mouse brain and other tissues. Proc Natl Acad Sci USA 1997;94:7001–7005.
13. Murakami T, Yamashita T, Iida M, Kuwajima M, Shima K: A short form of leptin receptor performs signal transduction. Biochem Biophys Res Commun 1997;231:26–29.
14. White DW, Kuropatwinski KK, Devos R, Baumann H, Tartaglia LA: Leptin receptor (OB-R) signaling. J Biol Chem 1997;272:4065–4071.
15. Wolf G, Hamann A, Han DC, Helmchen U, Thaiss F, Ziyadeh FN, Stahl RAK: Leptin stimulates proliferation and TGF-β expression in renal glomerular endothelial cells: potential role in glomerulosclerosis. Kidney Int 1999;56:860–872.
16. Elmquist JK, Elisa CF, Saper CB: From lesions to leptin: hypothalamic control of food intake and body weight. Neuron 1999;22:221–232.
17. Chehab FF, Mounzih K, Lu R, Lim ME: Early onset of reproductive function in normal female mice treated with leptin. Science 1997;275:88–90.
18. Morton NM, Emilsson V, Liu YL, Cawthorne MA: Leptin action in intestinal cells. J Biol Chem 1998;273:26194–26201.

19   Wang Y, Kuropatwinski KK, White DW, Hawley TS, Hawley RG, Tartaglia LA, Baumann H: Leptin receptor action in hepatic cells. J Biol Chem 1997;272:16216–16223.
20   Barzillai N, Wang J, Massilon D, Vuguin P, Hawkins M, Rossetti L: Leptin selectively decreases visceral adiposity and enhances insulin action. J Clin Invest 1997;100:3105–3110.
21   Emilsson Y, Liu YL, Cawthorne MA, Morton NM, Davenport M: Expression of the functional leptin receptor mRNA in pancreatic islets and direct inhibitory action of leptin on insulin secretion. Diabetes 1997;46:313–316.
22   Ducy P, Amling M, Takeda S, Priemel M, Schilling AF, Beil FT, Shen J, Vinson C, Rueger JM, Karsenty G: Leptin inhibits bone formation through a hypothalamic relay: a central control of bone mass. Cell 2000;100:197–207.
23   Malamaci F, Triipepi G, Zoccali G: Leptin in end-stage renal disease (ESRD): a link between fat mass, bone and the cardiovascular system. J Nephrol 2005;18:464–468.
24   Sierra-Honigmann MR, Nath AK, Murakami C, Garcia-Cardena G, Papapetropoulos A, Sessa WC, Madge LA, Schechner JS, Schwabb MB, Polverini PJ, Flores-Riveros JR: Biological action of leptin as an angiogenic factor. Science 1998;281:1683–1686.
25   Bouloumié A, Drexler HCA, Lafontan M, Busse R: Leptin, the product of Ob gene, promotes angiogenesis. Circ Res 1998;83:1059–1066.
26   Gainsford T, Willson TA, Metcalf D, Handman E, McFarlane C, Ng A, Nicola NA, Alexander WS, Hilton DJ: Leptin can induce proliferation, differentiation, and functional activation of hemopoietic cells. Proc Natl Acad Sci USA 1996;93:14564–14568.
27   Lord GM, Matarese G, Howard JK, Baker RJ, Bloom SR, Lechler RI: Leptin modulates the T-cell immune response and reverses starvation-induced immunosuppression. Nature 1998;394: 897–901.
28   Ottonello L, Gnerre P, Bertolotto M, Mancini M, Dapino P, Russo R, Garibotto G, Barreca T, Dallegri F: Leptin as a uremic toxin interferes with neutrophil chemotaxis. J Am Soc Nephrol 2004;15:2366–2372.
29   Hama H, Saito A, Takeda T, Tanuma A, Xie Y, Sato K, Kazama JJ, Gejyo F: Evidence indicating that renal tubular metabolism of leptin is mediated by megalin but not by the leptin receptors. Endocrinology 2004;145:3935–3940.
30   Nordfors L, Lönnqvist F, Heimbürger O, Danielsson A, Schalling M, Stenvinkel P: Low leptin gene expression and hyperleptinemia in chronic renal failure. Kidney Int 1998;54:1267–1275.
31   Wiesholzer M, Harm F, Hauser AC, Pribasnig A, Balcke P: Inappropriately high plasma leptin levels in obese haemodialysis patients can be reduced by high flux haemodialysis and haemodiafiltration. Clin Sci 1998;94:431–435.
32   Tsujimoto Y, Shoji T, Tabata T, Morita A, Emoto M, Nishizawa Y, Morii H: Leptin in peritoneal dialysate from continuous ambulatory peritoneal dialysis patients. Am J Kidney Dis 1999;34:832–838.
33   Briley LP, Szczech LA: Leptin and renal disease. Semin Dial 2006;16:54–59.
34   Mak RH, Cheung W, Cone RD, Marks DL: Leptin and inflammation-associated cachexia in chronic kidney disease. Kidney Int 2006;69:794–797.
35   Sharma K, Considine RV: The Ob protein (leptin) and the kidney. Kidney Int 1998;53:1483–1487.
36   Serradeil-Le Gal C, Raufaste D, Brossard G, Pouzet B, Marty E, Maffrand JP, Le Fur G: Characterization and localization of leptin receptors in the rat kidney. FEBS Lett 1997;404: 185–191.
37   Haynes WG, Morgan DA, Walsh SA, Mark SA, Sivitz WI: Receptor-mediated regional sympathetic nerve activation by leptin. J Clin Invest 1997;100:270–278.
38   Aizawa-Abe M, Ogawa Y, Masuzaki H, Ebihara K, Satoh N, Iwai H, Matsuoka N, Hayashi T, Hosoda K, Inoue G, Yoshimasa Y, Nakao K: Pathophysiological role of leptin in obesity-related hypertension. J Clin Invest 2000;105:1243–1252.
39   Haynes WG, Sivitz WI, Morgan DA, Walsh SA, Mark AL: Sympathetic and cardiorenal actions of leptin. Hypertension 1997;30:619–623.
40   Shek EW, Brands MW, Hall JE: Chronic leptin infusion increases arterial pressure. Hypertension 1998;31:409–414.
41   Dunbar JC, Lu H: Leptin-induced increase in sympathetic nervous and cardiovascular tone is mediated by proopiomelanocortin (POMC) products. Brain Res Bull 1999;50:215–221.

42 Jackson EK, Li P: Human leptin has natriuretic activity in the rat. Am J Physiol 1997;272: F333–F338.
43 Villarreal D, Reams G, Freeman RH, Taraben A: Renal effects of leptin in normotensive, hypertensive, and obese rats. Am J Physiol 1998;275:R2056–R2060.
44 Villlarreal D, Reams G, Freeman RH: Effects of renal denervation on the sodium excretory actions of leptin in hypertensive rats. Kidney Int 2000;58:989–994.
45 Orth SR, Amann K, Strojek K, Ritz E: Sympathetic overactivity and arterial hypertension in renal failure. Nephrol Dial Transplant 2001;16(suppl 1):67–69.
46 Wolf G, Helmchen U, Stahl RAK: Isoproterenol stimulates tubular DNA-replication in mice. Nephrol Dial Transplant 1996;11:2288–2292.
47 Wolf G, Han DC, Ziyadeh FN: Leptin and renal disease. Am J Kidney Dis 2002;39:1–11.
48 Han DC, Isono M, Chen S, Hong SW, Casaretto A, Wolf G, Ziyadeh FN: Leptin stimulates type I collagen production in *db/db* mesangial cells: role of increased glucose uptake and TGF-β type II receptor expression. Kidney Int 2001;59:1315–1323.
49 Ballermann BJ: A role for leptin in glomerulosclerosis? Kidney Int 1999;56:1154–1155.
50 Sharma K, Ziyadeh FN: The emerging role of transforming growth factor-β in kidney diseases. Am J Physiol 1994;266:F829–F842.
51 Suganami T, Mukoyama M, Mori K, Yokoi H, Koshikawa M, Sawai K, Hidaka S, Ebihara K, Tanaka T, Sugawara A, Kawachi H, Vinson C, Ogawa Y, Nakao K: Prevention and reversal of renal injury by leptin in a new mouse mdoel of diabetic nephropathy. FASEB J 2005;19:127–129, DOI:10.1096/fj.04–2183fje.
52 Lee MP, Orlov D, Sweeney G: Leptin induces rat glomerular mesangial cell hypertrophy but does not regulate hyperplasia or apoptosis. Int J Obes 2005;29:1395–1401.
53 Sahai A, Malladi P, Pan X, Paul R, Melin-Aldana H, Green RM, Whitington PF: Obese and diabetic *db/db* mice develop marked liver fibrosis in a mdoel of nonalcoholic steatohepatitis: role of short-form leptin receptors and osteopontin. Am J Physiol Gastrointest Liver Physiol 2004;287:G1035–G1043.
54 Bouloumié A, Marumo T, Lafontan M, Busse R: Leptin induces oxidative stress in human endothelial cells. FASEB J 1999;13:1231–1238.
55 Hannken T, Schroeder R, Stahl RAK, Wolf G: Angiotensin II-mediated expression of p27[Kip1] and induction of cellular hypertrophy in renal tubular cells depend on the generation of oxygen radicals. Kidney Int 1998;54:1923–1933.
56 Wolf G: Free radical production and angiotensin. Curr Hypertens Rep 2000;2:167–173.
57 Kasiske BL, Crosson T: Renal disease in patients with massive obesity. Arch Intern Med 1986;146:1105–1109.
58 Cohen AH: Massive obesity and the kidney. Am J Pathol 1975;81:117–130.
59 Praga M, Hernandez E, Herrero JC, Morales E, Revilla Y, Diaz-Gonzalez R, Rodicio JL: Influence of obesity on the appearance of proteinuria and renal insufficiency after unilateral nephrectomy. Kidney Int 2000;58:2111–2118.
60 Widjaja A, Si M, Horn R, Holman RR, Turner R, Brbant G: UKPDS 20- plasma leptin, obesity, and plasma insulin in type 2 diabetic subjects. J Clin Endo Metab 1997;82:654–657.
61 Fruehwald-Schultes B, Kern W, Beyer J, Forst T, Pfutzner A, Peters A: Elevated serum leptin concentrations in type 2 diabetic patients with microalbuminuria and macroalbuminuria. Metabolism 1999;48:1290–1293.

Gunter Wolf, MD, Professor of Medicine
Department of Internal Medicine III, Friedrich-Schiller-University
Erlanger Allee 101
DE–07740 Jena (Germany)
Tel. +49 3641 9324301, Fax +49 3641 9324302, E-Mail Gunter.Wolf@med.uni-jena.de

Obesity and Renal Hemodynamics

R.J. Bosma, J.A. Krikken, J.J. Homan van der Heide, P.E. de Jong, G.J. Navis

Department of Medicine, Division of Nephrology, University Medical Center Groningen, Groningen, The Netherlands

Abstract

Obesity is a risk factor for renal damage in native kidney disease and in renal transplant recipients. Obesity is associated with several renal risk factors such as hypertension and diabetes that may convey renal risk, but obesity is also associated with an unfavorable renal hemodynamic profile independent of these factors, and that may exert effects on renal damage as well. In animal models of obesity-associated renal damage, micro-puncture studies showed glomerular hypertension and hyperfiltration. In humans an elevated glomerular filtration rate has been demonstrated in several studies, sometimes associated with hyperperfusion as well, independent of blood pressure or the presence of diabetes. An elevated filtration fraction was found in several studies, consistent with glomerular hypertension. This renal hemodynamic profile resembles the hyperfiltration pattern in diabetes and is therefore assumed to be a pathogenetic factor in renal damage. Of note, the association between body mass index and renal hemodynamics is not limited to overt obesity or overweight, but is also present across the normal range, without a particular threshold. Multiple factors are assumed to contribute to these renal hemodynamic alterations, such as insulin resistance, the renin–angiotensin system and the tubulo-glomerular responses to increased proximal sodium reabsorption, and possibly also inappropriate activity of the sympathetic nervous system and increased leptin levels. Obesity has a high world-wide prevalence. On a population-basis, therefore, its contribution to long-term renal risk may be considerable, especially as it is usually clustered with risk factors like hypertension and insulin-resistance. In short-term studies the renal hemodynamic alterations in obesity and the associated proteinuria were reversible by weight loss, and renin–angiotensin system-blockade, respectively. These interventions are therefore likely to have the potential to limit the renal risks of obesity.

Copyright © 2006 S. Karger AG, Basel

Introduction

Obesity is a risk factor for renal damage in native kidney disease [1–3] and in renal transplant recipients [4–9]. Several factors that are usually associated with obesity, such as hypertension, insulin resistance and diabetes, are likely to be involved in the increased renal risk in obesity. However, also in the absence of these factors, obesity is associated with an unfavorable renal hemodynamic profile that may play a role in the susceptibility and progression of chronic renal damage. In this review, we will give an overview of the impact of obesity and overweight on renal hemodynamics, its possible underlying mechanisms and the implications for long-term renal risk.

Measurement of Renal Hemodynamics: Limitations for Obesity Studies

Current evidence on the role of renal hemodynamics as a factor in progressive renal function loss has mainly been derived from rat studies, where micropuncture allows direct assessment of glomerular flow and pressure. In remnant kidney models it has convincingly been shown that changes in renal hemodynamics leading to glomerular hypertension are important pathogenetic factors in progressive renal damage [10–14]. In animal models of obesity it has been shown that alterations in the glomerular micro-circulation resemble those after renal ablation, with glomerular hypertension and hyperfiltration at the single nephron level [15, 16], as will be discussed in more detail in the next paragraph.

In humans, on the other hand, glomerular flow and filtration pressure cannot directly been measured. Current knowledge on renal hemodynamics in man is therefore derived from indirect assessment of renal hemodynamics by clearance techniques. Accurate assessment of glomerular filtration rate (GFR) and effective renal plasma flow (ERPF) can be obtained by measuring the renal clearance of specific markers, such as ^{125}I-iothalamate, ^{99m}Tc-diethylenethiaminepenta-acetic acid, ^{51}Cr-ethylenediaminetetra-acetic acid, or inulin for GFR, and ^{131}I-hippurate or para-amino-hippuric acid for ERPF, respectively. When GFR and ERPF are measured simultaneously, filtration fraction (FF) can be calculated as the ratio from GFR and ERPF. FF can be a helpful parameter to interpret changes in GFR in terms of the underlying hemodynamic changes. When a rise in GFR is primarily related to hyperperfusion, it will be associated with a proportional rise in ERPF and thus unaltered FF. When a rise in GFR is primarily due to a change in filtration pressure, this will be apparent as a rise in FF. A rise in filtration pressure is assumed to be unfavorable in terms of long-term renal risk, as directly proven in animal studies. This assumption is supported by indirect evidence in human studies, demonstrating the prognostic impact of changes in FF during therapy for long-term renal outcome [17, 18]. For these reasons, FF is considered a surrogate parameter for glomerular hypertension in man.

The assessment of glomerular hyperfiltration by clearance studies, however, can be misleading, as clearance measurements reflect the total filtration rate of all nephrons together, so a normal GFR can reflect a normal GFR in a normal nephron number, but also hyperfiltration in a decreased number of nephrons. Only if total GFR is elevated, one can be sure that single nephron GFR is elevated as well.

Measurements of GFR and ERPF by specific tracers, albeit the golden standard, are expensive and time consuming. Therefore, for clinical and epidemiological purposes GFR is usually estimated from serum creatinine by renal function equations that include antropometric indices such as age, weight and gender to account for between-individual differences in muscle mass and the consequent differences in creatinine generation. Many equations are available, mostly empirically developed in populations with native kidney disease [19–36]. However, these equations are subject to bias inherent to their algorithm. The association between obesity and renal function can be confounded in particular, as in obesity a higher body weight is mainly due to a higher fat mass, whereas the factor body weight in the equation is assumed to reflect muscle mass. We evaluated this bias in transplant recipients, showing that, at higher body mass index (BMI), the MDRD equation progressively underestimated, and the Cockcroft-Gault equation progressively overestimated GFR [37]. In native kidney disease, in a relatively small population, the Cockcroft-Gault equation underestimated GFR at low BMI, and overestimated GFR at high BMI, whereas for the MDRD equation the BMI-dependent bias was of borderline significance, and the recently developed equation by Rule et al. [33] showed progressive overestimation of GFR at high BMI. [38] Thus, bias can erroneously suggest presence of hyperfiltration at high BMI! These shortcomings hamper the interpretation of epidemiological analyses of the association between obesity and renal function, and development of an equation that accounts for differences in body composition would be highly relevant for epidemiological analyses on obesity and renal function.

Another issue relevant to renal hemodynamic assessment in obesity is the normalization of data to body surface area (BSA). This applies to renal function equations as well as accurate renal function measurements. Renal function indices are usually expressed per 1.73 m^2 BSA, as standard reference body size. However, excess weight does not only lead to a rise in BMI, but also to a rise in BSA. Thus, when an individual gains weight, BSA-normalized kidney function would 'decrease', in the absence of a true change in renal function. Likewise, BSA-normalization may confound comparisons between obese subjects and non-obese control populations, as illustrated by Schmieder et al. [39] who found a lower BSA-corrected ERPF in obese subjects compared to lean subjects, whereas height-corrected and uncorrected ERPF were not different. Similar findings, also for GFR, were shown by Anastasio et al. [40]. For these reasons it has been proposed that normalization for height would be more appropriate, albeit no consensus exists as yet [41]. At any rate, FF, the ratio of GFR and ERPF, is not affected by the adopted method of normalization, and can be interpreted to reflect true differences between obese and non-obese subjects.

Impact of Obesity on Renal Hemodynamics

Several experimental studies addressed the impact of obesity on renal hemodynamics. Multiple studies were done in obese Zucker rats, a genetic model of obesity associated with slowly progressive renal damage, proteinuria and focal glomerulosclerosis. Micro-puncture studies [15, 16, 42, 43] revealed an elevated GFR, which was attributable to an increased single nephron plasma flow rate as well as glomerular transcapillary hydraulic pressure. The elevated GFR, however, was not a prerequisite for the development of structural damage in this model [42, 43]. Also, in diet-induced animal models of obesity, an elevated GFR is usually documented as reviewed elsewhere [44].

Human studies on the impact of obesity and overweight on renal hemodynamics are summarized in table 1. To avoid the BMI-dependent bias inherent to renal function equations, only studies using accurate clearance methods are shown here. It shows that an elevated GFR was found in most, but not all studies. Frequently, but not always, this was accompanied by an elevated ERPF and/or an elevated FF. From measuring dextran sieving curves, and entering these data in a theoretical model on the determinants of filtration, Chagnac et al. [2] could attribute the elevated GFR in severe obesity to an increase in transcapillary hydraulic pressure, ΔP, supporting the presence of glomerular hypertension. A more recent study of this group is of special interest as well, as it provides renal hemodynamic data before and after weight loss by bariatric surgery [45]. A massive weight loss of 48 kg in nine morbidly obese, non-diabetic subjects over a period of 12–17 months, with a decrease in mean BMI from 48 to 32 kg/m^2, resulted in a decrease in GFR of 145 ± 14 to 110 ± 7 ml/min (vs. 92 ± 4 ml/min in controls), with corresponding decreases in ERPF, FF, albuminuria, and blood pressure, along with improvement in insulin resistance.

For the five studies that analyzed both GFR and renal perfusion [2, 40, 41, 45, 46] in obese subjects and controls an overview of the mean values for GFR, ERPF and FF is given in figure 1, ranked by BMI (from left to right). For the two studies that provide data for hypertensives separately [41, 46] data are given by a break-up by presence or absence of hypertension. The overview shows that in the studies reporting a difference between obese and non-obese subjects, the difference is usually in the same direction, namely hyperfiltration

Table 1. Human studies on overweight and obesity and renal hemodynamics

Reference	n	Population	BMI	GFR	ERPF	Normalization for	Renal hemodynamics in obese subjects vs. non-obese controls
Reisin et al. [104]	42	HT	31	–	Hippurate	Height	↑ RBF in normotensive and hypertensive obese
Ribstein et al. [41]	60	HC/HT obese, non-obese	32	Tc-DTPA	Hippurate	Height	↑ GFR and ↑ ERPF in obesity, ↑ FF in HT obese only
Scaglione et al. [46]	64	HC/HT	34	Tc-DTPA	Hippurate	–	ERPF↓ and ↑ FF in centrally obese HT
Porter and Hollenberg [48]	45	Healthy moderately obese	29	–	Xenon	Kidney volume	↑ RBF in obese, on low sodium diet only
Anastasio et al. [40]	20	Severely overweight, normotensive	47	Insulin	Hippurate	BSA	No differences
Chagnac et al. [2]	12	Severely obese with IR	44	Insulin	Hippurate	–	↑ GFR; ↑ ERPF; ↑ FF
Chagnac et al. [45]	8	Severe obesity/IR, weight reduction	48	Insulin	Hippurate	–	↑ GFR; ↑ ERPF; ↑ FF correction by weight loss
Bosma et al. [49]	102	Healthy	24	Iothalamate	Hippurate	BSA and height	↑ FF with higher BMI

BMI = Body mass index; BSA = body surface area; ERPF = effective renal plasma flow; GFR = glomerular filtration rate; HC = healthy controls; HT = hypertensive subjects; IR = insulin resistance; n = number of overweight/obese patients; RBF = renal blood flow.

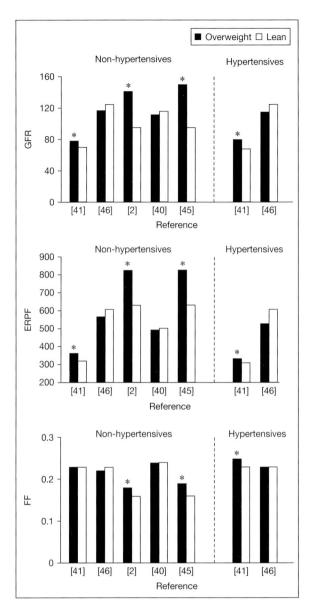

Fig. 1. Five studies [2, 40, 41, 45, 46] showing the impact of overweight and obesity on GFR, ERPF and FF, respectively. Bars left of dashed line show non-hypertensive subjects, bars right of dashed line show the hypertensive subpopulation of studies 1 and 2. For reference [46], data of the centrally and peripherally obese are presented together. *$p < 0.05$.

(elevated GFR), hyperperfusion (elevated ERPF) or elevated glomerular pressure (FF). In two of the studies an effect of obesity was present only under specific conditions. Ribstein et al. [41] found that obesity was associated with altered renal hemodynamics in hypertensive, but not in normotensive subjects, and Scaglione et al. [46] found that FF was elevated (and renal plasma flow reduced) in hypertensives with central obesity (presumably reflecting insulin resistance) only, but not in normotensive subjects with central or peripheral obesity, or in hypertensives with peripheral obesity. These data illustrate the relevance of the interaction between obesity/overweight and its frequent co-morbid conditions such as hypertension and insulin resistance for renal hemodynamics. The presence of hypertension may result in differences in ERPF in particular as in established hypertension usually a decreased ERPF is found, with a preserved GFR, and consequently and elevated FF [47]. Thus, as to FF the effects of hypertension and obesity are in the same direction, but as to ERPF their effects are opposite. Moreover, sodium intake may be relevant as well; Porter and Hollenberg [48] found an elevated renal plasma flow in moderately obese, healthy hypertensives as compared to normotensive controls during a severe dietary sodium restriction, but not during a liberal sodium diet. The mechanism underlying this effect of sodium status has not been established.

Overweight and obesity can modulate renal hemodynamics also in the absence of co-morbid conditions, as apparent from the data in the non-hypertensives in figure 1 and from data from our own group [49]. We found a relationship between a higher BMI and a higher FF in a population of normal subjects, well-documented to be healthy and with a BMI not exceeding $30 \, kg/m^2$ (figure 2). This demonstrates, first, that excess weight is associated with altered renal hemodynamics even in the absence of overt obesity without an apparent lower threshold, and second that the impact of weight excess on glomerular hemodynamics does not depend on presence of co-morbid conditions. Nevertheless, as apparent from the above studies, when hypertension and/or insulin resistance are concomitantly present, their renal manifestations may interact with those of the excess weight, and thus account for some of the discrepancies between studies.

It should be noted that many other factors can modulate the effects of overweight and obesity on renal hemodynamics as well. These include age [50], sodium intake, drug use, and possibly also the severity of obesity, although the overview in figure 1 does not support the latter. The respective effects of age and BMI on GFR are illustrated in figure 3, showing cross-sectional data in healthy subjects. BMI-associated hyperfiltration was only present in young adults, but in older subjects GFR was similar for those with a BMI above or below the median. This could suggest a steeper age-related decline in subjects with higher BMI, but also that BMI-associated hyperfiltration in young adults is relatively innocent in terms of renal risk at an older age. However, these are

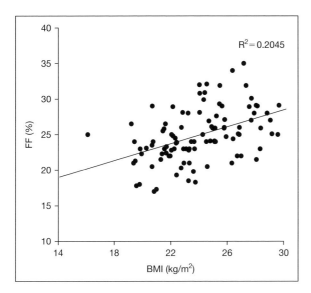

Fig. 2. Scatterplot depicting the univariate correlation between BMI and FF in healthy subjects. Adapted from reference [49].

cross-sectional data, in selected healthy subjects with a BMI not exceeding 30 kg/m². A proper interpretation in terms of presence of absence of long-term renal risk induced by hyperfiltration would therefore require longitudinal data, also in subjects with the usual co-morbid conditions.

It should be noted moreover, that renal hemodynamic status in adults and older subjects also depends to a considerable extent on their prior medical history as regards longstanding hypertension and cardiovascular disease and the resulting presence of subtle renal end organ damage [50, 51]. Obviously, this is all the more true for populations with an underlying primary renal condition. Whether BMI modulates renal hemodynamics in patients with a primary renal disorder would be of great interest, but has not been investigated so far. Recently, in transplant recipients, preliminary data support the impact of BMI on GFR and FF, with a higher GFR and FF at 1 year after transplantation in recipients with a higher BMI [52].

Mechanisms Underlying the Obesity-Related Changes in Renal Hemodynamics

Together, the above data suggest that obesity and overweight are associated with an altered afferent–efferent glomerular vasomotor balance. A decreased

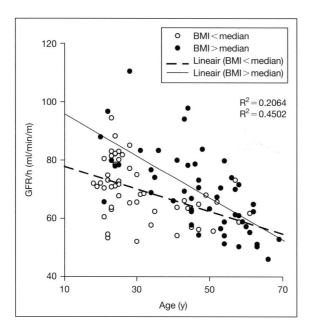

Fig. 3. Scatterplot showing the association between age and GFR in a healthy study population with a break-up by median BMI (median: 23.9 kg/m^2), showing a more steep decline with age in subjects with higher BMI, mainly due to a higher GFR at younger age. Note that GFR is presented in ml/min per meter of height.

afferent arteriolar tone is present, that allows a larger transmission of the systemic arterial pressure to the glomerular capillary bed [53] and an increase in filtration as well as perfusion. As apparent from an elevated FF, when combined with a relative increase of efferent arteriolar tone the rise in filtration is more prominent than the rise in perfusion. The elevated glomerular transcapillary hydraulic pressure as calculated by Chagnac et al. [2] is in line with this assumption. Extracellular volume expansion, due to the obesity-associated increase in sodium reabsorption, can also be expected to contribute to renal hyperperfusion and hyperfiltration. Furthermore, glomerulomegaly, as documented in obesity-related focal glomerulosclerosis [54], should be considered as a possible factor involved in the hyperfiltration and hyperperfusion of obesity. However, data on glomerular morphology are available only from subjects in whom proteinuria and renal function impairment warrant a renal biopsy, and no data are available to substantiate this assumption for subjects with less severe renal involvement.

Insulin Resistance

Excess weight gain, especially when associated with visceral obesity, leads to glucose intolerance, insulin resistance and a compensatory hyperinsulinemia [55], which has been associated with sodium retention as well as increased GFR [2, 56]. Dengel et al. [57] demonstrated a close correlation between insulin resistance and renal hemodynamics in obese, mildly hypertensive, older subjects, in whom a lower glucose disposal rate during hyperinsulinemic euglycemic clamping correlated with a higher GFR and FF. This association was blunted by a high sodium diet, i.e. a condition of suppression of the renin–angiotensin system. The mechanism underlying the link between insulin resistance and renal hemodynamic alterations in obesity has not been elucidated yet. Hyperinsulinemia has been suggested to be involved by its stimulating effects on sympathetic nervous system activity and vasoconstrictor effects [58–61]. However, experimental hyperinsulinemia does not lead to elevated FF, neither in acute nor in chronic experiments [56, 62, 63]. Whereas, the effects of exogenous insulin in an experimental set-up may not be completely similar to those of endogenous hyperinsulinemia, this renders a direct effect of high insulin levels less likely, and indirect effects are more likely.

Indirect evidence from epidemiological data suggests that the effects of insulin resistance, as estimated from presence of a central fat distribution, can dissociate from those of overweight. In the general population-based Prevend cohort, obesity was associated with hyperfiltration, whereas a central fat distribution was associated with a greater risk for mild renal function impairment not only in obese, but also in lean subjects [64]. Thus, the renal risks of obesity may be related more closely to the concomitant presence of insulin resistance rather than with the obesity as such. Whether this relates to specific renal hemodynamic factors has not been established.

Renin–Angiotensin System

Several lines of evidence suggest the renin–angiotensin system (RAS) to be involved in the renal hemodynamic changes in obesity. Circulating components of the RAS are elevated in obesity in experimental animals [65] and in humans. In this respect, activation of the RAS in adipose tissue has been proposed to provide a link between obesity and hypertension [66] by production of angiotensinogen – the substrate for renin – by adipose tissue [67]. The resulting generation of the effector hormone angiotensin II can elicit hypertension by its vasoconstrictor effects on the peripheral vascular bed and by promoting tubular sodium reabsorption. Moreover, in the kidney it will elicit efferent vasoconstriction, as manifest from a rise in FF. Several studies reported correlations between plasma angiotensinogen concentrations, blood pressure and BMI [68–70]. Moreover [71], an effective weight loss program reduced circulating angiotensinogen, plasma renin activity, aldosterone and ACE activity, as well as adipose tissue expression of angiotensinogen. In humans, Scaglione et al. [46] found significantly higher plasma renin levels in centrally obese normotensive subjects compared to peripheral obese and lean subjects. Ruano et al. [72] retrospectively studied 100 morbidly obese patients, split up for central and peripheral obesity. In centrally obese patients plasma renin activity, aldosterone and ACE levels were elevated, and associated with sodium retention, potassium loss and high insulin levels. After gastric bypass these abnormal hormone levels tended to normalize. Interestingly, Hopkins et al. [73] showed a blunted renal vascular response to angiotensin II infusion in subjects with higher BMI, suggesting increased intrarenal RAS activity as well. In line with this, a study in mice showed that obesity was associated with a tissue-specific increase in ACE activity in the kidney [74], but otherwise data on the intrarenal RAS are scarce.

The renal impact of increased activity of the RAS can also be inferred from the renal hemodynamic response to RAS-blockade. The renal vasodilator responses to RAS blockade were shown to be directly proportional to BMI in two studies. In 100 healthy, normotensive, predominantly overweight and obese subjects [75] a highly significant relationship was found between the age- and plasma renin activity-adjusted BMI and the renal plasma flow response to captopril, supporting the assumption that a BMI-associated higher intrarenal RAS activity is involved in renal vascular tone in overweight and obesity. A similar relationship was found by Price et al. [76] in 12 type II diabetic patients with a wide range of BMI, in whom the rise in ERPF elicited by angiotensin-receptor blocker irbesartan was directly proportional to BMI, with BMI accounting for 50% of the variation in the renal hemodynamic response. These BMI-dependent renal effects of RAS-blockade also suggest that RAS-blockade may be a suitable tool to reduce the renal risks of obesity – assuming that the renal hemodynamic profile indeed reflects a long-term risk.

In accord with this assumption, RAS-blockade was shown to reduce proteinuria in obese proteinuric patients. Interestingly, in a small study weight loss was as effective as RAS-blockade in reducing proteinuria [77]. Whether the antiproteinuric effect of weight loss relates to renal hemodynamic changes has not been established, but theoretically, the reduction in GFR and FF that would be anticipated to occur with weight loss can be expected to lead to reduction of proteinuria.

Interaction between Renal Sodium Handling and
Renal Hemodynamics in Obesity
Altered tubular sodium handling can affect renal hemodynamics by the tubuloglomerular feedback (TGF) mechanism. TGF regulates glomerular hemodynamics by modifying afferent arteriolar tone in response to changes in

NaCl delivery to the macula densa. A lower distal NaCl supply elicits a decrease in afferent tone, thus eliciting a rise in glomerular filtration pressure and filtration rate. The ensuing rise in filtered load then restores NaCl supply to the macula densa towards its original level [78]. Obesity is well-established to be associated with increased sodium reabsorption [79–81], presumably at the level of Henle's loop. Increased TGF activity can thus contribute to the afferent vasodilation observed in obesity.

The impaired sodium excretion in obesity is related to an impaired pressure natriuresis response, in particular in central obesity [82–84], and has been attributed to increased activity of the RAS, of the sympathetic nervous system [75, 85–87], hyperinsulinemia [88], as well as to compression of the kidneys by excess retroperitoneal adipose tissue [89]. The impaired sodium excretion of obesity can be considered to act as a two-edged sword. First, it contributes to volume expansion and obesity hypertension. The latter is well-established to be sensitive to sodium intake, and interestingly, the sodium-sensitivity of blood pressure is attenuated by weight-loss [87]. Second, by its effect on afferent tone it impairs the glomerular protection against elevated systemic pressure. This provides a very plausible candidate mechanism for the renal susceptibility in obesity hypertension, although for obvious reasons no human data are available to support this assumption.

The combination of sodium-sensitive blood pressure and an unfavorable renal hemodynamic profile is not unique to obesity. It is also observed in sodium-sensitive black hypertensives [90], non-modulating hypertension, and sodium-sensitive hypertensives with an unfavorable cardiovascular risk profile and micro-albuminuria [91–94], conditions that appear to share an elevated cardiovascular risk as well, and that also display a considerable overlap (fig. 4). The maladaptation to sodium in these conditions may be worthwhile to consider for its clinical implications, as excess sodium not only increases blood pressure, but can also induce adverse renal effects, with a further rise in glomerular pressure and micro-albuminuria [95] as demonstrated in high risk sodium-sensitive hypertensives. Whereas in obesity the main dietary measures should obviously be aimed at weight reduction, moderate dietary sodium restriction may be a useful adjunct to improve the obesity-associated risk profile. As the studies by Porter and Hollenberg [48] and Dengel et al. [57], however, suggest that the renal hemodynamic abnormalities of obesity can be unmasked, or promoted by sodium restriction, the renal hemodynamic effects of dietary sodium restriction in obesity warrant further study.

Leptin and the Sympathetic Nervous System

Other mechanisms suggested to be involved in the renal effects of obesity are increased activity of the sympathetic nervous system, and elevated leptin

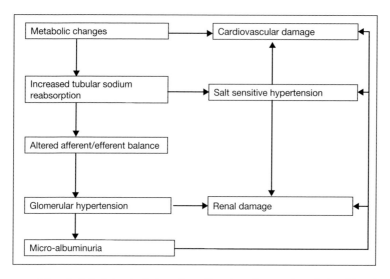

Fig. 4. Schedule depicting the alleged pathogenetic relationships between renal sodium handling, renal hemodynamics, sodium-sensitive hypertension, and cardiovascular and renal end-organ damage in obesity and other conditions.

levels. Increased activity of the sympathetic nervous system is well-documented in obesity, which includes an increased renal norepinephrin spill-over indicating increased intrarenal sympathetic activity [79, 96, 97], and studies in experimental animals have shown its involvement in the sodium-retention and hypertension of obesity [79]. However, its role in the obesity-associated changes in renal hemodynamics is not well-documented. Leptin, an adipocyte-derived hormone that is involved in the regulation of weight by effects on appetite and energy expenditure, leads to sympathic coactivation of the kidneys and adrenal glands, suggesting that the obesity-associated increase in sympathetic nerve activity could be due in part to these sympathetic effects of leptin [96, 97]. Human obesity is often associated with leptin resistance, providing sympathetic underactivity with subsequent positive energy balance and weight gain. However, leptin resistance does not lead to sympathetic underactivity in the kidneys. Rahmouni et al. [98] examined the role of leptin in hypertension associated with diet-induced obesity in mice and demonstrated a preservation of the renal sympathetic but also of arterial pressure response to leptin, which could be a candidate mechanism for the elevated cardiovascular risk in obesity. The other way round leptin deficiency, as in the ob/ob mouse, leads to decreased arterial pressure, despite severe obesity [99]. Whereas these results support a role for leptin in obesity hypertension, however, leptin does not seem to exert a clear-cut effect on renal hemodynamics [100].

Implications of Altered Renal Hemodynamics in Obesity

By analogy with diabetic hyperfiltration and with remnant kidney models in experimental animals [10–13], the renal hemodynamic profile with glomerular hyperfiltration and hypertension in obesity is assumed to be a pathogenetic factor in the susceptibility to progressive renal function loss by hypertensive glomerular capillary damage and the ensuing increased albumin leakage [2]. In micro-puncture studies in obese Zucker rats the elevated glomerular pressure precedes the development of focal sclerosis [16], but in other studies glomerular hyperfiltration was not a prerequisite for the development of structural renal damage [42, 43], suggesting that renal hemodynamic changes are not a crucial pathogenetic factor in this model. Moreover, RAS blockade protects against focal sclerosis in this model without affecting renal hemodynamics [16], so the pathogenetic role of the renal hemodynamics in obesity-associated renal damage is not straightforward, and presumably involves interaction with other obesity-associated factors.

In human, in spite of the recognition of obesity as an increasingly important renal risk factor [1, 101], no longitudinal data are present to document a pathogenetic role of renal hemodynamics in the elevated renal risk of obesity. Nevertheless, it is logical to assume that the obesity-associated renal hemodynamic profile increases the renal susceptibility to hypertensive damage, by combined effects of afferent vasodilation with impaired protection against elevated systemic arterial pressure, and an elevated efferent vascular tone that further increases glomerular pressure. Considering the usual co-existence of obesity and hypertension this alleged mechanism warrants further exploration. It is supported by the association of obesity, glomerular hyperfiltration and albuminuria in many experimental and clinical studies [2, 102]. The potential of RAS-blockade to alleviate the renal hemodynamic consequences of obesity, along with a reduction in blood pressure, suggests that RAS-blockade may be a fruitful intervention to reduce the long-term renal risks of obesity, an assumption that is supported by its beneficial effects on obesity-associated proteinuria [77].

Conclusion

Obesity and overweight are associated with altered renal hemodynamics, usually characterized by elevated filtration that can also be associated with elevated perfusion, elevated filtration pressure, or both. By analogy with diabetes and other hyperfiltration conditions, this renal hemodynamic profile is assumed to contribute to the elevated renal risk in obesity, in interaction with other obesity-associated factors like hypertension and insulin-resistance, although direct evidence for a pathogenetic role of renal hemodynamics in man is lacking. In short-term studies, the renal hemodynamic alterations in obesity and the associated proteinuria were reversible by weight loss, and by RAS-blockade, respectively. These interventions are therefore likely to have the potential to limit the renal risks of obesity. The effects of novel weight loss regimens based on cannabinoid CB1 receptor antagonism [103] on renal hemodynamics and renal risk remains to be established.

References

1 Bonnet F, Deprele C, Sassolas A, Moulin P, Alamartine E, Berthezene F, et al: Excessive body weight as a new independent risk factor for clinical and pathological progression in primary IgA nephritis. Am J Kidney Dis 2001;37:720–727.
2 Chagnac A, Weinstein T, Korzets A, Ramadan E, Hirsch J, Gafter U: Glomerular hemodynamics in severe obesity. Am J Physiol Renal Physiol 2000;278:F817–F822.
3 Kasiske BL, Napier J: Glomerular sclerosis in patients with massive obesity. Am J Nephrol 1985;5:45–50.
4 de Vries AP, Bakker SJ, van Son WJ, van der Heide JJ, Ploeg RJ, The HT, et al: Metabolic syndrome is associated with impaired long-term renal allograft function; not all component criteria contribute equally. Am J Transplant 2004;4:1675–1683.
5 el Agroudy AE, Wafa EW, Gheith OE, Shehab el-Dein AB, Ghoneim MA: Weight gain after renal transplantation is a risk factor for patient and graft outcome. Transplantation 2004;77:1381–1385.
6 Halme L, Eklund B, Kyllonen L, Salmela K: Is obesity still a risk factor in renal transplantation? Transpl Int 1997;10:284–288.
7 Meier-Kriesche HU, Vaghela M, Thambuganipalle R, Friedman G, Jacobs M, Kaplan B: The effect of body mass index on long-term renal allograft survival. Transplantation 1999;68: 1294–1297.
8 Meier-Kriesche HU, Arndorfer JA, Kaplan B: The impact of body mass index on renal transplant outcomes: a significant independent risk factor for graft failure and patient death. Transplantation 2002;73:70–74.
9 Mitsnefes MM, Khoury P, McEnery PT: Body mass index and allograft function in pediatric renal transplantation. Pediatr Nephrol 2002;17:535–539.
10 Brenner BM: Hemodynamically mediated glomerular injury and the progressive nature of kidney disease. Kidney Int 1983;23:647–655.
11 Hostetter TH, Olson JL, Rennke HG, Venkatachalam MA, Brenner BM: Hyperfiltration in remnant nephrons: a potentially adverse response to renal ablation. Am J Physiol 1981;241:F85–F93.
12 Yoshida Y, Fogo A, Shiraga H, Glick AD, Ichikawa I: Serial micropuncture analysis of single nephron function in subtotal renal ablation. Kidney Int 1988;33:855–867.
13 Yoshioka T, Shiraga H, Yoshida Y, Fogo A, Glick AD, Deen WM, et al: 'Intact nephrons' as the primary origin of proteinuria in chronic renal disease. Study in the rat model of subtotal nephrectomy. J Clin Invest 1988;82:1614–1623.
14 Hall RL, Wilke WL, Fettman MJ: Captopril slows the progression of chronic renal disease in partially nephrectomized rats. Toxicol Appl Pharmacol 1985;80:517–526.
15 Park SK, Kang SK: Renal function and hemodynamic study in obese Zucker rats. Korean J Intern Med 1995;10:48–53.
16 Schmitz PG, O'Donnell MP, Kasiske BL, Katz SA, Keane WF: Renal injury in obese Zucker rats: glomerular hemodynamic alterations and effects of enalapril. Am J Physiol 1992;263 (pt 2):F496–F502.
17 Apperloo AJ, de Zeeuw D, de Jong PE: A short-term antihypertensive treatment-induced fall in glomerular filtration rate predicts long-term stability of renal function. Kidney Int 1997;51:793–797.

18 Hansen HP, Rossing P, Tarnow L, Nielsen FS, Jensen BR, Parving HH: Increased glomerular filtration rate after withdrawal of long-term antihypertensive treatment in diabetic nephropathy. Kidney Int 1995;47:1726–1731.
19 Agarwal R, Nicar MJ: A comparative analysis of formulas to predict creatinine clearance (abstract). J Am Soc Nephrol 1994;5:386.
20 Baracskay D, Jarjoura D, Cugino A, Blend D, Rutecki GW, Whittier FC: Geriatric renal function: estimating glomerular filtration in an ambulatory elderly population. Clin Nephrol 1997;47:222–228.
21 Bjornsson TD: Use of serum creatinine concentrations to determine renal function. Clin Pharmacokinet 1979;4:200–222.
22 Cockcroft DW, Gault MH: Prediction of creatinine clearance from serum creatinine. Nephron 1976;16:31–41.
23 Edwards KDG, Whyte HM: Plasma creatinine level and creatinine clearance as tests of renal function. Aust Ann Med 1959;8:218–224.
24 Gates GF: Creatinine clearance estimation from serum creatinine values: an analysis of three mathematical models of glomerular function. Am J Kidney Dis 1985;5:199–205.
25 Hull JH, Hak LJ, Koch GG, Wargin WA, Chi SL, Mattocks AM: Influence of range of renal function and liver disease on predictability of creatinine clearance. Clin Pharmacol Ther 1981;29:516–521.
26 Jelliffe RW: Estimation of creatinine clearance when urine cannot be collected. Lancet 1971;1:975–976.
27 Jelliffe RW: Letter: Creatinine clearance: bedside estimate. Ann Intern Med 1973;79:604–605.
28 Levey AS, Bosch JP, Lewis JB, Greene T, Rogers N, Roth D: A more accurate method to estimate glomerular filtration rate from serum creatinine: a new prediction equation. Modification of Diet in Renal Disease Study Group. Ann Intern Med 1999;130:461–470.
29 Mawer GE, Lucas SB, Knowles BR, Stirland RM: Computer-assisted prescribing of kanamycin for patients with renal insufficiency. Lancet 1972;1:12–15.
30 Nankivell BJ, Gruenewald SM, Allen RD, Chapman JR: Predicting glomerular filtration rate after kidney transplantation. Transplantation 1995;59:1683–1689.
31 Nankivell BJ, Chapman JR, Allen RD: Predicting glomerular filtration rate after simultaneous pancreas and kidney transplantation. Clin Transplant 1995;9:129–134.
32 Nguyen HT, Shannon AG, Coates PA, Owens DR: Estimation of glomerular filtration rate in type II (non-insulin dependent) diabetes mellitus patients. IMA J Math Appl Med Biol 1997;14:151–160.
33 Rule AD, Gussak HM, Pond GR, Bergstralh EJ, Stegall MD, Cosio FG, et al: Measured and estimated GFR in healthy potential kidney donors. Am J Kidney Dis 2004;43:112–119.
34 Salazar DE, Corcoran GB: Predicting creatinine clearance and renal drug clearance in obese patients from estimated fat-free body mass. Am J Med 1988;84:1053–1060.
35 Toto RD, Kirk KA, Coresh J, Jones C, Appel L, Wright J, et al: Evaluation of serum creatinine for estimating glomerular filtration rate in African Americans with hypertensive nephrosclerosis: results from the African-American Study of Kidney Disease and Hypertension (AASK) Pilot Study. J Am Soc Nephrol 1997;8:279–287.
36 Walser M, Drew HH, Guldan JL: Prediction of glomerular filtration rate from serum creatinine concentration in advanced chronic renal failure. Kidney Int 1993;44:1145–1148.
37 Bosma RJ, Doorenbos CR, Stegeman CA, van der Heide JJ, Navis G: Predictive performance of renal function equations in renal transplant recipients: an analysis of patient factors in bias. Am J Transplant 2005;5:2193–2203.
38 Cirillo M, Anastasio P, De Santo NG: Relationship of gender, age, and body mass index to errors in predicted kidney function. Nephrol Dial Transplant 2005;20:1791–1798.
39 Schmieder RE, Beil AH, Weihprecht H, Messerli FH: How should renal hemodynamic data be indexed in obesity? J Am Soc Nephrol 1995;5:1709–1713.
40 Anastasio P, Spitali L, Frangiosa A, Molino D, Stellato D, Cirillo E, et al: Glomerular filtration rate in severely overweight normotensive humans. Am J Kidney Dis 2000;35:1144–1148.
41 Ribstein J, du CG, Mimran A: Combined renal effects of overweight and hypertension. Hypertension 1995;26:610–615.

42 Kasiske BL, Cleary MP, O'Donnell MP, Keane WF: Effects of genetic obesity on renal structure and function in the Zucker rat. J Lab Clin Med 1985;106:598–604.
43 O'Donnell MP, Kasiske BL, Cleary MP, Keane WF: Effects of genetic obesity on renal structure and function in the Zucker rat. II. Micropuncture studies. J Lab Clin Med 1985;106:605–610.
44 Hall JE, Hildebrandt DA, Kuo J: Obesity hypertension: role of leptin and sympathetic nervous system. Am J Hypertens 2001;14(pt 2):103S–115S.
45 Chagnac A, Weinstein T, Herman M, Hirsh J, Gafter U, Ori Y: The effects of weight loss on renal function in patients with severe obesity. J Am Soc Nephrol 2003;14:1480–1486.
46 Scaglione R, Ganguzza A, Corrao S, Parrinello G, Merlino G, Dichiara MA, et al: Central obesity and hypertension: pathophysiologic role of renal haemodynamics and function. Int J Obes Relat Metab Disord 1995;19:403–409.
47 van Paassen P, de Zeeuw D, Navis G, de Jong PE: Does the renin-angiotensin system determine the renal and systemic hemodynamic response to sodium in patients with essential hypertension? Hypertension 1996;27:202–208.
48 Porter LE, Hollenberg NK: Obesity, salt intake, and renal perfusion in healthy humans. Hypertension 1998;32:144–148.
49 Bosma RJ, van der Heide JJ, Oosterop EJ, de Jong PE, Navis G: Body mass index is associated with altered renal hemodynamics in non-obese healthy subjects. Kidney Int 2004;65:259–265.
50 Lindeman RD, Tobin J, Shock NW: Longitudinal studies on the rate of decline in renal function with age. J Am Geriatr Soc 1985;33:278–285.
51 Fliser D, Franek E, Joest M, Block S, Mutschler E, Ritz E: Renal function in the elderly: impact of hypertension and cardiac function. Kidney Int 1997;51:1196–1204.
52 Bosma RJ, Kwakernaak AJ, Homan van der Heide JJ, Jong de PE, Navis G: Does a high body mass index induce glomerular hyperfiltration in renal transplant patients? Am J Transplant 2006.
53 Srivastava T: Nondiabetic consequences of obesity on kidney. Pediatr Nephrol 2006;21:463–470.
54 Praga M, Hernandez E, Morales E, Campos AP, Valero MA, Martinez MA, et al: Clinical features and long-term outcome of obesity-associated focal segmental glomerulosclerosis. Nephrol Dial Transplant 2001;16:1790–1798.
55 Landsberg L, Krieger DR: Obesity, metabolism, and the sympathetic nervous system. Am J Hypertens 1989;2(pt 2):125S–132S.
56 ter Maaten JC, Bakker SJ, Serne EH, ter Wee PM, Donker AJ, Gans RO: Insulin's acute effects on glomerular filtration rate correlate with insulin sensitivity whereas insulin's acute effects on proximal tubular sodium reabsorption correlation with salt sensitivity in normal subjects. Nephrol Dial Transplant 1999;14:2357–2363.
57 Dengel DR, Goldberg AP, Mayuga RS, Kairis GM, Weir MR: Insulin resistance, elevated glomerular filtration fraction, and renal injury. Hypertension 1996;28:127–132.
58 Landsberg L: Insulin resistance, energy balance and sympathetic nervous system activity. Clin Exp Hypertens A 1990;12:817–830.
59 Minaker KL, Rowe JW, Tonino R, Pallotta JA: Influence of age on clearance of insulin in man. Diabetes 1982;31:851–855.
60 Welle S, Lilavivathana U, Campbell RG: Increased plasma norepinephrine concentrations and metabolic rates following glucose ingestion in man. Metabolism 1980;29:806–809.
61 Rowe JW, Young JB, Minaker KL, Stevens AL, Pallotta J, Landsberg L: Effect of insulin and glucose infusions on sympathetic nervous system activity in normal man. Diabetes 1981;30:219–225.
62 Hall JE, Brands MW, Mizelle HL, Gaillard CA, Hildebrandt DA: Chronic intrarenal hyperinsulinemia does not cause hypertension. Am J Physiol 1991;260(pt 2):F663–F669.
63 Tucker BJ, Anderson CM, Thies RS, Collins RC, Blantz RC: Glomerular hemodynamic alterations during acute hyperinsulinemia in normal and diabetic rats. Kidney Int 1992;42:1160–1168.
64 Pinto-Sietsma SJ, Navis G, Janssen WM, de Zeeuw D, Gans RO, de Jong PE: A central body fat distribution is related to renal function impairment, even in lean subjects. Am J Kidney Dis 2003;41:733–741.
65 Henegar JR, Bigler SA, Henegar LK, Tyagi SC, Hall JE: Functional and structural changes in the kidney in the early stages of obesity. J Am Soc Nephrol 2001;12:1211–1217.
66 Engeli S, Sharma AM: Role of adipose tissue for cardiovascular-renal regulation in health and disease. Horm Metab Res 2000;32:485–499.

67 Ailhaud G, Fukamizu A, Massiera F, Negrel R, Saint-Marc P, Teboul M: Angiotensinogen, angiotensin II and adipose tissue development. Int J Obes Relat Metab Disord 2000;24 (suppl 4):S33–S35.
68 Pratt JH, Ambrosius WT, Tewksbury DA, Wagner MA, Zhou L, Hanna MP: Serum angiotensinogen concentration in relation to gonadal hormones, body size, and genotype in growing young people. Hypertension 1998;32:875–879.
69 Rotimi C, Cooper R, Ogunbiyi O, Morrison L, Ladipo M, Tewksbury D, et al: Hypertension, serum angiotensinogen, and molecular variants of the angiotensinogen gene among Nigerians. Circulation 1997;95:2348–2350.
70 Schorr U, Blaschke K, Turan S, Distler A, Sharma AM: Relationship between angiotensinogen, leptin and blood pressure levels in young normotensive men. J Hypertens 1998;16:1475–1480.
71 Engeli S, Bohnke J, Gorzelniak K, Janke J, Schling P, Bader M, et al: Weight loss and the renin-angiotensin-aldosterone system. Hypertension 2005;45:356–362.
72 Ruano M, Silvestre V, Castro R, Garcia-Lescun MC, Rodriguez A, Marco A, et al: Morbid obesity, hypertensive disease and the renin-angiotensin-aldosterone axis. Obes Surg 2005;15:670–676.
73 Hopkins PN, Lifton RP, Hollenberg NK, Jeunemaitre X, Hallouin MC, Skuppin J, et al: Blunted renal vascular response to angiotensin II is associated with a common variant of the angiotensinogen gene and obesity. J Hypertens 1996;14:199–207.
74 Barton M, Carmona R, Morawietz H, d'Uscio LV, Goettsch W, Hillen H, et al: Obesity is associated with tissue-specific activation of renal angiotensin-converting enzyme in vivo: evidence for a regulatory role of endothelin. Hypertension 2000;35(pt 2):329–336.
75 Ahmed SB, Fisher ND, Stevanovic R, Hollenberg NK: Body mass index and angiotensin-dependent control of the renal circulation in healthy humans. Hypertension 2005;46:1316–1320.
76 Price DA, Lansang MC, Osei SY, Fisher ND, Laffel LM, Hollenberg NK: Type 2 diabetes, obesity, and the renal response to blocking the renin system with irbesartan. Diabetes Med 2002;19:858–861.
77 Praga M, Hernandez E, Andres A, Leon M, Ruilope LM, Rodicio JL: Effects of body-weight loss and captopril treatment on proteinuria associated with obesity. Nephron 1995;70:35–41.
78 Schnermann J, Briggs JP: The macula densa is worth its salt. J Clin Invest 1999;104:1007–1009.
79 Hall JE: The kidney, hypertension, and obesity. Hypertension 2003;41(pt 2):625–633.
80 Kikuchi K, Iimura O, Yamaji I, Shibata S, Nishimura M, Aoki K, et al: The pathophysiological role of water-sodium balance and renal dopaminergic activity in overweight patients with essential hypertension. Am J Hypertens 1988;1:31–37.
81 Waki M, Kral JG, Mazariegos M, Wang J, Pierson RN Jr, Heymsfield SB: Relative expansion of extracellular fluid in obese vs. nonobese women. Am J Physiol 1991;261(pt 1):E199–E203.
82 Hall JE, Brands MW, Dixon WN, Smith MJ Jr: Obesity-induced hypertension. Renal function and systemic hemodynamics. Hypertension 1993;22:292–299.
83 Rocchini AP, Key J, Bondie D, Chico R, Moorehead C, Katch V, et al: The effect of weight loss on the sensitivity of blood pressure to sodium in obese adolescents. N Engl J Med 1989;321:580–585.
84 Ahmed SB, Fisher ND, Stevanovic R, Hollenberg NK: Body mass index and angiotensin-dependent control of the renal circulation in healthy humans. Hypertension 2005;46:1316–1320.
85 Alonso-Galicia M, Brands MW, Zappe DH, Hall JE: Hypertension in obese Zucker rats. Role of angiotensin II and adrenergic activity. Hypertension 1996;28:1047–1054.
86 Hall JE, Henegar JR, Dwyer TM, Liu J, da Silva AA, Kuo JJ, et al: Is obesity a major cause of chronic kidney disease? Adv Ren Replace Ther 2004;11:41–54.
87 Wofford MR, Hall JE: Pathophysiology and treatment of obesity hypertension. Curr Pharm Des 2004;10:3621–3637.
88 Quinones-Galvan A, Ferrannini E: Renal effects of insulin in man. J Nephrol 1997;10:188–191.
89 Hall JE, Kuo JJ, da Silva AA, de Paula RB, Liu J, Tallam L: Obesity-associated hypertension and kidney disease. Curr Opin Nephrol Hypertens 2003;12:195–200.
90 Aviv A, Hollenberg NK, Weder AB: Sodium glomerulopathy: tubuloglomerular feedback and renal injury in African Americans. Kidney Int 2004;65:361–368.
91 Campese VM, Parise M, Karubian F, Bigazzi R: Abnormal renal hemodynamics in black salt-sensitive patients with hypertension. Hypertension 1991;18:805–812.

92 Chiolero A, Maillard M, Nussberger J, Brunner HR, Burnier M: Proximal sodium reabsorption: an independent determinant of blood pressure response to salt. Hypertension 2000;36:631–637.
93 Hollenberg NK, Moore T, Shoback D, Redgrave J, Rabinowe S, Williams GH: Abnormal renal sodium handling in essential hypertension. Relation to failure of renal and adrenal modulation of responses to angiotensin II. Am J Med 1986;81:412–418.
94 Hollenberg NK: Sodium and the kidney: the non-modulator concept. Nephrol Dial Transplant 2001;16(suppl 6):38–39.
95 Bigazzi R, Bianchi S, Baldari D, Sgherri G, Baldari G, Campese VM: Microalbuminuria in salt-sensitive patients. A marker for renal and cardiovascular risk factors. Hypertension 1994;23: 195–199.
96 Rahmouni K, Haynes G: Leptin and the central neural mechanisms of obesity hypertension. Drugs Today (Barc) 2002;38:807–817.
97 Vaz M, Jennings G, Turner A, Cox H, Lambert G, Esler M: Regional sympathetic nervous activity and oxygen consumption in obese normotensive human subjects. Circulation 1997;96:3423–3429.
98 Rahmouni K, Morgan DA, Morgan GM, Mark AL, Haynes WG: Role of selective leptin resistance in diet-induced obesity hypertension. Diabetes 2005;54:2012–2018.
99 Beltowski J, Jochem J, Wojcicka G, Zwirska-Korczala K: Influence of intravenously administered leptin on nitric oxide production, renal hemodynamics and renal function in the rat. Regul Pept 2004;120:59–67.
100 Brenner BM: Hemodynamically mediated glomerular injury and the progressive nature of kidney disease. Kidney Int 1983;23:647–655.
101 Kambham N, Markowitz GS, Valeri AM, Lin J, D'Agati VD: Obesity-related glomerulopathy: an emerging epidemic. Kidney Int 2001;59:1498–1509.
102 de Jong PE, Verhave JC, Pinto-Sietsma SJ, Hillege HL: Obesity and target organ damage: the kidney. Int J Obes Relat Metab Disord 2002;26(suppl 4):S21–S24.
103 Boyd ST, Fremming BA: Update on rimonabant – a selective cannabinoid CB1 antagonist (May). Ann Pharmacother 2006;40:994.
104 Reisin E, Messerli FG, Ventura HO, Frohlich ED: Renal haemodynamic studies in obesity hypertension. J Hypertens 1987;5:397–400.

G.J. Navis
Department of Medicine, University Medical Center
PO Box 30001
NL–9700 RB Groningen (The Netherlands)
Tel. +31 503612955, Fax +31 503619310, E-Mail g.j.navis@int.umcg.nl

Insulin Resistance and Renal Disease

Danilo Fliser, Jan T. Kielstein, Jan Menne

Division of Nephrology, Department of Internal Medicine,
Hannover Medical School, Hannover, Germany

Abstract

In non-diabetic patients with (primary) kidney diseases a syndrome of insulin resistance can be diagnosed even before the onset of impaired renal function and uremia. It is presenting with hyperinsulinemia, glucose intolerance, hyperglycemia and dyslipidemia. As a result, patients with kidney diseases own the same metabolic cardiovascular risk factors as patient with the classic metabolic syndrome. Thus, this renal insulin resistance syndrome may not only contribute to the excessive cardiovascular risk of patients with kidney diseases, but may also help to explain why even a minor impairment of renal function is a significant and independent cardiovascular risk factor in the general population.

Copyright © 2006 S. Karger AG, Basel

The Syndrome of Insulin Resistance in Renal Patients

Insulin resistance describes a clinical condition where there is a reduced biological effect for any given blood concentration of insulin. In other words, insulin resistance is present when a normal insulin blood concentration produces a less than normal biological response. Insulin resistance can be physiological (e.g. in pregnancy), or pathological. It may occur as a primary phenomenon or secondary to other disorders. The most common clinical condition of insulin resistance is the classic metabolic syndrome, which is also characterized by hyperglycemia, hypertension, obesity and dyslipidemia (e.g. low HDL-cholesterol and high triglycerides), and which is accompanied by cardiovascular sequels from atherosclerotic vascular disease [1]. Among the many other clinical conditions presenting with insulin resistance, uremia is of particular interest for nephrologists (table 1) [2]. An impaired glucose tolerance in patients with renal failure has been recognized almost 100 years ago [3], but only after the euglycemic clamp technique had become available, diminished insulin-stimulated

Table 1. Physiological and pathological conditions associated with insulin resistance

Type of insulin resistance	Clinical condition
Primary	Type 2 diabetes
Secondary	Uremia, starvation, liver chirrhosis, acromegaly, Cushing's disease, polycystic ovary syndrome, pheocromocytoma
Physiological	Pregnancy, puberty, old age, stress
Insulin receptor mutations	Leprachaunism, Rabson–Mendenhall syndrome
Insulin receptor autoantibodies	Ataxia telangiectasia, systemic autoimmune disease, myeloma
Other syndromes	Prader–Willi, Lawrence Moon Biedl, Laron dwarfism, etc.

glucose uptake (as a direct measure of insulin resistance) could be documented [4]. Moreover, complex disturbances of insulin secretion and other features of the classic metabolic syndrome such as glucose intolerance and dyslipidemia accompany insulin resistance in uremic patients [5, 6].

Do Kidney Diseases Invariably Cause Insulin Resistance?

Several explanations for the presence of insulin resistance in uremia have been proposed such as deficiency of vitamin D, secondary hyperparathyroidism, lack of erythropoietin and/or anemia, metabolic acidosis, and putative uremic toxins [6–10]. Evidence in support for each of these possibilities came from studies in laboratory animals as well as patients with kidney diseases, but most clinical studies were small and restricted to patients with terminal renal failure. However, results from several recent clinical studies have clearly documented that insulin resistance is present already in patients with mild degrees of renal dysfunction or even in patients with apparently normal renal function, i.e. normal glomerular filtration rate [11–14]. In these patients, complications of uremia such as secondary hyperparathyroidism or anemia could not have account for the presence of insulin resistance. Therefore, it was proposed that kidney disease *per se* causes a syndrome of insulin resistance.

We could recently confirm this assumption in a large prospective study in 227 non-diabetic patients with different types of primary kidney disease [15], in whom we assessed insulin sensitivity using the Homeostasis Model Assessment of Insulin Resistance (HOMA-IR) [2, 16]. We found significant insulin resistance in renal patients and, in addition, significantly higher fasting plasma

Table 2. Insulin resistance (HOMA-IR) and plasma insulin and glucose concentrations in 227 non-diabetic patients with primary kidney diseases, and in 76 healthy subjects of similar age, gender and body mass index (BMI) distribution

	Healthy subjects	Renal patients GFR (in ml/min/1.73 m^2)		
		>90	45–90	<90
N	76	73	77	77
GFR, ml/min/1.73 m^2	–	119 ± 28	65 ± 13	28 ± 11
HOMA-IR	1.39 ± 0.06	3.32 ± 0.26*	4.04 ± 0.56*	3.35 ± 0.33*
Insulin, mU/l	6.5 ± 0.3	13.1 ± 0.8*	15.3 ± 1.5*	13.2 ± 0.9*
Glucose, mg/dl	86 ± 1	98 ± 2*	99 ± 2*	98 ± 2*

Patients were divided in 3 groups according to their glomerular filtration rate (GFR). *= p<0.01 vs. healthy subjects (data are presented as mean ± SEM).

insulin (i.e. hyperinsulinemia) and glucose (i.e. hyperglycemia) levels as compared with healthy subjects matched with respect to age, gender and body mass index (BMI). Importantly, these metabolic disorders were present even in patients in whom measured glomerular filtration rate (iothalamate clearance) was within the normal range (table 2). Moreover, as in prediabetic patients with the metabolic syndrome [1, 17], insulin resistance (i.e. HOMA-IR) in renal patients was closely correlated with higher BMI (r = 0.477; p < 0.01) and serum triglyceride concentrations (r = 0.384; p < 0.01) [15]. The association of higher BMI with insulin resistance, hyperinsulinemia, hyperglycemia and hypertriglyceridemia in patients with kidney disease is of interest, because it suggests that this deleterious combination explains, at least in part, the excessive cardiovascular risk of renal patients. Interestingly, insulin resistance was present even in patients matched with respect to BMI to healthy subjects, so that overweight seems to be only an aggravating factor of an underlying (renal?) metabolic disorder [15]. Collectively, these findings support the notion that non-diabetic (primary) kidney diseases are associated with a syndrome of insulin resistance, and underline the currently not widely appreciated role of the kidney in glucose homeostasis.

Renal Glucose Metabolism

Past and more recent experimental studies and studies in humans have revealed that the kidney is an important organ of glucose homeostasis [18–23].

For example, it has been known for decades that renal gluconeogenesis contributes substantially to the maintanance of normal blood glucose levels [18], but only after sophisticated isotopic techniques for measurement of net balance glucose turnover had become available, the contribution of renal glucose metabolism to whole body glucose turnover could be exactly quantified [19–23]. These studies revealed that kidneys are as important as the liver for total postabsorptive gluconeogenesis in normal humans [20–23]. Furthermore, it has been shown that lactate is the dominant precursor for renal gluconeogenesis, pointing to the kidney as an important organ for lactate disposal [18, 20, 23]. An elegant proof that the kidneys are a major site of extrahepatic glucose production have provided Battezzati et al. [24], who studied anhepatic patients prior to liver transplantation. They could clearly demonstrate that the rate of extrahepatic gluconeogenesis in these patients was similar to that of postabsorptive healthy subjects. In other words, during the (short) anhepatic phase increased renal gluconeogenesis maintained normal blood glucose levels. Similar results have been obtained already 60 years ago in hepatectomized and nephrectomized rabbits [25].

In face of the important role of the kidney in glucose homeostasis, the question arises how kidney diseases may affect (renal) glucose metabolism. Metabolic studies in patients with type 2 diabetes mellitus have revealed that renal gluconeogenesis – similarly as hepatic gluconeogenesis – is not suppressed by insulin to a similar extend as in healthy individuals [21]. Thus, increased total body gluconeogenesis and hyperglycemia in diabetic patients result from inappropriately high hepatic and renal glucose production. This observation points to the presence of insulin resistance on the level of renal glucose metabolism. Such information is not available for patients with kidney dysfunction and further experimental and clinical studies are warranted to elucidate the action of insulin in the diseased kidney. Furthermore, it has also been shown that in healty volunteers, epinephrine significantly stimulates renal gluconeogenenesis from lactate and glycerol by increasing substrate availability and the gluconeogenic efficiency of the kidney [19, 23]. In patients with kidney diseases increased sympathetic activity is observed early on in the course of the disease [26, 27], and one might speculate that higher sympathetic activity could contribute to increased renal gluconeogensis and hyperglycemia in renal patients.

Insulin Resistance and Cardiovascular Events in Patients with Kidney Disease

The finding of insulin resistance (and other disturbances of glucose metabolism) in patients with kidney diseases did not attract much attention until a

Table 3. Key features of the classic and the renal metabolic syndrome

Classic	Renal
Insulin resistance/hyperinsulinemia	Insulin resistance/hyperinsulinemia
Glucose intolerance/hyperglycemia	Glucose intolerance/hyperglycemia
Hypertriglyceridemia and low HDL-C	Hypertriglyceridemia and low HDL-C
Abdominal ('central') obesity	Abdominal ('central') obesity (?)
Hypertension	Hypertension
Microalbuminuria	Microalbuminuria
Renal disease (?)	Renal disease
Salt sensitivity	Salt sensitivity
Hyperuricemia	Hyperuricemia
Endothelial dysfunction	Endothelial dysfunction
Reduced vascular compliance	Reduced vascular compliance
Left ventricular hypertrophy	Left ventricular hypertrophy

Both syndromes have striking similarities.

recent study indicated that insulin resistance is related to cardiovascular events in patients with terminal renal failure [28]. Moreover, in our cohort of 227 non-diabetic patients with chronic kidney diseases several components of the insulin resistance syndrome were not only associated with prevalent cardiovascular events, but were also significantly related to cardiovascular events during a 7 year follow-up [15]. Thus, in patients with kidney diseases metabolic (and other) cardiovascular risk factors alike those found in patients with the classic metabolic syndrome are present (table 3), and there is little doubt that these two populations are characterized by considerably cardiovascular risk [29]. This may also help to explain why even a minor impairment of renal function has been recently recognized as a significant and independent cardiovascular risk factor in the general population [30, 31]. It is therefore of interest that we found a significant correlation between insulin resistance on the one hand, and the use of anti-hypertensive drugs on the other hand in our cohort of renal patients [15]. Particularly the use of diuretics was associated with hyperinsulinemia, hyperglycemia and increased HOMA-IR. This finding is reminiscent to results of intervention studies with (higher dose) diuretic treatment in patients with essential hypertension [32], and could result from a 'diabetogenic' action of diuretics in renal patients as well, particularly when higher doses are administered. An alternative explanation could be that the need for aggressive anti-hypertensive treatment – including diuretics – is highest in patients in whom the renal insulin resistance syndrome is most pronounced.

In addition to insulin resistance and hyperglycemia, we also found markedly lower adiponectin plasma concentrations in renal patients with prevalent and incident cardiovascular events [15], similar to observations in individuals with normal renal function [33], and in patients with terminal renal failure [34]. Adiponectin is a recently discovered adipokine that plays an important pathophysiological role in patients with impaired glucose homeostasis [35]. It is synthesized and secreted by adipose tissue and exhibits potent anti-inflammatory and anti-atherosclerotic properties. Adiponectin is a component of a novel signaling network between adipocytes, insulin-sensitive tissues and vasculature [35]. Our results, therefore support the notion that adiponectin is a vasoprotective factor, and that in patients with renal dysfunction hypo-adiponectemia is a cardiovascular risk factor [15, 34].

Insulin Resistance and Progression

Another important consideration when discussing the role of insulin resistance in patients with kidney diseases is the question, whether it affects progression. In this respect, results of recent large population-based prospective studies are of interest in which insulin resistance predicted the risk for chronic kidney disease [36–38]. Chen et al. [36, 37] have examined a large representative sample of U.S. adults who participated in the third National Health and Nutrition Examination Survey (NHANES III). They were able to demonstrate a strong, significant, and dose–response relationship among insulin resistance and risk of chronic kidney disease among non-diabetic participants. This relationship was independent of age, gender, race, and other potential risk factors for chronic kidney disease, such as blood pressure, obesity, total cholesterol, education, physical activity, cigarette smoking and use of non-steroidal anti-inflammatory drugs. Similar findings were also reported by Kurella et al. [38], who studied data from participants in the ARIC Study, i.e. a large, prospective, community-based cohort. They demonstrated that the metabolic syndrome is associated with an increased risk for incident chronic kidney disease, defined as progression to a glomerular filtration rate below 60 ml/min/1.73 m^2 over a 9 year period. In this analysis, the risk was also independent of several potential confounding factors such as age, gender, race, hypertension, etc. Moreover, there were graded relations among the number of clinical traits of the metabolic syndrome, HOMA-IR, and fasting insulin levels and the risk for chronic kidney disease [38]. In contrast to these findings from community-based cohorts, we could not identify insulin resistance and accompanying metabolic disturbances as significant progression promoters in our cohort of non-diabetic patients with primary kidney diseases [39]. Nevertheless, it is plausible to assume that the classical metabolic syndrome may cause renal dysfunction in prediabetic patients through the joined action of hyperinsulinemia, hyperglycemia, hypertension, hyperlipidemia, obesity, etc. [29, 40–42]. However, a complementary explanation would be that some of these patients have unrecognized (primary) kidney diseases. Further experimental and clinical research efforts are needed in order to clarify this issue.

References

1 Reilly MP, Rader DJ: The metabolic syndrome: more than the sum of its parts? Circulation 2003;108:1546–1551.
2 Wallace TM, Matthews DR: The assessment of insulin resistance in man. Diabet Med 2002;19:527–534.
3 Neubauer E: Über Hyperglykämie bei Hochdrucknephritis und die Beziehung zwischen Glykämie und Glucosurie beim Diabetes mellitus. Biochem Z 1910;25:285–289.
4 DeFronzo RA, Alverstrand A, Smith D, Hendler R, Hendler E, Wahren J: Insulin resistance in uremia. J Clin Invest 1981;67:563–572.
5 Feneberg R, Sparber M, Veldhuis JD, Mehls O, Ritz E, Schaefer F: Altered temporal organization of plasma insulin oscillations in chronic renal failure. J Clin Endocrinol Metab 2002;87:1965–1973.
6 Mak RHK, DeFronzo R: Glucose and insulin metabolism in uremia. Nephron 1992;61:377–382.
7 Mak RHK: Intravenous 1,25 dihydroxycholecalciferol corrects glucose intolerance in hemodialysis patients. Kidney Int 1992;41:1049–1054.
8 Lu KC, Shieh SD, Lin SH, Chyr SH, Lin YF, Diang LK, Sheu WHH, Ding YA: Hyperparathyroidism, glucose tolerance and platelet intracellular free calcium in chronic renal failure. QJM 1994;87:359–365.
9 Kautzky-Willer A, Pacini G, Barnas U, Ludvik B, Streli C, Graf H, Prager R: Intravenous calcitriol normalizes insulin sensitivity in uremic patients. Kidney Int 1995;47:200–206.
10 Spaia S, Pangalos M, Askepidis N, Pazarloglou M, Mavropoulou E, Theodoridis S, Dimitrakopoulos K, Milionis A, Vayonas G: Effect of short-term rHuEPO treatment on insulin resistance in haemodialysis patients. Nephron 2000;84:320–325.
11 Eidemak I, Feldt-Rasmussen B, Kanstrup IL, Nielsen SL, Schmitz O, Strandgaard S: Insulin resistance and hyperinsulinemia in mild to moderate progressive chronic renal failure and its association with aerobic work capacity. Diabetologia 1995;38:565–572.
12 Vareesangthip K, Tong P, Wilkinson R, Thomas TH: Insulin resistance in adult polycystic kidney disease. Kidney Int 1997;52:503–508.
13 Fliser D, Pacini G, Engelleiter R, Kautzky-Willer A, Franek E, Ritz E: Insulin resistance and hyperinsulinaemia are present already in patients with incipient renal disease. Kidney Int 1998;53:1243–1247.
14 Kato Y, Hayashi M, Ohno Y, Suzawa T, Sasaki T, Saruta T: Mild renal dysfunction is associated with insulin resistance in chronic glomerulonephritis. Clin Nephrol 2000;54:366–373.
15 Becker B, Kronenberg F, Kielstein JT, Haller H, Morath C, Ritz E, Fliser D: Renal insulin resistance syndrome, adiponectin and cardiovascular events in patients with kidney diseases: the Mild to Moderate Kidney Disease (MMKD) Study. J Am Soc Nephrol 2005;16:1091–1098.
16 Bonora E, Targher G, Alberiche M, Bonadonna RC, Saggiani F, Zenere MB, Monauni T, Muggeo M: Homeostasis model assessment closely mirrors the glucose clamp technique in the assessment of insulin sensitivity: studies in subjects with various degrees of glucose tolerance and insulin sensitivity. Diabetes Care 2000;23:57–63.
17 Asia Pacific Cohort Studies Colloboration: Body mass index and cardiovascular disease in the Asia-Pacific Region: an overview of 33 cohorts involving 310 000 participants. Int J Epidemiol 2004;33:751–758.

18 Wirthensohn G, Guder WG: Renal substrate metabolism. Physiol Rev 1986;66:469–497.
19 Stumvoll M, Chintalapudi U, Perriello G, Welle S, Gutierrez O, Gerich J: Uptake and release of glucose by the human kidney. Postabsorptive rates and responses to epinephrine. J Clin Invest 1995;96:2528–2533.
20 Stumvoll M, Meyer C, Perriello G, Kreider M, Welle S, Gerich J: Human kidney and liver gluconeogenesis: evidence for organ substrate selectivity. Am J Physiol 1998;274:E817–E26.
21 Meyer C, Stumvoll M, Nadkarni V, Dostou J, Mitrakou A, Gerich J: Abnormal renal and hepatic glucose metabolism in type 2 diabetes mellitus. J Clin Invest 1998;102:619–624.
22 Cersosimo E, Garlick P, Ferretti J: Insulin regulation of renal glucose metabolism in humans. Am J Physiol 1999;276:E78–E84.
23 Meyer C, Stumvoll M, Welle S, Woerle HJ, Haymond M, Gerich J: Relative importance of liver, kidney, and substrates in epinephrine-induced increased gluconeogenesis in humans. Am J Physiol 2003;285:E819–E26.
24 Battezzati A, Caumo A, Martino F, Sereni LP, Coppa J, Romito R, Ammatuna M, Regalia E, Matthews DE, Mazzaferro V, Luzi L: Nonhepatic glucose production in humans. Am J Physiol 2004;86:E129–E135.
25 Bergman H, Drury DR: The relationship of kidney function to the glucose utilization of the extra abdominal tissues. Am J Physiol 1938;124:279–286.
26 Ishii M, Ikeda T, Takagi M, Sugimoto T, Atarashi K, Igari T, Uehara Y, Matsuoka H, Hirata Y, Kimura K, Takeda T, Murao S: Elevated plasma catecholamines in hypertensives with primary glomerular diseases. Hypertension 1983;5:545–551.
27 Converse RL, Jacobsen TN, Toto RD, Jost CMT, Cosentino F, Tarazi FF, Victor GG: Sympathetic overactivity in patients with chronic renal failure. N Engl J Med 1992;327:1912–1918.
28 Shinohara K, Shoji T, Emoto M, Tahara H, Koyama H, Ishimura E, Miki T, Tabata T, Nishizawa Y: Insulin resistance as an independent predictor of cardiovascular mortality in patients with end-stage renal disease. J Am Soc Nephrol 2002;13:1894–1900.
29 El-Atat FA, Stas SN, McFarlane SI, Sowers JR: The relationship between hyperinsulinemia, hypertension and progressive renal disease. J Am Soc Nephrol 2004;15:2816–2827.
30 Sarnak MJ, Levey AS, Schoolwerth AC, Coresh J, Culleton B, Hamm LL, McCullough PA, Kasiske BL, Kelepouris E, Klag MJ, Parfrey P, Pfeffer M, Raij L, Spinosa DJ, Wilson PW; American Heart Association Councils on Kidney in Cardiovascular Disease, High Blood Pressure Research, Clinical Cardiology, and Epidemiology and Prevention: Kidney disease as a risk factor for development of cardiovascular disease: a statement from the American Heart Association Councils on Kidney in Cardiovascular Disease, High Blood Pressure Research, Clinical Cardiology, and Epidemiology and Prevention. Circulation 2003;108:2154–2169.
31 Mann JFE, Gerstein HC, Dulau-Florea I, Lonn E: Cardiovascular risk in patients with mild renal insufficiency. Kidney Int 2003;63(suppl 84):S192–S196.
32 Opie LH, Schall R: Old antihypertensives and new diabetes. J Hypertens 2004;22:1453–1458.
33 Pischon T, Girman CJ, Hotamisligil GS, Rifai N, Hu FB, Rimm EB: Plasma adiponectin levels and risk of myocardial infarction in men. JAMA 2004;291:1730–1737.
34 Zoccali C, Mallamaci F, Tripepi G, Benedetto FA, Cutrupi S, Parlongo S, Malatino LS, Bonanno G, Seminara G, Rapisarda F, Fatuzzo P, Buemi M, Nicocia G, Tanaka S, Ouchi N, Kihara S, Funahashi T, Matsuzawa Y: Adiponectin, metabolic risk factors, and cardiovascular events among patients with end-stage renal disease. J Am Soc Nephrol 2002;13:134–141.
35 Goldstein BJ, Scalia R: Adiponectin: A novel adipokine linking adipocytes and vascular function. J Clin Endocrinol Metab 2004;89:2563–8256.
36 Chen J, Muntner P, Hamm LL, Fonseca V, Batuman V, Whelton PK, He J: Insulin resistance and risk of chronic kidney disease in nondiabetic US adults. J Am Soc Nephrol 2003;14:469–477.
37 Chen J, Muntner P, Hamm LL, Jones DW, Batuman V, Fonseca V, Whelton PK, He J: The metabolic syndrome and chronic kidney disease in U.S. adults. Ann Intern Med 2004;140:167–174.
38 Kurella M, Lo JC, Chertow GM: Metabolic syndrome and the risk for chronic kidney disease among nondiabetic adults. J Am Soc Nephrol 2005;16:2134–2140.
39 Fliser D, Kronenberg F, Kielstein JT, Morath C, Bode-Boger SM, Haller H, Ritz E: Asymmetric dimethylarginine (ADMA) and progression of chronic kidney disease: the Mild to Moderate Kidney Disease (MMKD) Study. J Am Soc Nephrol 2005;16:2456–2461.

40 Iseki K, Ikemiya Y, Kinjo K, Inoue T, Iseki C, Takishita S: Body mass index and the risk of development of end-stage renal disease in a screened cohort. Kidney Int 2004;65:1870–1876.
41 Schelling JR, Sedor JR: The metabolic syndrome as a risk factor for chronic kidney disease: more than a fat chance? J Am Soc Nephrol 2004;15:2773–2774.
42 Bagby SP: Obesity-initiated metabolic syndrome and the kidney: a recipe for chronic kidney disease? J Am Soc Nephrol 2004;15:2775–2791.

Danilo Fliser, MD
Associate Professor of Medicine
Department of Internal Medicine, Hannover Medical School
Carl-Neuberg-Strasse 1
DE–30625 Hannover (Germany)
Tel. +49 511 532 6319, Fax +49 511 55 23 66, E-Mail fliser.danilo@mh-hannover.de

Treatment of Obesity: A Challenging Task

Arya M. Sharma, Gianluca Iacobellis

Department of Medicine, Cardiovascular Obesity Research and Management at the Michael G. deGroote School of Medicine, Hamilton, Canada

Abstract

Background: Obesity is a chronic and heterogeneous medical condition. Weight loss is clearly the most desirable goal in obese subjects management. Successful obesity treatment should be defined as long-term weight loss maintenance. In this chapter, we briefly review the current evidences regarding the treatment of obesity. **Methods:** We searched MEDLINE and PubMed for original articles published between 1995 and 2006, focusing on obesity treatment. The search terms we used, alone or in combination, were 'obesity', 'lifestyle changes', 'diet', 'exercise', 'pharmacological treatment', 'surgical treatment'. **Conclusions:** The conventional management of obese patients involves weight reduction with lifestyle changes, including dietary therapy and increased physical activity, or a combined approach with lifestyle changes and pharmacological or surgical interventions. Exercise appears crucial in the successful maintenance of weight loss and in fostering cardiovascular health in obese patients. Some anti-obesity drugs, such as sibutramine and orlistat, have been shown to induce a significant weight loss and long-term weight loss maintenance. Surgical therapy is often necessary in morbidly obese patients and generally results in more significant and long-lasting weight loss than other treatments.

Copyright © 2006 S. Karger AG, Basel

Introduction

Obesity is a chronic and heterogeneous medical condition. The prevalence of obesity in the United States and worldwide is dramatically increasing [1]. According to data from the National Health and Nutrition Examination Survey (NHANES) for 1999–2000, almost two-thirds of US adults are overweight (defined as body mass index (BMI) 25–29.9 kg/m^2), and of these almost one-third are obese (BMI ≥ 30 kg/m^2) [2]. While short-term weight loss is rather easy to achieve and has demonstrated positive effects on associated comorbidities such as hypertension or diabetes, successful obesity treatment is defined as long-term weight-loss maintenance. Thus, most weight-loss treatments do not qualify as obesity treatment.

The conventional management of obese patients involves weight reduction with lifestyle changes, including dietary therapy and exercise, or a combined approach with lifestyle changes and pharmacological or surgical interventions. Weight loss has multiple other beneficial effects, including the following: lowering blood pressure, lowering serum cholesterol and triglycerides; raising high-density lipoprotein cholesterol, lowering glucose and reducing insulin resistance, proinflammatory and prothrombotic factors.

Lifestyle Intervention

The primary intervention in obese people is to achieve sustained lifestyle changes. The difficulty in achieving this goal is a common experience in the clinical practice. A combination of diet, physical activity, and behavior therapy, including self-monitoring, goal setting and social support may result in moderate weight loss, but often fail to sustain long-term weight-loss maintenance.

Behavioral Treatment

Lifestyle and behavior modification can be facilitated by counseling from a dietician, exercise specialist, or behavioral therapist who has weight-management experience. The goal of behavioral treatment is to help obese patients identify and modify their eating habits, activity level, and thinking that contribute to their excess weight. Behavioral treatment typically yields a 9–10% reduction in body weight during the first 6 months of treatment [3]. As with other chronic diseases, the challenge lies in sustaining such lifestyle changes in the long-term.

Dietary Intervention

Reduced caloric intake is one of the cornerstones of obesity treatment. Obesity is due to an imbalance between the amount of food energy consumed and the amount of energy expended through resting metabolism; the absorption, metabolism, and storage of food; and physical activity. The ideal goal to achieve weight loss is to induce a sustained reduction of energy intake below that of energy expenditure. In fact, the rate of weight loss on a given caloric intake should be related to the rate of energy expenditure, although studies have shown that the total amount of weight lost does not directly correlate with the degree of the energy restriction [4]. However, chronic caloric restriction diminishes metabolic rate because of reduced lean body mass. In addition, the reduction in metabolic rate with food restriction slows the rate of weight loss. Several different diet plans in terms of calorie content, specific food content and form have been proposed. Nevertheless, there are no studies demonstrating convincingly that one type of diet is more effective or safer than others. The degree of energy restriction of the diet seems to have not a significant impact on long-term weight-loss maintenance. Very low energy diets (e.g., <900 kcal/day), usually offered as liquid meal replacements may be appropriate for short-term treatment of obesity in selected patients. They are most commonly used for short periods to induce more rapid weight loss, improve comorbidities, and provide patients with positive feedback. A very low energy diet consisting of 45–70 g high-quality protein, 30–50 g carbohydrate, and ~2 g fat per day, as well as supplements of vitamins, minerals, and trace elements, appears to be safe in selected patients under medical supervision. Low-calorie diets, >800 kcal/day, are applicable to most obese patients. A diet rich in fruits, vegetables, low-fat dairy products, whole grains, and other low-glycemic index carbohydrates may promote weight loss and is preferable to low-fat diets in which large amounts of simple carbohydrates are substituted for fats. Some have advocated diets with protein replacement of simple carbohydrates in an effort to minimize insulin production. However, these diets have not been shown to be more effective in maintaining weight loss, and the possible long-term consequences of maintaining a lower body weight at the expense of consuming more saturated fat are unknown. Both low-fat and low-carbohydrate diets have been shown to induce weight loss and reduce obesity-related comorbidities [5]. A body of evidences suggests that not only total energy intake influence the development of obesity, but the proportion of dietary energy from fat can also play a role. A recent meta-analysis also suggests that reducing dietary fat is a component of successful weight maintenance [6]. Low-carbohydrate diets cause greater short-term (up to 6 months) weight loss than low-fat diets, but the long-term clinical safety and efficacy of these diets is unknown. High-protein weight-loss diets have become also popular. Nevertheless, there is some evidence that a sustained high-protein diet can adversely affect renal function, especially in obese diabetic patients with chronic kidney disease. High-protein diets should be avoided if possible and low-protein diets (<10% of total energy) should be prescribed to induce weight loss in these patients [7]. Increased consumption of dietary fiber could also play a role in the management of obesity through its ability to control body weight evolution through its effect on satiety. Prospective studies have also suggested that substituting whole grain for refined grain products may lower the risk of overweight and obesity [8]. Meal replacements are another valid alternative dietary strategy in the treatment of obesity and have been shown to aid maintenance of weight loss in one long-term uncontrolled study [9].

Exercise Treatment

Observational evidence support the idea that even small amounts of regular physical activity can reduce all-cause, and particularly, cardiovascular mortality by 20–30% [10]. Exercise increases energy expenditure, stimulates fat loss and also produces gains in lean muscle mass that may affect the absolute amount of weight lost. Physical activity has been reported to provide a favorable effect on body fat distribution. It has also been suggested that the amount of lean body mass gained through exercise was related to exercise intensity rather than volume. A recent Roundtable Consensus Statement from the American College of Sports Medicine (ACSM) reported a 'moderately strong' relationship between physical inactivity and the risk of developing obesity [11]. Exercise should be considered a crucial component of weight management, although current evidence suggests exercise alone is only minimally effective for weight loss in obese subjects. In fact, several randomized controlled trials analyzed the effects of aerobic exercise without diet intervention on weight loss and concluded that it is not enough to have any substantial effect on weight loss without a concurrent modification in diet. In contrast, regular exercise appears crucial in the prevention of weight gain and successful maintenance of weight loss. Subjects having a combined diet and exercise treatment maintained weight loss better and longer than the. There are clinical evidences of better weight maintenance in subjects having a combined diet and exercise treatment compared with subjects in the diet only. Most studies examining the role of exercise in the management of obesity have focused on aerobic exercise. However, a few studies showed that resistance training has a favorable effect on body composition and metabolic profile in both overweight and obese populations. An exercise prescription in obese subjects should include a weekly energy expenditure of at least 4,200 kcal/week, the equivalent of 30 min or 2 miles of brisk walking 5 times per week.

Pharmacological Intervention

The Food and Drug Administration-approved indications for drug therapy are a BMI between 27.0 and 29.9 in patients with an obesity-related medical complication or a BMI > 30. Orlistat and sibutramine are currently approved in most countries for the long-term pharmacological treatment of obesity. Both have been shown to induce a 5 ±10% weight loss that can be largely maintained under treatment in studies lasting up to 2 years.

Sibutramine leads to weight loss by enhancing satiety and increasing basal metabolism rate [12]. Sibutramine inhibits reuptake of the biogenic amine neurotransmitters norepinephrine and serotonin in the central and peripheral nervous system. Treatment with sibutramine resulted in clinically significant weight loss during both short- and long-term therapy in obese adults [13–17]. Sibutramine therapy causes much greater weight loss when given in combination with behavior therapy and meal replacements. Sibutramine can induce a slight increase in heart rate, and has been reported induce a significant increase in blood pressure in upto $1\pm3\%$ of participants in clinical trials [18]. Sibutramine treatment is unlikely to elicit a critical increase in blood pressure even in hypertensive patients [19] and new data suggests that this compound may in fact have central sympatholytic properties [20]. Nevertheless, it is advised that blood pressure and heart rate to be monitored. The use of sibutramine in patients with impaired renal function has not been reported. Orlistat is a reversible lipases inhibitor for obesity management that acts by inhibiting the absorption of dietary fats [21]. It acts in the lumen of the stomach and small intestine by forming a covalent bond with the active serine site of gastric and pancreatic lipases. The inactivated enzymes are unavailable to hydrolyze dietary fat in the form of triglycerides into absorbable free fatty acids and monoglycerides. Orlistat inhibits activity of pancreatic and gastric lipases, blocks gastrointestinal uptake of approximately 30% of ingested fat at the dose of 360 mg/daily (120 mg three times a day). Several studies have evaluated the efficacy and tolerability of orlistat [21–27]. Orlistat administered for 2 years promotes weight loss, maintenance of this weight loss and minimizes weight regain. Additionally, orlistat therapy improves lipid profile, blood pressure, and quality of life [21–27]. Orlistat leads to predictable gastrointestinal effects related to its mode of action, which were generally mild, transient, and self-limiting and usually occurred early during treatment. Orlistat has been shown to reduce the bioavailability of fat-soluble vitamins and of cyclosporin A [28] – a finding that may be relevant for the use of this agent in renal transplant patients.

There is a growing interest in the development of newer anti-obesity agents. Among these Rimonabant has shown encouraging results. Rimonabant is a selective CB1 endocannabinoid receptor antagonist indicated for the treatment of obesity [29]. The endocannabinoid system consists of several endogenous lipids, including anandamide and 2-arachidonoyl-glycerol (2-AG), and constitutes a retrograde signaling system, which modulates neurotransmitter release and synaptic plasticity [30–31]. Specific cannabinoid receptors (CB [1]) are widely distributed in the central nervous system as well as other tissues. CB [1] receptor antagonists/inverse agonists reduce food intake and suppress operant responding for food rewards. Hence, endocannabinoids provide the first example of a retrograde signaling system, which is strongly implicated in the control of food intake. Rimonabant works by blocking endogenous cannabinoid binding to neuronal CB1 receptors. Activation of these receptors by endogenous cannabinoids, such as anadamide, increases appetite. Rimonabant at a dose of 20 mg has been shown to reduce body weight and improve cardiovascular risk factors in obese patients (The Rimonabant in Obesity, RIO studies) [32–34]. It is estimated that a significant proportion of the positive metabolic effects of rimonabant are mediated by peripheral weight-loss independent effects on other organs like the liver, adipocytes or skeletal muscle.

Surgical Intervention

Surgical therapy provides the most effective modality for treating severe obesity. Surgical treatment of obesity generally results in more significant and long-lasting weight loss than other treatments. Bariatric surgery also improves or resolves most comorbidities associated with severe obesity, such as type 2 diabetes and hypertension. The indications for bariatric surgery were established at a National Institutes of Health Consensus Conference held in 1991 [35]. It was recommended that suitable patients for bariatric surgery be selected using the following criteria: (1) BMI $> 40 \text{ kg/m}^2$ or BMI $> 35 \text{ kg/m}^2$ with significant obesity-related comorbidities; (2) age between 16 and 65 years; (3) acceptable operative risks; (4) repeated failures of other non-surgical therapeutic approaches; (4) psychologically stable patient with realistic expectations; (5) well informed and motivated patient; (6) commitment to prolonged lifestyle changes; (7) supportive family/social environment; (8) commitment to long-term follow-up; (9) absence of alcoholism, other addictions, or major psychopathology. A follow-up NIH consensus meeting was held recently in June 2004, and new recommendations will be available in the near future. Bariatric surgical techniques share two fundamental features: intestinal malabsorption and gastric restriction. Malabsorptive operations shorten the functional length of the intestinal surface for nutrient absorption, while restrictive procedures decrease food intake by creating a small neogastric pouch and the outlet. Malabsorptive procedures include gastric bypass and biliopancreatic diversion with or without an additional duodenal switch. Restrictive procedures include vertical banded gastroplasty and adjustable gastric banding. Roux-en-Y gastric bypass, which is a combination of a restrictive and bypass procedure is widely considered the current gold standard for bariatric surgery. The most common bariatric surgical procedures performed in Europe and United States are the Roux-en-Y gastric bypass and the laparoscopic adjustable gastric band procedures. Laparoscopic gastric banding represents the least invasive among frequently performed bariatric procedures. Several studies reported significant weight loss and improvement of comorbid conditions in morbid obese subjects who underwent a laparoscopic adjustable gastric band procedure [36]. The use of bariatric surgery has been described in patients with renal failure, both before and following renal transplantation [37, 38]. Both gastric bypass and laparoscopic adjustable gastric band procedures have been reported to be safe and effective for achieving significant long-term weight loss and improvement of comorbidity in patients with renal failure on dialysis, in preparation for transplantation, or after renal transplantation [37, 38].

Conclusions

The treatment of obesity requires a multidisciplinary approach, which ideally includes clinical and psychological supervision, along with skilled counseling in nutrition, exercise, and behavior modification. The conventional management of obese patients involves weight reduction with lifestyle changes, including dietary therapy and increased physical activity, or a combined approach with lifestyle changes and pharmacological or surgical interventions. Exercise appears crucial in the successful maintenance of weight loss and in fostering cardiovascular health in obese patients. Some anti-obesity drugs, such as sibutramine and orlistat, have been shown to induce a significant weight loss and long-term weight loss maintenance. Surgical therapy is often necessary in morbidly obese patients and generally results in more significant and long-lasting weight loss than other treatments.

References

1 Grundy SM, Brewer HB Jr, Cleeman JI, Smith SC Jr, Lenfant C, for the Conference Participants: Definition of metabolic syndrome: report of the National Heart, Lung, and Blood Institute/American Heart Association conference on scientific issues related to definition. Circulation 2004;109: 433–438.
2 United States Department of Health and Human Services: National Institutes of Health, National Heart, Lung, and Blood Institute, Clinical guidelines on the identification, evaluation, and treatment of overweight and obesity in adults: the evidence report: Bethesda, MD, National Institutes of Health, NIH Publication No. 98–4083, 1998.
3 Foster G: The behavioral approach to treating obesity. Am Heart J 2006;151:625–627.
4 Pruitt JD, Bensimhon D, Kraus WE: Nutrition as a contributor and treatment option for overweight and obesity. Am Heart J 2006;151:628–632.
5 Klein S: Clinical trial experience with fat-restricted vs. carbohydrate-restricted weight-loss diets. Obes Res 2004;12(suppl 2):141S–144S.
6 Astrup A, Ryan L, Grunwald GK, et al: The role of dietary fat in body fatness: evidence from a preliminary meta-analysis of ad libitum low-fat dietary intervention studies, Br J Nutr 2000;83: S25–S32.
7 Friedman AN: High-protein diets: potential effects on the kidney in renal health and disease. Am J Kidney Dis 2004;44:950–962.
8 Bazzano LA, Song Y, Bubes V, Good CK, Manson JE, Liu S: Dietary intake of whole and refined grain breakfast cereals and weight gain in men. Obes Res 2005;13:1952–1960.
9 Keogh JB, Clifton M: The role of meal replacements in obesity treatment. Obes Rev 2005;6: 229–234.

10 Bensimhon DR, Kraus WE, Donahue MP: Obesity and physical activity: a review. Am Heart J 2006;151:598–603.
11 American College of Sports Medicine: position stand on fitness: the recommended quantity and quality of exercise for developing and maintaining cardiorespiratory and muscular fitness, and flexibility in healthy adults. Med Sci Sports Exerc 1998;30:975–991.
12 Ryan DH, Kaiser P, Bray GA: Sibutramine: a novel new agent for obesity treatment. Obes Res 1995;3:553S–559S.
13 Hanotin C, Thomas F, Jones SP: Efficacy and tolerability of Sibutramine in obese patients: a dose-ranging study. Int J Obes Relat Metab Disord 1998;22:32–38.
14 Bray GA, Ryan DH, Gordon D: A double-blind randomized placebo-controlled trial of Sibutramine. Obes Res 1996;4:263–270.
15 Lean ME: Sibutramine – a review of clinical efficacy. Int J Obes Relat Metab Disord 1997;21:S30–S36.
16 Apfelbaum M, Vague P, Ziegler O, et al: Long-term maintenance of weight loss after a very-low-calorie diet: a randomized blinded trial of the efficacy and tolerability of sibutramine. Am J Med 1999;106:179–184.
17 James WP, Astrup A, Finer N, et al: Effect of sibutramine on weight maintenance after weight loss: a randomized trial. STORM Study Group. Sibutramine trial of obesity reduction and maintenance. Lancet 2000;356:2119–2125.
18 Pischon T, Sharma AM: Recent developments in the treatment of obesity-related hypertension. Curr Opin Nephrol Hypertens 2002;11:497–502.
19 Jordan J, Scholze J, Matiba B, Wirth A, Hauner H, Sharma AM: Influence of Sibutramine on blood pressure: evidence from placebo-controlled trials. Int J Obes (Lond) 2005;29:509–516.
20 Birkenfeld AL, Schroeder C, Boschmann M, Tank J, Franke G, Luft FC, Biaggioni I, Sharma AM, Jordan J: Paradoxical effect of sibutramine on autonomic cardiovascular regulation. Circulation 2002;106:2459–2465.
21 Drent ML, Larsson I, William-Olsson T, et al: Orlistat (Ro 18–0647), a lipase inhibitor, in the treatment of human obesity: a multiple dose study. Int J Obes Relat Metab Disord 1995;19:221–226.
22 Davidson MH, Hauptman J, DiGirolamo M, Foreyt JP, Halsted CH, Heber D, Heimburger DC, Lucas CP, Robbins DC, Chung J, Heymsfield SB: Weight control and risk factor reduction in obese subjects treated for 2 years with orlistat: a randomized controlled trial. JAMA 1999;281:235–242.
23 Sjostrom L, Rissanen A, Andersen T, Boldrin M, Golay A, Koppeschaar HPF, Krempf M: Randomised placebo-controlled trial of orlistat for weight loss and prevention of weight regain in obese patients. Lancet 1998;352:167–172.
24 O'Meara S, Riemsma R, Shirran L, Mather L, ter Riet G: A systematic review of the clinical effectiveness of orlistat used for the management of obesity. Obes Rev 2004;5:51–68.
25 Rossner S, Sjostrom L, Noack R, Meinders AE, Noseda G: Weight loss, weight maintenance, and improved cardiovascular risk factors after 2 years treatment with orlistat for obesity. European Orlistat Obesity Study Group. Obes Res 2000;8:49–61.
26 Chanoine JP, Hampl S, Jensen C, Boldrin M, Hauptman J: Effect of orlistat on weight and body composition in obese adolescents: a randomized controlled trial. JAMA 2005;293:2873–2883.
27 Krempf M, Louvet JP, Allanic H, Miloradovich T, Joubert JM, Attali JR: Weight reduction and long-term maintenance after 18 months treatment with orlistat for obesity. Int J Obes Relat Metab Disord 2003;27:591–597.
28 Iacobellis G: Drug-Drug Interactions in the Metabolic Syndrome. Nova Publishers, 2006.
29 Boyd ST, Fremming BA: Rimonabant – a selective CB1 antagonist. Ann Pharmacother 2005;39:684–690.
30 Cooper SJ: Endocannabinoids and food consumption: comparisons with benzodiazepine and opioid palatability-dependent appetite. Eur J Pharmacol 2004;500:37–49.
31 Kirkham TC, Williams CM: Endocannabinoid receptor antagonists: potential for obesity treatment.Treat Endocrinol 2004;3:345–360.
32 Despres JP, Golay A, Sjostrom L; Rimonabant in Obesity-Lipids Study Group: Effects of rimonabant on metabolic risk factors in overweight patients with dyslipidemia. N Engl J Med 2005;353:2121–2134.

33 Van Gaal LF, Rissanen AM, Scheen AJ, Ziegler O, Rossner S; RIO-Europe Study Group: Effects of the cannabinoid-1 receptor blocker rimonabant on weight reduction and cardiovascular risk factors in overweight patients: 1-year experience from the RIO-Europe study. Lancet 2005;365: 1389–1397.
34 Pi-Sunyer FX, Aronne LJ, Heshmati HM, Devin J, Rosenstock J; RIO-North America Study Group: Effect of rimonabant, a cannabinoid-1 receptor blocker, on weight and cardiometabolic risk factors in overweight or obese patients: RIO-North America: a randomized controlled trial. JAMA 2006;295:761–775.
35 National Institutes of Health Conference: Gastrointestinal surgery for severe obesity. Consensus Development Conference Panel. Ann Intern Med 1991;115:956–961.
36 Schneider BE, Mun EC: Surgical management of morbid obesity. Diabetes Care 2005;28: 475–480.
37 Alexander JW, Goodman HR, Gersin K, Cardi M, Austin J, Goel S, Safdar S, Huang S, Woodle ES: Gastric bypass in morbidly obese patients with chronic renal failure and kidney transplant. Transplantation 2004;78:469–474.
38 Newcombe V, Blanch A, Slater GH, Szold A, Fielding GA: Laparoscopic adjustable gastric banding prior to renal transplantation. Obes Surg 2005;15:567–570.

Arya M. Sharma, MD, FRCPC
Canada Research Chair in Cardiovascular Obesity Research and Management
McMaster University, Hamilton General Hospital
237 Barton Street East
Hamilton, ON, L8L 2X2 (Canada)
Tel. +1 905 527 4322, Ext. 46806, Fax +1 905 525 2260
E-Mail sharma@ccc.mcmaster.ca, gianluca@ccc.mcmaster.ca

Weight Loss and Proteinuria

Manuel Praga, Enrique Morales

Servicio de Nefrología, Hospital Universitario 12 de Octubre, Madrid, Spain

Abstract

The level of proteinuria is one of the most determinant risk factors for the progression of proteinuric renal diseases. Several studies have shown that weight loss, either induced by low calorie diets, physical exercise, or bariatric surgery is accompanied by an important antiproteinuric effect. Reduction in proteinuria is already observed after a few weeks from the onset of weight loss and it is evident even in patients with modest weight losses. The magnitude of body weight loss and proteinuria decrease show a significant correlation. Reduction in proteinuria by weight loss has been described in obesity-induced glomerulopathy, obese diabetic patients and overweight or obese patients with different types of chronic proteinuric nephropathies. Attenuation of the obesity-induced glomerular hyperfiltration, decrease in the activity of renin–angiotensin–aldosterone system and modifications in the serum concentrations of adipocyte-derived cytokine are likely to be involved in the anti-proteinuric effect of weight loss, together with a better control of blood pressure, and improvement of serum lipid profile and insulin sensitivity.

Copyright © 2006 S. Karger AG, Basel

Introduction

Obesity is a well-known cause of proteinuria. Nevertheless, only a minority of obese patients appears to develop significant proteinuria, by reasons that are only partially understood. The alarming worldwide epidemy of obesity has reinforced the interest in obesity-related proteinuria, either as a clinical expression of the glomerular damage induced by obesity (obesity-related glomerulopathy) or as a superimposed clinical factor accelerating the progression of other proteinuric renal diseases. In spite of this generalized interest in the renal consequences of obesity, the beneficial effects of weight loss on renal diseases is an issue just starting to emerge in experimental and clinical studies.

Proteinuria and Progression of Renal Damage

An impressive number of clinical studies performed throughout the last two decades have clearly established that the level of proteinuria is one of the most determinant risk factors for the progression of both diabetic and non-diabetic renal diseases [1]. Consequently, any therapeutic measure that reduces proteinuria will induce a favorable influence on the long-term outcome of these diseases. The most compelling evidence to support this assumption comes from studies using ACE inhibitors or AT1-receptor antagonists (ARA) in chronic proteinuric nephropathies: virtually all the studies agree that the renoprotection offered by these drugs is closely related to their anti-proteinuric effect [1–5]. Proteinuria, thus, should not be considered a passive marker of the severity of glomerular damage, but an active contributor to the progression of renal diseases.

Microalbuminuria and Proteinuria in Overweight and Obese Patients

The presence of severe proteinuria as a complication of massive obesity has been well-described by several clinical studies [6]. Other studies have shown a high prevalence of microalbuminuria among diabetic and non-diabetic obese patients [7]. The degree of proteinuria is very variable, but it can reach the nephrotic range (more than 3.5 g/day) in a significant number of cases [8, 9]. The clinical expression of obesity-related proteinuria is surprisingly silent: by reasons poorly investigated, obese patients do not develop hypoproteinemia, hypoalbuminemia, edema, or the other typical findings of nephrotic syndrome even in the presence of massive proteinuria levels [8, 9]. This characteristic lack of clinical expression is very useful for the differential diagnosis with other proteinuric renal diseases (idiopathic focal and segmental glomerulosclerosis [FSG], membranous nephropathy, minimal change disease) that can also affect obese people. However, it carries a dangerous consequence: the lack of edema ends up very frequently with a delay in the detection of proteinuria. The long-term prognosis of FSG associated to obesity is serious: a significant proportion of patients can progress to end-stage renal failure, mainly those already showing any degree of renal insufficiency at the onset of follow-up [8].

No specific studies have been conducted to investigate the influence of the level of proteinuria on the progression of renal damage among patients with obesity-related glomerulopathy. However, the available data suggest that the relationship between proteinuria level and progression of renal damage could also been applied to this entity: reported patients with obesity-related FSG and a progression into end-stage renal failure usually show very high levels of proteinuria, whereas those showing only mild proteinuria or microalbuminuria usually exhibit an stable renal function [8, 9].

Weight Loss and Reduction in Proteinuria or Albuminuria

Some studies have reported a significant decrease of proteinuria and microalbuminuria in obese diabetic patients submitted to hypocaloric diets. In 1984, Vasquez et al. studied 24 obese (BMI: $39.1 \pm 1.7\,kg/m^2$) patients with type 2 diabetes and mild urinary excretions of proteins and albumin. They were treated with low calorie diets. After a mean weight loss of $13.6 \pm 1.6\%$, both albuminuria and proteinuria showed a significant reduction, 50% higher than the initial values. Metabolic control of diabetes and blood pressure improved significantly, but proteinuria reduction showed only a significant correlation with the initial levels of urinary proteins [10].

After this initial description, other studies have confirmed these findings. Solerte et al. [11] studied 24 type 1 and type 2 diabetic patients with obesity and overt nephropathy after 12 months on hypocaloric diet (1,410 kcal/day without changes of protein:carbohydrate ratio). In parallel with a significant weight loss (BMI decreased from 33 ± 1.6 to $26 \pm 1.8\,kg/m^2$) proteinuria showed an important reduction, from $1,280 \pm 511$ to $623 \pm 307\,mg/24\,h$. Blood pressure and serum total cholesterol and triglycerides also showed a significant decrease coincidental with weight loss. In another study also enrolling obese diabetic patients with overt nephropathy (urinary protein $3.3 \pm 2.6\,g/24\,h$), a low calorie diet was administered to 22 patients for 4 week. A mean body weight loss of $6.2 \pm 3\,kg$ was accompanied by significant decreases of blood pressure, urinary protein, serum creatinine and blood urea nitrogen [12]. Some clinical reports exemplify how dramatically an important weight loss can induce disappearance of microalbuminuria and improvement of metabolic syndrome in obese type 2 patients submitted to hypocaloric diets [13].

Reduction in proteinuria induced by weight loss has also been reported in patients with obesity-induced proteinuria without the concomitant presence of diabetes. Praga et al. [14] studied 17 obese non-diabetic patients showing proteinuria persistently higher than $1\,g/24\,h$. Obesity-related proteinuria was established as the most likely clinical diagnosis, taking into account the absence of other identifiable renal diseases, the slowing progression of proteinuria in parallel with obesity and the characteristic clinical picture of absence of hypoalbuminemia and edema in those patients with nephrotic-range proteinuria [8, 9]. A renal biopsy had been obtained in five cases: FSG lesions were observed in two, the remaining showing minimal glomerular changes (in two) and mild mesangial proliferation (in one). They were randomized into two different therapeutic schemes: nine patients were instructed to follow a hypocaloric diet (1,000–1,400 kcal/day) without protein restriction or to captopril treatment without changes in their dietary habits. Follow up was 12 months in both groups. In the group of patients submitted to hypocaloric diet, BMI decreased from 37.1 ± 3 to $32.6 \pm 3.2 \, kg/m^2$. At the end of study proteinuria had decreased from 2.9 ± 1.7 to $0.4 \pm 0.6 \, g/24\,h$. All the patients showed a reduction in proteinuria, ranging between 52 and 100% of the initial values. A significant correlation between weight loss and proteinuria reduction was observed. In the group of patients treated with captopril, weight loss and BMI remained stable during follow up but proteinuria also showed a significant decrease, similar to that of patients with calorie restriction: it decreased from 3.4 ± 1.7 to $0.7 \pm 1 \, g/24\,h$ (40–100% of the basal values). Blood pressure and serum lipids decreased throughout the study in both groups; however, no correlation between reduction in proteinuria and blood pressure changes or variations in serum lipids was found.

Whether weight loss can induce a proteinuria reduction in patients affected by different types of proteinuric renal diseases was assessed by Morales et al. [15] in a prospective randomized study. Overweight or obese patients (BMI $> 27 \, kg/m^2$) with persistent proteinuria higher than $1 \, g/24\,h$ and serum creatinine lower than $2 \, mg/dl$ were randomly assigned either to follow a low calorie diet or to maintain their usual dietary intake for 5 months. Patients assigned to the low calorie group were prescribed a diet with an energy reduction of 500 kcal with respect to their usual diet. Protein content was 1–1.2 g/kg/day. Of the thirty patients included, type 2 overt diabetic nephropathy has been diagnosed in 14, chronic glomerulonephritis of different types in 7, hypertensive nephrosclerosis in 5, and reflux nephropathy in 4. Patients submitted to low calorie diet lost $4.1 \pm 3\%$ of their baseline weight. Mean BMI significantly decreased from 33 ± 3.5 to $31.6 \pm 3.2 \, kg/m^2$. In spite of this rather modest weight loss, mean proteinuria showed a significant decrease from 2.8 ± 1.4 to $1.9 \pm 1.4 \, g/24\,h$. Interestingly, reduction in proteinuria was already significant ($26.4 \pm 30\%$ of baseline values) after only 1 month of hypocaloric diet, when patients had lost $2.8 \pm 2.1\%$ of the baseline weight. Decrease in weight loss and proteinuria showed a significant correlation. Restricting the analysis to those patients who had achieved a weight loss greater than 3% of baseline values, a more important proteinuria decrease was demonstrated, equivalent to $50 \pm 21\%$ of the initial values. A significant increase of HDL cholesterol together with a tendency to a decrease in LDL cholesterol was observed in the diet group. Renal function (serum creatinine and Cockcroft-Gault creatinine clearance) remained stable.

By contrast, patients maintained in their usual dietary habits showed significant increases of weight and BMI and proteinuria tended to increase. Serum creatinine and estimated creatinine clearance showed a significant deterioration in this control group. Between-group comparisons confirmed the significant and divergent differences in the evolution of weight, BMI, proteinuria, and serum lipids during the study. The subanalysis of the 14 patients with type 2 diabetic nephropathy included in the study reproduced the global results: significant proteinuria decrease in those patients submitted to low calorie diets, in comparison with diabetics assigned to the control group. Once more, the magnitude of proteinuria reduction was related to the percentage of weight loss.

Bariatric Surgery in Patients with Proteinuria

The number of patients undergoing surgical procedures to induce drastic weight loss (bariatric surgery) is increasing wordlwide, since at present these surgical interventions are the most effective methods to achieve important and sustained weight loss in patients with morbid obesity [16]. Reductions in body weight higher than 30% can be obtained and sustained on the long-term. Recent studies have reported a dramatic reduction in the level of proteinuria in patients submitted to bariatric surgery, coinciding with drastic weight loss [17, 18]. Palomar et al. [19] reported on 35 patients with morbid obesity in whom a biliopancreatic diversion was performed. Excess weight loss was 67% at 1 year after surgical intervention. Microalbuminuria decreased and proteinuria disappeared in parallel with weight loss. In addition, the remaining cardiovascular risk factors common among these patients (hyperlipidemia, hypertension, diabetes) showed a marked improvement. The beneficial effect of bariatric surgery on obesity-induced proteinuria appears to be independent of the type of surgical intervention, since the same benefits have been reported with gastric bypass and gastric banding.

Pharmacotherapy and Lifestyle Modifications

Pharmacotherapy for obesity is one of the most rapidly developing fields in the last years [20]. Recent studies have demonstrated that available drugs for the therapy of obesity (sibutramine, rimonabant, orlistat) increase weight loss in obese patients undergoing hypocaloric diets, with a satisfactory risk profile [20–22]. Lifestyle modifications (regular physical exercise, avoiding sedentarism) are another fundamental approach to obtain sustained weight losses. Importantly, these pharmacological and non-pharmacological measures induce additive effects [21]. An important improvement of metabolic risk factors (dyslipidemia, insulin resistance) has been observed after weight loss in patients treated with anti-obesity drugs and lifestyle modifications [20–22]. However, their specific influence on microalbuminuria, proteinuria or histological lesions of obesity-induced glomerulopathy has not been investigated.

Mechanisms Involved in Weight Loss–Induced Proteinuria Reduction. Influence on the Renin–Angiotensin–Aldosterone System

Weight loss either induced by diet, physical exercise or pharmacotherapy is accompanied by metabolic changes that in other settings have demonstrated a favorable influence on proteinuric renal diseases: better control of blood pressure, improvement of serum lipid profile (HDL cholesterol increase, LDL cholesterol and triglyceride decrease) improvement of insulin sensitivity and better glycemic control in diabetic patients [20–23]. Certainly, the reduction in proteinuria that follows weight loss can be explained at least partially by the sum of these favorable changes. Nevertheless, in the few clinical studies that have investigated these issues [14, 15] the reduction in proteinuria was independent of changes in blood pressure or serum lipids, at least in short-term analysis.

Weight loss induces important changes in several cytokines and hormones produced by adipocytes. One of these hormones is leptin, whose serum levels are increased in obese patients and decline significantly during weight loss [24]. Experimental studies have shown that infusion of leptin into rats induced glomerular endothelial cell proliferation and enhanced expression of TGF-β and type IV collagen. These changes were accompanied by proteinuria and the appearance of progressive glomerulosclerosis [25]. Consequently, modifications in the serum concentrations of leptin, adiponectin and other substances produced by adipocytes are likely to play an important role in the anti-proteinuric effect of weight loss.

Almost all experimental and clinical studies agree that obesity induces glomerular hyperfiltration (preglomerular vasodilation, increase of hydrostatic pressure in glomerular capillaries, increase of filtration fraction) [26] by reasons that are only partially known. Bariatric surgery in obese subjects is followed by a sharp amelioration of glomerular hyperfiltration [27]. An illustrative example of the importance of hyperfiltration-induced obesity is the detrimental role that the presence of overweight or obesity plays in the long-term outcome of patients with remnant kidney and renal agenesis, clinical paradigms of hyperfiltration nephropathies [28]. Since proteinuria is a central manifestation of glomerular hyperfiltration, the amelioration of glomerular hemodynamics that accompanies weight loss is very likely to play an important role in the reduction in proteinuria.

Blockade of the renin–angiotensin–aldosterone system (RAAS) is currently the most powerful therapeutic strategy to slow or halt the progression of diabetic and non-diabetic proteinuric nephropathies. The effectiveness of ACE inhibitors or ARA in these diseases is determined by their capacity to lower proteinuria [1–5]. Very interestingly, several studies have conclusively demonstrated that adipocytes can produce all the components of RAAS and that obesity enhances their synthesis [29, 30]. Moderate weight loss (-5% of baseline values) by low calorie diet is followed by an important decrease of angiotensinogen levels, renin, aldosterone and ACE activity in plasma and adipose tissue [31]. In this regard, the reduction in proteinuria that follows weight loss can be attributed at least partially to a blockade of RAAS, similar to that induced by pharmacologic methods (ACE inhibitors and ARA). In fact, both the rapidity and the magnitude of proteinuria reduction in patients losing body weight are very reminiscent of the anti-proteinuric effects of these drugs. The anti-proteinuric effects of weight loss and ACE inhibitors or ARA could be additive. On the other hand, a hardly investigated question is whether obesity represents a condition of partial resistance to the anti-proteinuric and renoprotective actions of drugs that blockade RAAS [8]. Recent studies have shown that plasma aldosterone levels are elevated in obese hypertensive patients, particularly those with visceral or abdominal obesity [32] and that this elevation is relatively independent of plasma renin activity. Human adipocytes secrete factors partially characterized that directly induce the synthesis of aldosterone adrenal glands [33]. These data link the obesity-related hyperaldosteronism with the recently renewed interest in the anti-proteinuric, renoprotective and cardioprotective effects of spironolactone and other antagonists of aldosterone [34].

References

1. Ruggenenti P, Perna A, Mosconi L, et al: Urinary protein excretion rate is the best independent predictor of ESRF in non-diabetic proteinuric chronic nephropathies. Kidney Int 1998;53:1209–1216.
2. de Zeeuw D, Remuzzi G, Parving HH, et al: Proteinuria, a target for renoprotection in patients with type 2 diabetic nephropathy: lessons from RENAAL. Kidney Int 2004;65:2309–2320.
3. Jafar TH, Stark PC, Schmid CH, et al: Proteinuria as a modifiable risk factor for the progression of non-diabetic renal diseases. Kidney Int 2001;60:1131–1140.
4. Peterson JC, Adler S, Burkart JM, et al: Blood pressure control, proteinuria, and the progression of renal disease: The Modification in Diet in Renal Disease Study. Ann Intern Med 1995;123:754–762.
5. Praga M, Morales E: Renal damage associated with proteinuria. Kidney Int 2002;62(suppl 82):S42–S46.
6. Kambham N, Markowitz G, Valeri AM, et al: Obesity-related glomerulopathy: An emerging epidemic. Kidney Int 2001;59:1498–1509.
7. Valensi P, Assayag M, Busby M, Paries J, Lormeau B, Attali JR: Microalbuminuria in obese patients with or without hypertension. Int J Obes Relat Metab Disord 1996;20:574–579.

8   Praga M, Hernández E, Morales E, et al: Clinical features and long-term outcome of obesity-associated focal segmental glomerulosclerosis. Nephrol Dial Transplant 2001;16:1790–1798.
9   Praga M, Morales E, Herrero JC, et al: Absence of hypoalbuminemia despite massive proteinuria in focal segmental glomerulosclerosis secondary to hyperfiltration. Am J Kidney Dis 1999;33: 52–58.
10 Vasquez B, Flock EV, Savage PJ, et al: Sustained reduction of proteinuria in Type 2 (non-insulin dependent) diabetes following diet-induced reduction of hyperglucemia. Diabetologia 1984;26: 127–133.
11 Solerte SB, Fioravanti M, Schifino N, Ferrari E: Effects of diet-therapy on urinary protein excretion, albuminuria and renal hemodynamic function in obese diabetic patients with overt nephropathy. Int J Obes 1989;13:203–211.
12 Saiki A, Nagayama D, Ohhira M, et al: Effect of weight loss using formula diet on renal function in obese patients with diabetic nephropathy. Int J Obes 2005;29:1115–1120.
13 Fredrickson SK, Fedrro TJ, Schutrumpf AC: Disappearance of microalbuminuria in a patient with type 2 diabetes and the metabolic syndrome in the setting of an intense exercise and dietary program with sustained weight reduction. Diabetes Care 2004;27:1754–1755.
14 Praga M, Hernández E, Andrés A, León M, Ruilope LM, Rodicio JL: Effects of body-weight loss and captopril treatment on proteinuria associated with obesity. Nephron 1995;70:35–41.
15 Morales E, Valero MA, León M, Hernández E, Praga M: Beneficial effects of weight loss in overweight patients with chronic proteinuric nephropathies. Am J Kidney Dis 2003;41:319–327.
16 Brolin RE: Bariatric surgery and long-term control of morbid obesity. J Am Med Assoc 2002;288: 2793–2796.
17 Izzedine H, Coupaye M, Reach I, Deray G: Gastric bypass and resolution of proteinuria in an obese diabetic patient. Diabetes Med 2005;22:1761–1762.
18 Soto FC, Higa-Sansone G, Copley JB, et al: Renal failure, glomerulonephritis and morbid obesity: improvement after rapid weight loss following laparoscopic gastric bypass. Obes Surg 2005;15: 137–140.
19 Palomar R, Fernández-Fresnedo G, Domínguez-Díez A, et al: Effects of weight loss alter biliopancreatic diversión on metabolism and cardiovascular profile. Obes Surg 2005;15:794–798.
20 Yanovski SZ, Yanovski JA: Obesity. N Engl J Med 2002;346:591–602.
21 Wadden TA, Berkowitz RI, Womble LG, et al: Randomized trial of lifestyle modification and pharmacotherapy for obesity. N Engl J Med 2005;353:2111–2120.
22 Despres JP, Golay A, Sjöström L: Effects of rimonabant on metabolic risk factors in overweight patients with dyslipemia. N Engl J Med 2005;353:2121–2134.
23 Haffner S, Taegtmeyer H: Epidemic obesity and the metabolic syndrome. Circulation 2003;108: 1541–1545.
24 Verrotti A, Basciani F, de Simone M, Morgese G, Chiarelli F: Serum leptin changes during weight loss in obese diabetic subjects with and without microalbuminuria. Diabetes Nutr Metab 2001;14: 283–287.
25 Wolf G, Hamann A, Han CD, et al: Leptin stimulates proliferation and TGF-β expression in renal glomerular endothelial cells: potential role in glomerulosclerosis. Kidney Int 1999;56:860–872.
26 Henegar JR, Bigler SA, Henegar LK, Tyagi SC, Hall JE: Functional and structural changes in the kidney in the early stages of obesity. J Am Soc Nephrol 2001;12:1211–1217.
27 Chagnac A, Weinstein T, Herman M, Hirsh J, Gafter U, Ori Y: The effects of weight loss on renal function in patients with severe obesity. J Am Soc Nephrol 2003;14:1480–1486.
28 González E, Gutiérrez E, Morales E, et al: Factors influencing the progression of renal damage in patients with unilateral renal agenesis and remnant kidney. Kidney Int 2005;68:263–270.
29 Hall JE: The kidney, hypertension, and obesity. Hypertension 2003;41:625–633.
30 El-Atat FA, Stas SN, McFarlane SI, Sowers JR: The relationship between hyperinsulinemia, hypertension and progressive renal disease. J Am Soc Nephrol 2004;15:2816–2827.
31 Engeli S, Bohnke J, Gorzelniak K, et al: Weight loss and the renin-angiotensin-aldosterone system. Hypertension 2005;45:356–362.
32 De Paula RB, Da Silva AA, Hall JE: Aldosterone antagonism attenuates obesity-induced hypertension and glomerular hyperfiltration. Hypertension 2004;43:41–47.

33 Ehrhart-Bornstein M, Lamounier-Zepter V, Schraven A, et al: Human adipocytes secrete mineralocorticoid-releasing factors. Proc Natl Acad Sci USA 2003;10:14211–14216.
34 Hostetter TH, Ibrahim HN: Aldosterone in chronic kidney and cardiac disease. J Am Soc Nephrol 2003;14:2395–2401.

Dr. Manuel Praga
Servicio de Nefrología, Hospital Universitario 12 de Octubre
Avenida de Córdoba s/n
ES–28041 Madrid (Spain)
Tel. +34 91 390 8383, E-Mail mpragat@senefro.org

Treatment of Arterial Hypertension in Obese Patients

Ulrich O. Wenzel, Christian Krebs

III Medizinische Klinik, Universitätsklinikum Hamburg Eppendorf, Hamburg, Germany

Abstract

The prevalence of obesity is steadily increasing. Hypertension is one of the most common co-morbidities of obesity and significantly contributes to morbidity and mortality. Most obese hypertensive patients require antihypertensive drug treatment. However, current guidelines do not give specific recommendations for antihypertensive therapy of obese hypertensive patient. Some antihypertensive agents may have unwanted effects on the metabolic and hemodynamic abnormalities that link obesity and hypertension. Due to the lack of guidelines, this chapter provides recommendations for or against each class of antihypertensive agents mostly based on subjective criteria and pathophysiologic assumptions. Diuretics and beta-blockers are reported to reduce insulin sensitivity and increase lipid levels, whereas calcium antagonists are metabolically neutral and ACE-inhibitors as well as angiotensin receptor blockers increase insulin sensitivity. Sodium retention plays a central role in the development of obesity-related hypertension. Therefore, treatment with an ACE-inhibitors or a diuretic should be considered as first-line antihypertensive drug therapy in obesity–hypertension.

Copyright © 2006 S. Karger AG, Basel

Introduction

Obesity is a significant risk factor for hypertension and the cardiovascular sequel of hypertension. Several studies have demonstrated that an increase in weight raises blood pressure, whereas weight loss reduces blood pressure levels. Blood pressure increases in a linear manner over the whole range of body mass index or waist circumference or body mass index. An increase of 4.5 cm for men, respectively 2.5 cm for women in waist circumference or an increase of 1.7 kg/m^2 for men, respectively 1.25 kg/m^2 for woman in body mass index corresponds to an increase in blood pressure of 1 mm Hg [1]. Blood pressure levels are consistently higher in obese patients. Prevalence of hypertension is higher in obese patients compared to normal weight patients and the rate of good blood pressure control decreases with increasing obesity [2]. Most obese hypertensive patients require antihypertensive drug treatment. Despite years of investigation, our fundamental and clinical knowledge of obesity–hypertension is relatively meager and certainly inadequate [3]. Adding to this frustration is the inadequacy of the treatment of obesity–hypertension. Some classes of antihypertensive agents may have unwanted effects on the metabolic and hemodynamic abnormalities that link obesity and hypertension. The increasing prevalence of hypertension in obese patients and the low control rates in overweight and obese patients document the challenge that hypertension control in obese patients imposes on the physician. This chapter summarizes a series of comprehensive discussion about the theoretical reasons for the differential use of the major classes of antihypertensive agents in the pharmacological management of obesity–hypertension [3–8].

Guidelines

Guidelines for the management of hypertension provide specific recommendations for a variety of special populations, including black patients, patients with heart failure, pregnancy, coronary heart disease, diabetes mellitus, chronic kidney disease or the elderly [9]. However, the guidelines do not give specific recommendations for antihypertensive therapy of obese hypertensive patient. This is quite surprising since obviously most physicians treat more obese hypertensive patients than, e.g. pregnant hypertensive patients. Most guidelines obviously do not recognize obese patients as special population and only recommend that these patients need to lose weight [7, 10]. While indeed weight loss may be the most effective measure for the treatment of hypertension in obese patients, in practice, long-term outcome of weight management programs for obesity are generally poor. It should be recognized that hypertensive obese patients loosing weight need significantly fewer antihypertensive agents than those with stable obesity. Therefore, weight loss not only by itself lowers blood pressure but appears to be a useful tool in blood pressure management in patients who require medication to control their blood pressure [11]. Even the WHO technical report on prevention and management of the global epidemic of obesity does not spell out any specific recommendation or guidelines for antihypertensive therapy in obesity–hypertension [7, 12].

This lack of specific recommendations may also come from the paucity of data from prospective intervention studies for obese hypertensive patients. There are no outcome of studies on the management of hypertension in obese hypertensive patients, moreover, they are often even excluded from intervention trials. Obese patients may have been included in some of the intervention trials, but no sub-analyses for these patients have been presented yet [7, 10]. Clinical trials are needed to determine the most effective antihypertensive drugs for obese hypertensive patients.

Because of the lack of clear evidence for drawing up recommendations or guidelines for antihypertensive treatment of obesity–hypertension clinical practice is frequently based on subjective criteria. Only general recommendations can be given for or against various antihypertensive agents, based on mechanistic or pathophysiologic assumptions [10]. Each class of agents and the rationale for selection of each for initial therapy will be discussed.

Obesity–Hypertension

Obesity–hypertension is characterized hemodynamically by expanded intravascular volume associated with increased cardiopulmonary volume and cardiac output and a decrease rather than an increase in total peripheral resistance compared to lean hypertensive patients [13]. Since obese patients have sodium retention, even normal levels of renin activity must be considered as inappropriate elevated [5]. Activation of the renin–angiotensin system in adipose tissue may represent an important link between obesity and hypertension. Hypertension in obese patients is also associated with increased inflammation, left ventricular hypertrophy, sleep apnea, endothelial dysfunction, renal hyperfiltration and microalbuminuria. Obese patients often have metabolic abnormalities ranging from hyperinsulinemia to full-blown type 2 diabetes. In addition, obesity per se may cause end-organ damage [5]. Although it is incompletely understood by which mechanisms obesity alters blood pressure and renal function, three factors seem to be of particular importance: (1) activation of the renin–angiotensin system, (2) increased sympathetic activity, and (3) structural changes in the kidney itself [14].

Blood Pressure

There are currently no specific treatments goals for obese hypertensives. Probably these goals should be similar to those recommended for other high risk patients, including patients with diabetes (130/80 mm Hg).

Diuretics

As discussed earlier, sodium and volume retention play a central role in the development of obesity-related hypertension. Diuretics reduce blood pressure levels by decreasing intravascular volume and cardiac output and should thus be beneficial in obese patients. Therefore, diuretics merit a useful place to begin when considering antihypertensive drug treatment in obese patients [3]. Moreover, diuretics have been shown to decrease cardiovascular morbidity and mortality in patients with arterial hypertension [15]. However, used at higher doses, diuretics may stimulate the renin–angiotensin system, increase sympathetic nerve activity, and promote insulin resistance and dyslipidemia. Diuretic-based antihypertensive therapy is associated with the development of diabetes but with improved clinical outcomes [16]. It has been proposed that the duration of clinical trials like ALLHAT has been too short to detect the adverse effects of diabetes. During a mean follow-up of 14 years, for patients enrolled in the systolic hypertension in the elderly program (SHEP), patients given chlorthalidone who developed new onset diabetes had no increase in cardiovascular or total mortality rates, whereas there was an increased morbidity and mortality in those who developed diabetes while on placebo. The SHEP data may reflect a more benign course of diabetes induced by a diuretic than for diabetes that develops spontaneously [17]. It should be noted that thiazides are more potent to lower blood pressure than loop diuretics in hypertensive patients and should be preferred. Loop diuretics are indicated if the diuretic response to thiazides is inadequate and in patients with heart failure or decreased renal function. The diuretic dose should begin, e.g. with chlorthalidone or hydrochlorothiazide 12.5 or 25 mg. Diuretics are of particular value in the elderly patients. Also black patients with hypertension may be more volume dependent than white patients and may respond better to diuretics. Due to high sodium consumption, blockade of the renin–angiotensin system by itself does often not lower blood pressure very efficient. Therefore, low dose diuretics are excellent in combination with ACE-inhibitors and angiotensin receptor blockers (ARBs) as shown in the 'Birmingham Hypertension Square' (fig. 1) [18]. In clinical practice, it is almost impossible to control blood pressure in obese patients without the use of diuretics [4]. Thus, diuretics are most frequently needed to control blood pressure in obese patients and play a central role in combination with other agents in the management of obese hypertensives [6].

Beta-Blockers

Beta-blockers have mixed effects. Beta-blockers have been widely used in the treatment of hypertension and are recommended as first-line drugs in hypertension guidelines [9]. Beta-blockers have been shown to decrease cardiovascular morbidity and mortality in patients with arterial hypertension. Furthermore, beta-blockers work very well as the first line therapy for young hypertensive patients. Young hypertensive patients have a high sympathetic tone and

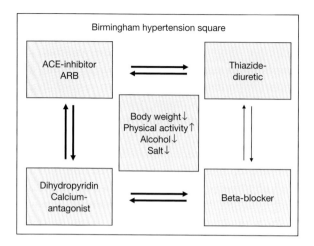

Fig. 1. The so called 'Birmingham Hypertension Square' shows the optimum choice of add-in drugs in the management of resistant hypertension. ACE-inhibitors and ARBs work very well in combination with thiazide diuretics or dihydropyridine calcium antagonists. The dihydropyridine calcium antagonists combine well with beta-blockers. Non-pharmacological treatment options are body weight reduction, increased physical activity as well as decrease in dietary sodium intake and avoidance of excess alcohol ingestion.

increased cardiac output. Increased sympathetic activity may play a role in obesity–hypertension since beta-blockers are more effective in lowering blood pressure in obese than in lean hypertensives. However, the initial use of beta-blockers in the elderly is less well established. Beta-blocker therapy is associated with an initial weight gain in the first month [19]. This may be caused by changes in energy metabolism. Beta-blocker treatment reduces energy expenditure by 4–9% [19]. Obese hypertensives are advised to exercise regularly for weight loss. However, at the same time beta-blockers may negatively influence exercise capacity. This has to be considered, when prescribing these drugs to obese patients [19]. Since weight reduction is an important aspect of hypertension management it may be hampered by the use of a beta-blocker [19]. In addition, beta blockers may disturb carbohydrate and lipid metabolism leading to impaired glucose tolerance as well as increased lipid levels. Recently, the first choice role of beta-blockers in the treatment of arterial hypertension has been challenged. Meta-analysis data showed that in comparison with other antihypertensive drugs, the effect of beta-blockers is less than optimum, with a raised risk of stroke. Hence, it was strongly suggested that beta-blockers should not remain first choice in the treatment of primary hypertension [20]. The discussion is still ongoing and remains to be resolved.

However, at least in obese hypertensive patients, the rationale for the use of beta-blockers as first choice agent patients may be questionable. Beta-blockers should be given when they are specifically indicated, i.e. in obese patients with atrial fibrillation, ischemic heart disease, or heart failure. However, without these specific indications other antihypertensive drugs should be taken first to lower blood pressure in obese patients [19].

Calcium Antagonists

Calcium antagonists are very efficacious in lowering blood pressure. Calcium antagonists lower blood pressure through a decrease in peripheral vascular resistance. Dihydropyridine calcium antagonists also promote natriuresis and may therefore work well in obese patients. Calcium antagonists do not have adverse metabolic effects, this makes them suitable for obese hypertensive patients. However, smaller studies have shown that calcium antagonists may be less effective in obese individuals [21]. This may be caused by the fact that the peripheral resistance is often reduced in these patients. A recent trial with verapamil showed that this drug is effective in reducing blood pressure and heart rate in obese hypertensive patients [22]. The precise role of calcium antagonists is still rather controversial with respect to hypertension, especially in those patients with cardiac complications. However, probably similar to the above-mentioned discussion about the first choice use of beta-blockers in the management of uncomplicated hypertension, the controversy has been overstated [3]. The black patient and the older patient with hypertension have been shown to respond particularly well to calcium antagonists [23]. The recently published Anglo-Scandinavian Cardiac Outcomes Trial (ASCOT) compared the effect on cardiovascular events of combinations of beta-blocker with a thiazide versus long-acting calcium antagonist with an ACE-inhibitor The calcium antagonist-based regimen lowered blood pressure slightly better and prevented more major cardiovascular events and induced less diabetes than the beta-blocker-based regimen [24]. Thus, calcium antagonists especially in combination with ACE-inhibitors (fig. 1) are useful antihypertensive drugs for obese hypertensive patients.

ACE-Inhibitors

Recently, the 100th anniversary of the discovery of renin by Tigerstedt and Bergman and the 60th anniversary of the identification of angiotensin by Braun-Menendez and Page was celebrated [25]. Much insight has been gained

Table 1. Reduction of rate of new onset diabetes by ACE-inhibitors and ARBs

Trial	ACE-inhibitor/ARB	Comparison	Reduction rate (%)
ALLHAT	Lisinopril	Thiazid	30
ALPINE	Candesartan	Atenolol	87
ANBP2	Enalapril	Thiazid	32
CAPPP	Captopril	Thiazid, Beta-blocker	14
CHARM	Candesartan	Plazebo	19
LIFE	Losartan	Atenolol	25
PEACE	Trandolapril	Plazebo	17
SCOPE	Candesartan	Plazebo	25
VALUE	Valsartan	Amlodipin	23

ALLHAT = Antihypertensive and Lipid-Lowering Treatment to Prevent Heart Attack Trial; ALPINE = Antihypertensive Treatment and Lipid Profile in a North of Sweden Efficacy; ANBP2 = The Second Australian National Blood Pressure Study; CAPPP = Captopril Prevention Project; CHARM = Candesartan in Heart Failure Assessment of Reduction in Mortality and Morbidity; LIFE = Cardiovascular morbidity and mortality in the Losartan Intervention For Endpoint reduction in hypertension; PEACE = Prevention of Events with Angiotensin Converting Enzyme Inhibition; SCOPE = Study on Cognition and Prognosis in the Elderly; VALUE = Valsartan Antihypertensive Long-Term Use Evaluation.

from the synthesis of pharmacological entities that antagonize the biosynthesis and physiological response to the generation of angiotensin II. ACE-inhibitors block the conversion of angiotensin I to angiotensin II. ACE-inhibitors reduce morbidity and mortality in patients with myocardial infarction and heart failure. In addition, they slow progression of renal disease in patients with chronic renal disease [26, 27]. These data and the results of the HOPE study indicate that ACE-inhibitors confer vascular and renal protection that are independent of their systemic antihypertensive action [28]. Recent data show increased activity of the renin–angiotensin system in adipose tissue. The renin–angiotensin system in the adipose tissue has been implicated in the development of insulin resistance and type 2 diabetes. ACE-inhibitors do not promote metabolic abnormalities but can even improve insulin sensitivity and reduce the risk of new onset type 2 diabetes. Use of ACE-inhibitors has been associated in several studies with a lower rate of new diagnosis of diabetes as shown in table 1 [29]. One should use these agents in full doses before a second agent of another class is chosen. The coadministration of a diuretic may be extremely valuable especially in the patients who are more volume dependent. In conclusion, ACE-inhibitors are currently considered the most appropriate form of antihypertensive treatment for obese hypertensive patients [4–6].

Angiotensin Receptor Blockers

ARBs directly inhibit the binding of angiotensin II to the AT1 receptor and provide a more specific blockade of the renin–angiotensin system than ACE-inhibitors. Their tolerability, particularly the absence of cough makes them an alternative to ACE-inhibitors [6]. The most frequent indication for ARB use is for patients who have had side effects from ACE-inhibitors. ARBs may share some of the beneficial properties of ACE inhibition. ARBs significantly reduce the velocity of progression of diabetic nephropathy in type 2 diabetes [30]. The CROSS study (Candesartan Role on Obesity and on Sympathetic System) examined the antihypertensive, metabolic and neuroadrenergic effects of the ARB candesartan in comparison to a thiazide diuretic in obese hypertensive patients. Treatment with the ARB significantly improved insulin sensitivity and reduced muscle sympathetic nerve activity compared with the diuretic [31]. Similar to ACE-inhibitors treatment with ARBs reduces the rate of new onset of diabetes compared to other antihypertensive drugs (table 1) [32]. A good number of reports exists dealing with the combination of ACE-inhibitor and ARBs [33] and there is a certain theoretical rationale for the coexistent use of both the ACE-inhibitor and the ARB for an almost complete blockade of the renin–angiotensin system [3].

Glitazones bind to PPAR-γ and are, as so called 'insulin sensitizers', highly effective in control of insulin resistance and diabetes. The ARBs telmisartan and irbesartan as well as metabolites of losartan have a molecular structure that imparts partial agonist properties with the PPAR-γ molecule, which results in reductions in glucose and lipid metabolism (fig. 2) [34–36]. This implies that some ARBs could treat both the hemodynamic and metabolic aberrations seen in subjects with the metabolic syndrome, such as insulin resistance, glucose intolerance, and hypertension. Ongoing megatrials like ONTARGET (Ongoing Telmisartan Alone and in Combination with Ramipril Global Endpoint Trial), in which telmisartan is compared with ramipril, will show whether the PPAR-γ activity confers additional protection. Whether clinically significant differences in cardiovascular or renoprotection exist between ARBs and ACE inhibitors in patients with metabolic syndrome or obesity remains to be fully investigated in appropriate head-to-head studies. In summary, ARBs provide an important and interesting alternative to the use of ACE-inhibitors in hypertensive obese patients [6].

Renin Inhibition

As mentioned above, angiotensin II is strongly implicated in the development of cardiovascular and renal disease as well as obesity hypertension.

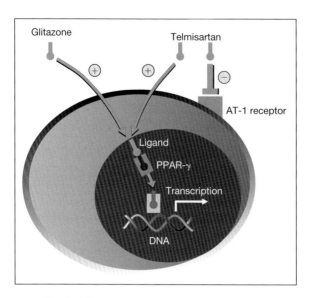

Fig. 2. Glitazone and telmisartan as well as irbesartan and metabolites of losartan are ligands for PPAR-γ. This activation occurs independently of inhibition of the AT-1 receptor by the ARBs. The activated receptor complex binds as transcription factor to DNA and induces transcription of genes, which improve insulin sensitivity. Modified from reference [41] with permission.

Therefore, the renin–angiotensin system is a highly successful pharmacologic target. However, compensatory increase in plasma renin levels may lead to adjustments in angiotensin II formation, and may present limitations for a complete blockade of the renin–angiotensin system [37]. In addition the recent discovery of a renin receptor and its role in a profibrotic response are challenging the view that the renin–angiotensin system can be completely blocked by ACE-inhibitors or ARBs [38, 39]. An orally effective renin inhibitor, aliskiren, is now available to address angiotensin production directly at its rate-limiting step [37]. Head-to-head comparison with ACE-inhibitors and ARBs will show whether the increase in plasma renin activity seen with these agents and signaling through the renin receptor has a pathogenic role in cardiovascular and renal disease. If this is the case the drug class of renin inhibitors may have a great potential [37].

Alpha-Blockers

Alpha-blockers have been associated with improved insulin sensitivity and lipid metabolism. Positive effects on insulin sensitivity have also been reported

Table 2. Summary of first choice use of antihypertensive drug classes in obesity hypertension

Antihypertensive drug class	Advantage	Disadvantage	Recommendation
ACE-inhibitors	Improve insulin sensitivity Decrease rate of new onset of diabetes mellitus Vascular and renal protection	Cough	First choice
ARBs	Similar to ACE-inhibitors Good tolerability PPAR-γ activity	Expensive	Alternative to ACE-inhibitor
Diuretics	Sodium and volume retention pathogenic in obesity hypertension Good in combination with other classes Not expensive	Decrease insulin sensitivity Disturb lipid metabolism	First choice (especially in combination with ACE-inhibitors)
Beta-blockers	Beneficial for patients with coronary heart disease, atrial fibrillation and heart failure Not expensive	Weight gain Decrease insulin sensitivity Disturb lipid metabolism Discussion about first choice treatment of hypertension ongoing	Not first choice but should be given when specifically indicated
Calcium antagonists	Metabolic neutral Good combination with ACE-inhibitor	Dihydropyridines less efficient in obesity-hypertension Discussion about first choice treatment of hypertensive patient with cardiac disease ongoing	Not first choice but good in combination with ACE-inhibitors
Alpha-blockers	Improve insulin sensitivity Improve lipid metabolism Beneficial in benign prostatic hyperplasia	Increases risk of congestive heart failure	No more first choice but acceptable in combination therapy or patients with prostatic hypertrophy

in obesity–hypertension [40]. In addition, alpha-blockers are used to treat benign prostatic hyperplasia. Noradrenaline acts on alpha1-adrenoceptors to contract the smooth muscle in the prostate and bladder and by opposing these actions, alpha-blockers are beneficial in prostatic hypertrophy.

A recent report from the 'Antihypertensive and Lipid-Lowering Treatment to Prevent Heart Attack Trial' (ALLHAT) has severely challenged the use of alpha-blocker for initial therapy in hypertension. Patients treated with doxazosine had a 25% increase in risk of cardiovascular disease events especially congestive heart failure in the ALLHAT trial compared to patients treated with the thiazide diuretic chlorthalidone [16]. This study very nicely underlines that it is difficult to rely on studies that show substantial advantages of alpha-blockers on surrogate parameters like insulin sensitivity or lipid levels. Consequently, these agents are used less today. Particular care must also be exercised in patients with postural hypotension with falls [3]. Therefore, alpha-blockers are no longer first line antihypertensive drugs [9]. However, they may be useful in combination with other antihypertensive drugs or in patients with benign prostatic hypertrophy.

Conclusion

Diuretics and beta-blockers are reported to reduce insulin sensitivity and increase lipid levels, whereas calcium antagonists are metabolically neutral and ACE-inhibitors as well as ARBs increase insulin sensitivity. Sodium and volume retention play a central role in the development of obesity-related hypertension. A summary of first choice treatment of obese hypertensives is shown in table 2. Treatment with an ACE-inhibitors or a diuretic should be considered as first-line antihypertensive drug therapy in obese patients. In most cases combination therapy of an ACE-inhibitor with low dose diuretics may be the first choice to lower blood pressure. If such a treatment does not lower blood pressure sufficiently, the addition or substitution of a calcium antagonist, an alpha- or beta-blocker may be considered. ARBs are indicated in patients who have had side effects from ACE-inhibitors. However, one should keep in mind that the beta-blocker can be associated with weight gain. Clearly, additional studies are needed to determine the long-term efficacy of antihypertensive drugs in obesity–hypertension.

References

1   Doll S, Paccaud F, Bovet P, Burnier M, Wietlisbach V: Body mass index, abdominal adiposity and blood pressure: consistency of their association across developing and developed countries. Int J Obes Relat Metab Disord 2002;26:48–57.

2   Bramlage P, Pittrow D, Wittchen HU, Kirch W, Boehler S, Lehnert H, Hoefler M, Unger T, Sharma AM: Hypertension in overweight and obese primary care patients is highly prevalent and poorly controlled. Am J Hypertens 2004;17:904–910.

3   Frohlich ED: Clinical management of the obese hypertensive patient. Cardiol Rev 2002;10:127–138.

4   Zanella MT, Kohlmann O Jr, Ribeiro AB: Treatment of obesity hypertension and diabetes syndrome. Hypertension 2001;38(pt 2):705–708.

5   Sharma AM: Is there a rationale for angiotensin blockade in the management of obesity hypertension? Hypertension 2004;44:12–19.

6   Sharma AM, Pischon T, Engeli S, Scholze J: Choice of drug treatment for obesity related hypertension: where is the evidence? J Hypertens 2001;19:667–674.

7   Sharma AM, Rossner S: Who cares about the obese hypertensive patient? J Intern Med 2002;251:369–371.

8   Dentali F, Sharma AM, Douketis JD: Management of hypertension in overweight and obese patients: a practical guide for clinicians. Curr Hypertens Rep 2005;7:330–336.

9   Chobanian AV, Bakris GL, Black HR, Cushman WC, Green LA, Izzo JL Jr, Jones DW, Materson BJ, Oparil S, Wright JT Jr, Roccella EJ: The Seventh Report of the Joint National Committee on Prevention, Detection, Evaluation, and Treatment of High Blood Pressure: the JNC 7 report. JAMA 2003;289:2560–2572.

10 Engeli S, Sharma AM: Emerging concepts in the pathophysiology and treatment of obesity-associated hypertension. Curr Opin Cardiol 2002;17:355–359.

11 Jones DW, Miller ME, Wofford MR, Anderson DC Jr, Cameron ME, Willoughby DL, Adair CT, King NS: The effect of weight loss intervention on antihypertensive medication requirements in the hypertension Optimal Treatment (HOT) study. Am J Hypertens 1999;12(pt 1–2):1175–1180.

12 Obesity: preventing and managing the global epidemic. Report of a WHO Consultation. World Health Organ Tech Rep Ser 2000;894:i–xii, 1–253.

13 Rocchini AP: Cardiovascular regulation in obesity-induced hypertension. Hypertension 1992;19(suppl):I56–160.

14 Wofford MR, Hall JE: Pathophysiology and treatment of obesity hypertension. Curr Pharm Des 2004;10:3621–3637.

15 Major outcomes in high-risk hypertensive patients randomized to angiotensin-converting enzyme inhibitor or calcium channel blocker vs diuretic: The Antihypertensive and Lipid-Lowering Treatment to Prevent Heart Attack Trial (ALLHAT). JAMA 2002;288:2981–2997.

16 Major cardiovascular events in hypertensive patients randomized to doxazosin vs chlorthalidone: the antihypertensive and lipid-lowering treatment to prevent heart attack trial (ALLHAT). ALLHAT Collaborative Research Group. JAMA 2000;283:1967–1975.

17 Kostis JB, Wilson AC, Freudenberger RS, Cosgrove NM, Pressel SL, Davis BR: Long-term effect of diuretic-based therapy on fatal outcomes in subjects with isolated systolic hypertension with and without diabetes. Am J Cardiol 2005;95:29–35.

18 Lip GY, Beevers M, Beevers DG: The 'Birmingham Hypertension Square' for the optimum choice of add-in drugs in the management of resistant hypertension. J Hum Hypertens 1998;12:761–763.

19 Sharma AM, Pischon T, Hardt S, Kunz I, Luft FC: Hypothesis: Beta-adrenergic receptor blockers and weight gain: a systematic analysis. Hypertension 2001;37:250–254.

20 Lindholm LH, Carlberg B, Samuelsson O: Should beta blockers remain first choice in the treatment of primary hypertension? A meta-analysis. Lancet, 2005;366:1545–1553.

21 Schmieder RE, Gatzka C, Schachinger H, Schobel H, Ruddel H: Obesity as a determinant for response to antihypertensive treatment. BMJ 1993;307:537–540.

22 White WB, Elliott WJ, Johnson MF, Black HR: Chronotherapeutic delivery of verapamil in obese versus non-obese patients with essential hypertension. J Hum Hypertens 2001;15:135–141.

23 Dickerson JE, Hingorani AD, Ashby MJ, Palmer CR, Brown MJ: Optimisation of antihypertensive treatment by crossover rotation of four major classes. Lancet 1999;353:2008–2013.

24 Dahlof B, Sever PS, Poulter NR, Wedel H, Beevers DG, Caulfield M, Collins R, Kjeldsen SE, Kristinsson A, McInnes GT, Mehlsen J, Nieminen M, O'Brien E, Ostergren J: Prevention of cardiovascular events with an antihypertensive regimen of amlodipine adding perindopril as required versus atenolol adding bendroflumethiazide as required, in the Anglo-Scandinavian Cardiac

Outcomes Trial-Blood Pressure Lowering Arm (ASCOT-BPLA): a multicentre randomised controlled trial. Lancet 2005;366:895–906.
25 Basso N, Terragno NA: History about the discovery of the renin-angiotensin system. Hypertension 2001;38:1246–1249.
26 Wenzel UO: Angiotensin-converting enzyme inhibitors and progression of renal disease: evidence from clinical studies. Contrib Nephrol 2001;135:200–211.
27 Wolf G, Butzmann U, Wenzel UO: The renin-angiotensin system and progression of renal disease: from hemodynamics to cell biology. Nephron Physiol 2003;93:P3–P13.
28 Yusuf S, Sleight P, Pogue J, Bosch J, Davies R, Dagenais G: Effects of an angiotensin-converting-enzyme inhibitor, ramipril, on cardiovascular events in high-risk patients. The Heart Outcomes Prevention Evaluation Study Investigators. N Engl J Med 2000;342:145–153.
29 Scheen AJ: Prevention of type 2 diabetes mellitus through inhibition of the Renin-Angiotensin system. Drugs 2004;64:2537–2565.
30 Brenner BM, Cooper ME, de Zeeuw D, Keane WF, Mitch WE, Parving HH, Remuzzi G, Snapinn SM, Zhang Z, Shahinfar S: Effects of losartan on renal and cardiovascular outcomes in patients with type 2 diabetes and nephropathy. N Engl J Med 2001;345:861–869.
31 Grassi G, Seravalle G, Dell'Oro R, Trevano FQ, Bombelli M, Scopelliti F, Facchini A, Mancia G: Comparative effects of candesartan and hydrochlorothiazide on blood pressure, insulin sensitivity, and sympathetic drive in obese hypertensive individuals: results of the CROSS study. J Hypertens 2003;21:1761–1769.
32 Dahlof B, Devereux RB, Kjeldsen SE, Julius S, Beevers G, de Faire U, Fyhrquist F, Ibsen H, Kristiansson K, Lederballe-Pedersen O, Lindholm LH, Nieminen MS, Omvik P, Oparil S, Wedel H: Cardiovascular morbidity and mortality in the Losartan Intervention For Endpoint reduction in hypertension study (LIFE): a randomised trial against atenolol. Lancet 2002;359:995–1003.
33 Wolf G, Ritz E: Combination therapy with ACE inhibitors and angiotensin II receptor blockers to halt progression of chronic renal disease: pathophysiology and indications. Kidney Int 2005;67:799–812.
34 Benson SC, Pershadsingh HA, Ho CI, Chittiboyina A, Desai P, Pravenec M, Qi N, Wang J, Avery MA, Kurtz TW: Identification of telmisartan as a unique angiotensin II receptor antagonist with selective PPARgamma-modulating activity. Hypertension 2004;43:993–1002.
35 Schupp M, Janke J, Clasen R, Unger T, Kintscher U: Angiotensin type 1 receptor blockers induce peroxisome proliferator-activated receptor-gamma activity. Circulation 2004;109:2054–2057.
36 Schupp M, Lee LD, Frost N, Umbreen S, Schmidt B, Unger T, Kintscher U: Regulation of peroxisome proliferator-activated receptor gamma activity by losartan metabolites. Hypertension 2006;47:586–589.
37 Müller DN, Luft FC: Direct renin inhibition with aliskiren in hypertension and target organ damage. Clin J Am Soc Nephrol 2006;1:221–228.
38 Huang Y, Wongamornthan S, Kasting J, McQuillan D, Owens RT, Yu L, Noble NA, Border W: Renin increases mesangial cell transforming growth factor-beta1 and matrix proteins through receptor-mediated, angiotensin II-independent mechanisms. Kidney Int 2006;69:105–113.
39 Nguyen G, Delarue F, Burckle C, Bouzhir L, Giller T, Sraer JD: Pivotal role of the renin/prorenin receptor in angiotensin II production and cellular responses to renin. J Clin Invest 2002;109:1417–1427.
40 Pollare T, Lithell H, Selinus I, Berne C: Application of prazosin is associated with an increase of insulin sensitivity in obese patients with hypertension. Diabetologia 1988;31:415–420.
41 Wenzel U, Wolf G: Blood pressure independent effects of antihypertensive agents. Internist 2005;46:548–556.

Ulrich O. Wenzel
III Medizinische Klinik, Universitätsklinikum Hamburg Eppendorf
Martinistrasse 52
DE–20246 Hamburg (Germany)
Tel. +49 40 42803 3908, Fax +49 40 42803 5186, E-Mail wenzel@uke.uni-hamburg.de

Bariatric Surgery

Michael Korenkov

Department of Visceral Surgery, University of Mainz, Mainz, Germany

Abstract

Background/Methods: Bariatric surgery today is the only effective therapy for morbid obesity. Commonly performed procedures include adjustable gastric banding (AGB) and vertical banded gastroplasty (VBG); variations of Roux-en-Y gastric bypass (RYGB), biliopancreatic diversion or duodenal switch (BPD) and mixed procedures. All these procedures can be performed by open surgery and more recently by laparoscopy. This review discusses key issues in the surgical management of morbid obesity. **Results:** The two most common bariatric procedures performed worldwide are laparoscopic AGB and laparoscopic RYGB. Controversy exists regarding the best surgical procedure. For example, gastric bypass is the procedure of choice in the United States, while most surgeons in Europe and Australia favor gastric banding. Weight loss decreased according to the procedures performed in following decreasing order: BPD, RYGB, VBG, AGB. **Conclusion:** Concerning the complications and quality of life there is no single operation for morbid obesity without drawbacks. According the currently opinion are gastric restrictive procedures (AGB, VBG) generally considered safe and quick to perform, but the long-term outcome and quality of life, especially with regard to eating patterns, have been questioned. On the other hand the long-term efficacy of AGB can be improved by the development of new band devices. More complex bariatric procedures, such as RYGB or BPD, have a greater potential for serious perioperative complications but are associated with good long-term outcome in terms of weight loss combined with less dietary restriction.

Copyright © 2006 S. Karger AG, Basel

Obesity is associated with multiple complications and related comorbidities that lead to both physical and psychological problems. It is estimated that there are 400,000 deaths attributable to obesity in the United States each year and obesity has been identified as the second most common cause of death after smoking from modifiable behavioral risk factors [1]. Unfortunately, the conservative approach to weight loss consisting of diet, exercise, and medication generally achieves no more than a 5–10% reduction in body weight and recidivism after such weight loss exceeds 90% within 5 years [2]. The disappointing results of these approaches have led to a burgeoning interest in bariatric surgery [3].

Obesity is defined as a body mass index (BMI = body weight in kilogram's divided by body height in meters squared); morbid obesity is defined as a BMI $\geq 40\,kg/m^2$ or a BMI $> 35\,kg/m^2$ with coexisting comorbid conditions, such as hypertension, atherosclerotic cardiovascular disease, diabetes mellitus, sleep apnea, weight-bearing osteoarthritis and others [4].

Indications for Surgery

Obesity surgery is considered appropriate for adult patients with either a BMI of 40 or more, or a BMI between 35 and 40 with obesity-related comorbidity such as type II diabetes mellitus, hypertension, cardiomyopathy, sleep apnea, asthma, pseudotumor cerebri, osteoarthritis and hyperlipidemia. Each patient should be treated conservatively for at least 5 years without success. These selection criteria have been laid down in March 1991 by the National Institutes of Health Consensus Development Panel and have subsequently been adopted by all major surgical and non-surgical societies [5]. Various contraindications must also be taken into account, although most have not been derived from firm clinical evidence. Severe mental or cognitive retardation are therefore generally considered absolute contraindications. Psychiatric disorders (psychotic, personality or affective disorders, alcoholism and/or drug abuse), lacks of compliance are considered as potential negative predictors for obesity surgery, but they were not found to be valid [6].

Operative Procedures

Bariatric surgery evolved in the 1950s from the jejunoileal bypass (fig. 1). This operation was widely performed during the 1960s and 1970s and was a highly effective weight-reduction operation, but was abandoned because of severe diarrhea, electrolyte imbalance, liver failure, nephrolithiasis, gas-bloat syndrome and other negative appearances [7].

The modern bariatric procedures presents as follow:
(1) Gastric restrictive operations [adjustable gastric banding (AGB); vertical banded gastroplasty (VBG)].
(2) Malabsorptive operations [biliopancreatic diversion (BPD); duodenal switch].
(3) Malabsorptive/restrictive operations [Roux-en-Y gastric bypass (RYGB)].
(4) Other operations (gastric pacemaker).

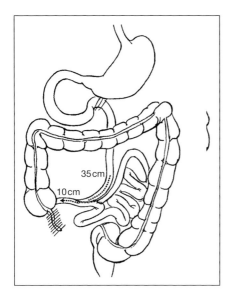

Fig. 1. Jejunoileal bypass.

The primary name in gastric restrictive surgery is Mason, who in association with Printen, performed the first restrictive procedure in 1971. They divided the stomach horizontally from lesser curvature to greater curvature, leaving a gastric conduit at the greater curvature [8]. This procedure was unsuccessful in maintaining weight loss. After the numerous of variations the VBG was established. In this procedure the upper stomach near the esophagus is stapled vertically for about 6 cm to create a smaller stomach pouch. The outlet from the pouch is restricted by a band or ring that slows the emptying of the food and thus creates the feeling of fullness (fig. 2). VBG is a non-adjustable restrictive procedure and nowadays is almost abandoned from the surgical repertoire.

AGB is the least invasive of the gastric restrictive procedures. The band is fitted around the upper part of the stomach, dividing it into two sections, the smaller of which is above the band and has a capacity of approximately 15–20 ml (pouch); the larger remaining part is below the band (fig. 3). The constriction is called a stoma. The inflatable band is connected with a tube to a subcutaneous port, which is used for the percutaneous introduction or removal of fluid to adjust the caliber of the gastric band.

The AGB makes it possible for the physician to alter the stoma diameter. Because the band system is adjustable, the stoma can be adjusted in a female patient to assure a normal pregnancy if this occurs.

BPD was developed by Scopinaro et al. [9] in Genoa. The small bowel is divided 250 cm proximal to the ileocecal valve, and a subtotal gastrectomy is

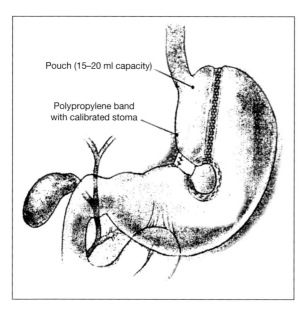

Fig. 2. Vertical banded gastroplasty (with permission from Brolin, JAMA, 2002).

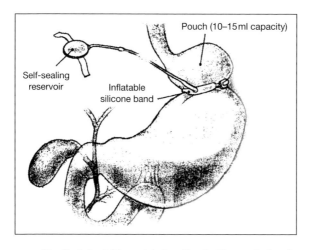

Fig. 3. Adjustable gastric banding (with permission from Brolin, JAMA, 2002).

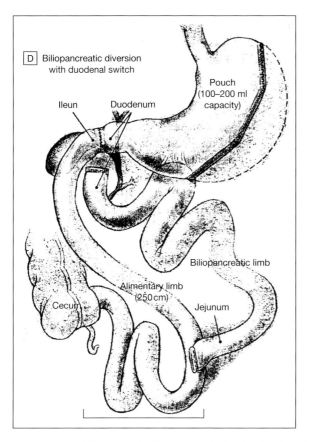

Fig. 4. Biliopancreatic diversion with duodenal switch (with permission from Brolin, JAMA, 2002).

performed. The distal (alimentary) limb is anastomosed to the proximal gastric pouch. The proximal limb (biliopancreatic conduit) is anastomosed to the side of the distal limb 50 cm proximal to the ileocecal valve (fig. 4). Absorption of nutrients occurs in this distal 50-cm common channel, with malabsorption of fats and starches. The BPD has been modified by Marceau and coworkers [10] in Quebec by a 'duodenal switch'. The greater curvature portion of the stomach is resected (leaving a tube), and the proximal duodenum is divided. At the point of division of the ileum the distal end is anastomosed to the proximal duodenum, and the proximal end is anastomosed to the side of the distal ileum.

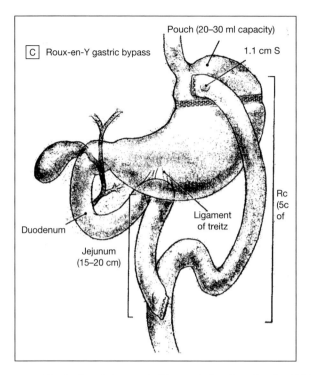

Fig. 5. Gastric bypass (with permission from Brolin, JAMA, 2002).

The gastric bypass procedure (RYGB) was published as a treatment for morbid obesity as early as 1967 by Mason and Ito [11]. Mason noted that a frequent sequelae of the Billroth II gastrectomy was weight loss. He thus developed the retrocolic loop gastric bypass to produce a small gastric reservoir, with early satiety and some 'dumping' to deter ingestion of sweets. Gastric bypass combines gastric restriction with a small amount of subclinical malabsorption. In gastric bypass, the upper stomach is completely closed off or transected, thereby excluding more than 95% of the stomach, all of the duodenum and about 50 cm of proximal jejunum from digestive continuity. The Roux-en-Y technique is currently the preferred method of gastric bypass (fig. 5).

Gastric pacemaker belongs to experimental procedures. In 1995, Cigaina et al. [12] discovered, when experimenting with pigs, that electrical stimulation of the stomach wall influenced the animal's eating habits. Animals whose stomach wall had been stimulated ate less. In this procedure, an electrode placed into the wall of the stomach between the incisura and the esophagogastric junction along the lesser curvature is connected by a wire lead to a pacer in a subcutaneous pocket, programmed to deliver a bipolar pulse to the stomach.

Preoperative Care

All patients should be evaluated for their medical history and undergo laboratory tests. Despite the lack of sound evidence in the obese, chest radiography, electrocardiography, spirometry, and abdominal ultrasonography are recommended for the evaluation of obesity-related comorbidity. Upper gastrointestinal endoscopy or upper GI series is advisable for all bariatric procedures, but is strongly recommended for gastric bypass patients [13]. All patients should be evaluated for psychological health, quality-of-life, possible personality disorders, social relationships, motivation, expectations and compliance [5]. Specifically important to obese patients is the evaluation of pulmonary function and obstructive sleep apnea.

Long-Term Outcomes of Effective Weight Loss and Comorbidity

Most common used criteria for the weight reduction after bariatric surgery is a loss of excess weight (the difference between actual weight and the ideal body weight for a given height). The estimation of the ideal body weight based on the Metropolian Tables for middle frame individuals [14]. Buchwald et al. [15] performed a systematic review and meta-analysis of 136 studies, which included for a total 22,094 patients. The meta-analysis has concentrated on the weight loss outcomes and on the impact of bariatric surgery on four selected obesity comorbidities: diabetes, hyperlipidemia, hypertension, and obstructive sleep apnea. The mean (95% confidence interval) percentage of excess weight loss was 61.2% (58.1–64.4%) for all patients; 47.5% (40.7–54.2%) for patients who underwent gastric banding; 61.6% (56.7–66.5%), gastric bypass; 68.2% (61.5–74.8%), gastroplasty; and 70.1% (66.3–73.9%), biliopancreatic diversion or duodenal switch. Diabetes was completely resolved in 76.8% of patients and resolved or improved in 86.6%. Hyperlipidemia improved in 70% or more of patients. Hypertension was resolved in 61.7% of patients and resolved or improved in 78.5%. Obstructive sleep apnea was resolved in 85.7% of patients and was resolved or improved in 83.6% of patients.

Choice of Procedure (Gastric Banding or Gastric Bypass?)

Ideally, the bariatric surgical procedure should: (1) present low-risk (mortality <1% and morbidity <10%), (2) give reduction >50% of the excess weight maintained for at least 5 years and benefiting >75% of the operated patients, (3) offer good quality of life with few side-effects, (4) be associated with a low rate of re-operations (<2% per year), and (5) be reversible and reproducible [16]. All four above-mentioned procedures (AGB, VBG, RYGB, BPD) can be effective in the treatment of morbid obesity and are performed by open surgery and more recently by laparoscopy. The two most common bariatric procedures performed worldwide are laparoscopic AGB and laparoscopic RYGB [17]. Controversy exists regarding the best surgical procedure. For example, gastric bypass is the procedure of choice in the United States, while most surgeons in Europe and Australia favor gastric banding. This discrepancy indicates that the choice of the procedure is driven by geographic factors and surgeon's skills rather than by medical evidence. According the currently opinion are restrictive procedures (AGB, VBG) generally considered safe and quick to perform, but the long-term outcome and quality of life, especially with regard to eating patterns, have been questioned. On the other hand, the long-term efficacy of AGB can be improved by the development of new band devices. Band-related complications include band slippage, leak, intolerance, infection, and migration as well as insufficient weight loss. The management of these complications includes: (1) band replacement for slippage, (2) band removal for infection, leak, and migration, (3) band removal plus RYGB for intolerance, and (4) the addition of BPD or band removal plus RYGB for insufficient weight loss. Evidence-based data for the choice of option selection are lacking.

It is common belief that certain predictors such as sweet eating or binging eating behavior and super-obesity (BMI > 50) can negative influent results of AGB or VBG. This hypothesis is controversial till now. Some authors demonstrated no influence of sweet eating and super-obesity on the postoperative weight reduction after laparoscopic AGB [18, 19].

VBG is a gastric restrictive procedure similar an AGB, but not so widespread as the last. Olbers et al. [20] compared the results of laparoscopic RYGB vs. laparoscopic VBG in a randomized clinical trial, which had 2-year follow-up time and 97.6% follow-up rate. RYGB and VBG were comparable in terms of operative safety and postoperative recovery, but weight reduction was greater after RYGB (84.4 vs. 59.8% at 2 years).

More complex bariatric procedures, such as RYGB or BPD, have a greater potential for serious perioperative complications but are associated with better long-term outcome in terms of weight loss combined with less dietary restriction [21, 22].

RYGB is currently a gold standard for most American bariatric surgeons. The standard GB includes a pouch volume of about 20 or 30 ml, an alimentary limb of at least 75 cm, and a biliary limb of at least 50 cm. Long limb distal gastric bypass seems to be preferable in superobese patients. RYGB can be performed without mortality [23]. Complications include stoma stenosis, gastric distension, anastomotic leakage, gastrointestinal hemorrhage, gastrojejunal ulcers and nutritional deficiencies as well as inadequate weight loss.

The BPD causes more malabsorption than RYGB. In its classic form, BPD consists of partial gastrectomy with a Roux-en-Y gastroenterostomy. In its duodenal switch form, a vertical sleeve gastrectomy is combined with a duodenoenterostomy. Few data have been published on limb length, but it is generally recommended that the common limb should measure more than 50 cm but less than 100 cm. Of note, no randomized trial to date has compared BPD with other procedures. BPD, however, can lead to massive weight loss: as much as 70% of the patient's initial excess weight [24]. The malabsorption created by this operation frequently leads to deficiencies in iron, calcium, and vitamins.

There is little consensus regarding the choice of bariatric surgical procedure. Although RYGB with standard or long limb or BPD may be the gold standard, AGB or VBG is frequently regarded as the procedure of first choice. In patients with failed AGB or VBG, RYGB or BPD may be offered. Some surgeons recommend AGB for patients with BMI < 50 and RYGB for patients with BMI > 50. Till now there is no convincing data about the impact of gastric pacemaker as a bariatric procedure.

Conclusion

Obesity is a serious medical problem that is increasing at an alarming rate. Significant obesity is associated with numerous comorbid conditions that respond to therapy directed toward a reduction of excess weight. There are no new non-surgical approaches that are effective in the long-term for treating morbid obesity. At this time, bariatric surgery offers the best treatment to produce sustained weight loss in patients who are morbidly obese. Weight loss decreased according to the procedures performed in following decreasing order: BPD, RYGB, VBG, AGB. Since currently there is no single operation for morbid obesity without drawbacks and in consequence of that no everywhere accepted gold standard bariatric operation. Evidence-based data for tailored patient selection for gastric restrictive or malabsorptive surgical procedures are currently lacking.

References

1 Mokdad AH, Marks JS, Stroup DF, Gerberding JL: Correction: actual causes of death in the United States, 2000. JAMA 2005;293:293–294.
2 Solomon CG, Dluhy RG: Bariatric surgery – quick fix or long-term solution? N Engl J Med 2004;351:2751–2753.
3 Sjostrom L, Lindroos AK, Peltonen M, Torgerson J, Bouchard C, Carlsson B, et al: Lifestyle, diabetes, and cardiovascular risk factors 10 years after bariatric surgery. N Engl J Med 2004;351: 2683–2693.
4 WHO: Obesity – preventing and managing the global epidemic. Report of a WHO Consultation on Obesity. Geneva, 1997, pp 3–5.
5 Sauerland S, Angrisani L, Belachew M, Chevallier JM, Favretti F, Finer N, et al: Obesity surgery: evidence-based guidelines of the European Association for Endoscopic Surgery (EAES). Surg Endosc 2005;19:200–221.
6 Black DW, Goldstein RB, Mason EE: Psychiatric diagnosis and weight loss following gastric surgery for obesity. Obes Surg 2003;13:746–751.
7 Deitel M: Overview of operations for morbid obesity. World J Surg 1998;22:913–918.
8 Printen KJ, Mason EE: Gastric surgery for relief of morbid obesity. Arch Surg 1973;106:428–431.
9 Scopinaro N, Gianetta E, Civalleri D, Bonalumi U, Bachi V: Bilio-pancreatic bypass for obesity: II. Initial experience in man. Br J Surg 1979;66:618–620.
10 Lagace M, Marceau P, Marceau S, Hould FS, Potvin M, Bourque RA, et al: Biliopancreatic diversion with a new type of gastrectomy: some previous conclusions revisited. Obes Surg 1995;5: 411–418.
11 Mason EE, Ito C: Gastric bypass in obesity. Surg Clin 1967;47:1345–1347.
12 Cigaina VV, Saggioro A, Rigo VV, Pinato G, Ischai S: Long-term effects of gastric pacing to reduce feed intake in swine. Obes Surg 1996;6:250–253.
13 Korenkov M, Sauerland S, Shah S, Junginger T: Is routine preoperative upper endoscopy in gastric banding patients really necessary? Obes Surg 2006;16:45–47.
14 Deitel M, Greenstein RJ: Recommendations for reporting weight loss. Obes Surg 2003;13: 159–160.
15 Buchwald H, Avidor Y, Braunwald E, Jensen MD, Pories W, Fahrbach K, et al: Bariatric surgery: a systematic review and meta-analysis. JAMA 2004;292:1724–1737.
16 Manterola C, Pineda V, Vial M, Losada H, Munoz S: Surgery for morbid obesity: selection of operation based on evidence from literature review. Obes Surg 2005;15:106–113.
17 Buchwald H, Williams SE: Bariatric surgery worldwide 2003. Obes Surg 2004;14:1157–1164.
18 Korenkov M, Kneist W, Heintz A, Junginger T: Laparoscopic gastric banding as a universal method for the treatment of patients with morbid obesity. Technical alternatives in laparoscopic placement of an adjustable gastric band: experience of two German university hospitals. Obes Surg 2004;14:1123–1127.
19 Mittermair RP, Aigner F, Nehoda H: Results and complications after laparoscopic adjustable gastric banding in super-obese patients, using the Swedish band. Obes Surg 2004;14:1327–1330.
20 Olbers T, Fagevik-Olsen M, Maleckas A, Lonroth H: Randomized clinical trial of laparoscopic Roux-en-Y gastric bypass versus laparoscopic vertical banded gastroplasty for obesity. Br J Surg 2005;92:557–562.
21 De Waele B, Lauwers M, Van Nieuwenhove Y, Delvaux G: Outpatient laparoscopic gastric banding: initial experience. Obes Surg 2004;14:1108–1110.
22 Jones DB, Provost DA, DeMaria EJ, Smith CD, Morgenstern L, Schirmer B: Optimal management of the morbidly obese patient SAGES appropriateness conference statement. Surg Endosc 2004;18: 1029–1037.

23 Obeid F, Falvo A, Dabideen H, Stocks J, Moore M, Wright M: Open Roux-en-Y gastric bypass in 925 patients without mortality. Am J Surg 2005;189:352–356.
24 Van Hee RH: Biliopancreatic diversion in the surgical treatment of morbid obesity. World J Surg 2004;28:435–444.

Privatdozent Dr. Michael Korenkov
Department of Visceral Surgery, University of Mainz
Langenbeckstrasse 1
DE–55101 Mainz (Germany)
Tel. +49 6131 177291, Fax +49 6131 176630, E-Mail korenkov@ach.klinik.uni-mainz.de

Author Index

Abrass, C.K. 106
Adamczak, M. 70
Axelsson, J. 165

Bosma, R.J. 184

Chudek, J. 70

de Jong, P.E. 184

Engeli, S. 122

Fliser, D. 203

Heimbürger, O. 165
Homan van der Heide, J.J. 184

Iacobellis, G. 212
Iseki, K. 42

Kalantar-Zadeh, K. 57
Kataoka, H. 91
Kielstein, J.T. 203
Kopple, J.D. 57
Korenkov, M. 243
Kramer, H. 1
Krebs, C. 230
Krikken, J.A. 184

Lord, G.M. 151

Meier-Kriesche, H.-U. 19
Menne, J. 203
Morales, E. 221
Morse, S. 135

Navis, G.J. 184
Nieszporek, T. 70

Praga, M. 221

Reisin, E. 135

Sharma, A.M. 212
Sharma, K. 91
Srinivas, T.R. 19
Stenvinkel, P. 165

Thakur, V. 135

Wenzel, U.O. 230
Więcek, A. 70
Wolf, G. VII, 175

Ziyadeh, F.N. 175

Subject Index

ABC transporters, lipid transport 112, 113
Adipokines, *see also* specific adipokines
   angiotensin II interactions 99, 100
   chronic kidney disease-cardiovascular
     disease interactions 101
   sources 71, 72, 166
   types 70, 71
Adiponectin
   chronic kidney disease levels and
     significance 76, 77, 97
   forms 75
   functions 76, 96, 97
   inflammatory response 169
   obesity effects on kidney 141
   receptors 76
   sources 85
   therapeutic targeting 77
Adipose tissue
   cell types 71
   endocrine function, *see* Adipokines;
     Adiponectin; Apelin; Leptin; Renin-
     angiotensin system; Resistin; Visfatin
   physiological functions 71
Adjustable gastric banding, *see* Bariatric
   surgery
Aldosterone, *see* Renin-angiotensin system
Aliskiren, obese hypertensive management
   238
Alpha-blockers, obese hypertensive
   management 238, 240
Angiotensin II
   adipokine interactions 99, 100

   functions 99
   inhibition interventions
     chronic kidney disease in obesity 13
     metabolic syndrome 127
     microalbuminuria resolution 142
     obese hypertensive management
       angiotensin-converting enzyme
         inhibitors 235, 236
       angiotensin receptor blockers 237
       renin inhibitors 238
     renal effects in obesity 143, 144, 197
Apelin
   expression regulation 79
   functions 79
   sources 79

Bariatric surgery
   approaches
     adjustable gastric banding 245
     biliopancreatic diversion 245
     gastric pacemaker 247, 248
     overview 217, 218
     popularity by technique 251
     Roux-en-Y gastric bypass 247
     selection of technique 250, 251
     vertical banded gastroplasty 244
   indications 244
   long-term outcomes 249, 250
   preoperative care 248, 249
   prospects for study 251, 252
   proteinuria and microalbuminuria
     reduction 225

Behavioral treatment, obesity management 213
Beta-blockers, obese hypertensive management 233–235
Biliopancreatic diversion, *see* Bariatric surgery

Calcium channel blockers, obese hypertensive management 235
CD36
  lipid metabolism in disease 113
  modified low-density lipoprotein binding 110
Cholesterol, *see also* Hyperlipidemia
  reverse transport 111, 112
  synthesis and uptake 110
Chronic kidney disease (CKD), *see also* specific diseases
  adiponectin levels 76, 77, 97
  cardiovascular disease risks 43, 165, 166
  epidemiology in obesity 2–5
  hemodialysis patients, *see* Hemodialysis
  inflammation
    cytokines from adipose tissue 167–170
    macrophage mediation 170
    systemic inflammation and outcomes 167, 168
  mortality 64, 165
  leptin levels 74
  obesity
    independent risk factor 7–10
    interventions
      hypertension control 12
      renin-angiotensin system inhibition 13
      weight loss 12, 13
    Japan study
      cardiovascular disease risk analysis 50, 51
      data sources 43–45
      end-stage renal disease and obesity 48, 49
      end-stage renal disease trends 49, 50
      general population analysis 46–48
      obesity trends 45, 46
      prevention 51
    sex differences 52
    shared risk factors 5–7
C-reactive protein (CRP), chronic kidney disease and cardiovascular outcomes 167

Diabetes
  epidemiology 135, 136
  kidney transplant recipients 24, 25
  obesity and chronic kidney disease 5–7
  tumor necrosis factor-α and insulin resistance role 80
  visfatin studies 78, 79
Diuretics, obese hypertensive management 232, 233

Effective renal plasma flow, *see* Renal blood flow
End-stage renal disease, *see* Chronic kidney disease
Exercise, obesity management 215

FAT, *see* CD36
Filtration fraction (FF)
  body surface area normalization 186, 187
  measurement 185
  obesity effects 187, 190, 191
Foam cell, mesangial cell conversion 115–117
Focal segmental glomerulosclerosis (FSGS), obesity association 10–12, 21, 137, 222
Free fatty acid (FFA), renal injury 141

Gastric bypass, *see* Bariatric surgery
Gastric pacemaker, weight loss studies 247, 248
Ghrelin, obesity effects on kidney 140
Glomerular filtration rate (GFR), obesity effects
  body surface area normalization 186, 187
  clinical implications 197
  measurement 185, 186
  mechanisms 191–196
  overview 137–139
  study analysis 187, 190, 191
Glucose, renal metabolism 205, 206

Hemodialysis (HD)
  mortality rate 58
  obesity benefits in survival
    Japan study 51, 52
    reverse epidemiology overview 58
    survival analysis 59–61, 170, 171
  plasminogen activator inhibitor-1 levels 98
Hemodynamics, see Filtration fraction; Glomerular filtration rate; Renal blood flow
Hyperlipidemia
  cardiovascular disease risks 106, 107
  chronic kidney disease risks 106–109
  kidney transplant recipients 24, 38
  reverse epidemiology in end-stage renal failure 64
Hypertension
  control and chronic kidney disease prevention 12
  kidney transplant recipients
    obesity risks 23
    prevalence 23
  management
    guidelines 231, 232
    obese hypertensives
      alpha-blockers 238, 240
      angiotensin-converting enzyme inhibitors 235, 236
      angiotensin receptor blockers 237
      beta-blockers 233–235
      blood pressure goals 232
      calcium channel blockers 235
      diuretics 232, 233
      drug selection 239, 240
      renin inhibitors 238
  obesity association and features 230–232
  obesity with chronic kidney disease 5–7
  weight loss response 142, 143

Infliximab, insulin resistance response 168
Insulin
  adipokine expression regulation
    plasminogen activator inhibitor-1 81
    tumor necrosis factor-α 80
  functions 92, 93
  renal handling 92

Insulin-like growth factor-1 (IGF-1), lipid metabolism role 109, 110, 115
Insulin resistance, see also Metabolic syndrome
  cardiovascular disease risks in kidney disease 206–208
  chronic kidney disease
    association 92, 93
    progression 208, 209
  classification 203, 204
  primary kidney disease association 204, 205
  renal glucose metabolism 205, 206
  renal hemodynamic effects 193
Interleukin-6 (IL-6)
  adipose expression 79
  chronic kidney disease levels 80, 167
  functions 79, 80
  peritoneal dialysis levels 168

Kidney transplantation
  diabetes in post-transplant period 25
  hyperlipidemia in post-transplant period 24, 38
  hypertension in post-transplant period
    obesity risks 23
    prevalence 23
  obesity in recipients
    cardiac events 24
    delayed graft function 23
    epidemiology 19, 20
    graft survival 25–34
    infection 33, 37
    management 36–38
    pathophysiology 21, 22
    perioperative complications 22
    survival analysis 26–34
    weight gain following transplant 34, 35

Leptin
  adverse effects 74
  body weight regulation 153
  chronic kidney disease levels and significance 74, 75, 94, 177
  elimination and catabolism 74, 75, 93, 94, 177
  expression 72, 152

Subject Index

257

Leptin (continued)
  fibrosis role 180, 181
  functions 72–74, 93, 152, 177
  immune system effects
    inflammatory immune response 158, 159
    innate immune response 156, 157
    mouse model studies 156
  inflammatory response 168, 169
  obesity effects on kidney 140
  ob/ob mouse 152, 155, 156, 175, 176
  peripheral effects 153, 154
  prospects for study 159, 160
  receptor
    circulating levels 152
    db/db mouse 155, 156, 176
    distribution 154, 155
    gene structure and forms 154, 176
    signaling 176
  renal effects 177–181
  renal hemodynamic effects in obesity 195, 196
  sex differences 152
  starvation signaling 153
  structure 152
Low birth weight, metabolic syndrome risks 139, 140
Low-density lipoprotein, see Hyperlipidemia

Macrophage, inflammation mediation 170
Mesangial cell
  foam cell formation 115–117
  leptin effects 178
  lipid metabolism 114, 115
  low-density lipoprotein response 114, 115
  sclerosis 115
Metabolic syndrome
  chronic kidney disease risks 136, 137
  insulin-like growth factor-1 role 109, 110, 115
  low birth weight risks 139, 140
  renin-angiotensin-aldosterone system dysregulation
    adipose tissue relationship with circulating renin-angiotensin-aldosterone system 125, 126

  clinical interventions 127
  local tissue renin-angiotensin systems 124, 125
  pathophysiology 127–129
  systemic renin-angiotensin-aldosterone system 123, 124
  weight loss effects 126, 130
Microalbuminuria
  bariatric surgery and reduction 225
  intervention goals 142
  obesity association 221–223
  weight loss and reduction 223–225

Obesity
  bariatric surgery, see Bariatric surgery
  chronic kidney disease shared risk factors 5–7
  definition 20, 243
  epidemiology and trends 1, 2, 20, 21, 58, 59, 135, 212
  hypertension, see Hypertension
  kidney transplant recipients, see Kidney transplantation
  management
    bariatric surgery, see Bariatric surgery
    behavioral treatment 213
    dietary intervention 213, 214
    exercise 215
    orlistat 216
    rimonabant 216, 217
    sibutramine 215, 216
  mortality 243
  proteinuria association, see Proteinuria
  renal effects
    blood flow 137, 187, 190, 191
    endocrine effects 140, 141
    excess excretory load 139, 140
    glomerular filtration rate 137–139, 187, 190, 191
    prevention and treatment 141–146
  renin-angiotensin-aldosterone system dysregulation
    adipose tissue relationship with circulating renin-angiotensin-aldosterone system 125, 126
    clinical interventions 127

local tissue renin-angiotensin systems
      124, 125
   pathophysiology 127–129
   systemic renin-angiotensin-aldosterone
      system 123, 124
   weight loss effects 126, 130
   reverse epidemiology, see Reverse
      epidemiology
Orlistat
   obesity management 216
   proteinuria and microalbuminuria
      reduction 225, 226

Peritoneal dialysis, interleukin-6 levels 168
Peroxisome proliferator-activated receptors
   (PPARs)
   agonists and renal protection in obesity
      144, 145
   lipid metabolism 112, 113
   PPAR-γ and CD36 expression regulation
      110
Plasminogen activator inhibitor-1 (PAI-1)
   adipose expression 80, 81
   expression regulation 81, 97, 98
   functions 81, 97
   hemodialysis patient levels 98
Proteinuria
   bariatric surgery and reduction 225
   obesity association 221–223
   obesity pharmacotherapy effects 225,
      226
   renal disease progression 222
   weight loss and reduction 223–225

Renal blood flow, obesity effects
   body surface area normalization 186, 187
   clinical implications 197
   effective renal plasma flow measurement
      185, 186
   mechanisms 191–196
   overview 137
   study analysis 187, 190, 191
Renin-angiotensin system
   adipose tissue 81
   components 122, 123
   dysregulation in obesity and metabolic
      syndrome

   adipose tissue relationship with
      circulating renin-angiotensin-
      aldosterone system 125, 126
   clinical interventions 127
   local tissue renin-angiotensin systems
      124, 125
   pathophysiology 127–129
   systemic renin-angiotensin-aldosterone
      system 123, 124
   weight loss effects 126, 130
   proteinuria and microalbuminuria
      reduction mechanisms in weight loss
      226, 227
   renal hemodynamic effects in obesity
      193, 194
   renin inhibitors in obese hypertensive
      management 238
Resistin
   functions 77, 78, 100
   inflammatory response 169, 170
   kidney pathophysiology 100, 101
   obesity effects on kidney 140, 141
   sources 77
Reverse epidemiology
   hemodialysis patients 59–61, 170,
      171
   Japan study 51, 52
   non-renal disease patterns 62, 63
   overview 58
   pathophysiology of obesity paradox
      63–65
   peritoneal dialysis patients 61, 62
   prospects for study 65, 66
   survival bias 64, 65
Rimonabant
   obesity management 216, 217
   proteinuria and microalbuminuria
      reduction 225, 226
Roux-en-Y gastric bypass, see Bariatric
   surgery

Sibutramine
   obesity management 215, 216
   proteinuria and microalbuminuria
      reduction 225, 226
Sodium handling, renal hemodynamic
   effects in obesity 194, 195

Sterol regulatory element binding protein (SREBP)
  cholesterol synthesis 110
  diabetes and kidney expression 108
Syndrome X, *see* Metabolic syndrome

T cell, leptin effects 157–160
Transforming growth factor-β (TGF-β)
  clearance 95, 96
  fibrosis role 180
  functions 95
  kidney pathophysiology 96
  leptin interactions in kidney 179, 180
Triglycerides, *see also* Hyperlipidemia
  metabolism 111
  toxicity 111
Tubular cells, lipid accumulation following injury 114
Tubuloglomerular feedback, renal hemodynamic effects in obesity 194, 195
Tumor necrosis factor-α (TNF-α)
  adipose expression 80, 167, 168
  insulin resistance role 80
  leptin interactions 156, 157
  therapeutic targeting 168

Vascular endothelial growth factor (VEGF)
  functions 98
  kidney pathophysiology 99
Vertical banded gastroplasty, *see* Bariatric surgery
Very low density lipoprotein (VLDL) receptors 111
Visfatin, function 78, 79

Weight loss
  children 143
  chronic kidney disease management 12, 13
  hypertension response 142, 143
  proteinuria and microalbuminuria reduction
    mechanisms 226, 227
    overview 223–225
  renal response in metabolic syndrome 126, 130